SECOND EDITION

Friction Science and Technology

FROM CONCEPTS to APPLICATIONS

SECOND EDITION

Friction Science and Technology

FROM CONCEPTS to APPLICATIONS

Peter J. Blau

Society of Tribologists
and Lubrication Engineers

CRC Press
Taylor & Francis Group
Boca Raton London New York

CRC Press is an imprint of the
Taylor & Francis Group, an **Informa** business

CRC Press
Taylor & Francis Group
6000 Broken Sound Parkway NW, Suite 300
Boca Raton, FL 33487-2742

First issued in paperback 2019

ISBN-13: 978-1-4200-5404-0 (hbk)
ISBN-13: 978-0-367-38666-5 (pbk)

Library of Congress Cataloging-in-Publication Data
Blau, P. J.
Friction science and technology : from concepts to applications / Peter J. Blau. -- 2nd ed.
p. cm.
Includes bibliographical references and index.
ISBN 978-1-4200-5404-0
1. Friction. I. Title.
TJ1075.B555 2008
621.8'9--dc22   2008018724

Visit the Taylor & Francis Web site at
http://www.taylorandfrancis.com

and the CRC Press Web site at
http://www.crcpress.com

Dedication

This book is dedicated to the memory of my parents: to my father, a principled, hardworking man who valued ethics and personal responsibility, and had a wonderful sense of humor; and to my mother, a small woman with a big heart, who opened my eyes to the richness of music and art.

One researcher had an addiction
To seeking the causes of friction;
He'd often confide,
Whilst watching things slide,
That he suffered that mental affliction.

Contents

Foreword

The first edition appeared in late 1995. Since that time, there have been many new developments in our understanding of friction. Examples of these are new ASTM standards for friction measurement, laser dimpled surfaces for friction control, friction of nanocomposites and alloys for light-weight bearings, and most importantly, leading edge research on friction at the molecular scale—perhaps the fastest growing aspect of the field.

This book begins with a thorough development of the history of thought on the subject of friction, which puts the book in context. This history provides grounding for the main goal of this book, which is to address the mechanics, materials, and applications-oriented aspects of friction and friction technology. As a result, this book does a fine job of comprehensively covering the subject. Key topic areas are mechanics-based treatments of friction, including typical problems and equations for estimating the effects of friction in simple machines; the wide range of devices that have been crafted to measure the magnitude of friction, some designed to simulate the behavior of engineering tribosystems; modeling of static and kinetic friction; the effects of tribosystem variables such as load, speed, temperature, surface texture, and vibration on frictional behavior, the result of which demonstrates how the same materials can exhibit much different frictional behavior when the contact conditions are changed; and the response of different types of material combinations to frictional contact.

I think the discussion on the same materials exhibiting different frictional behavior under differing contact conditions is particularly beneficial as so often in the past engineers would look up a material's inherent coefficient of friction in some handbook, apply that to a design, with the result of total mystification that the resultant friction is much different.

Subsequent chapters deal with run-in processes, which I found interesting as the importance of this is particularly acute in the bearings used in laser targeting and high-resolution photoimaging devices. There is also a useful chapter on lubrication by gases, liquids, and solids.

There is also an interesting chapter on the solid friction of materials. It covers a wide variety of combinations such as leather, wood, stone, metals, a variety of alloys, metallic glasses, ceramics, polymers, carbon-/diamondlike materials, ice structures, just to name a few.

A unique feature is the inclusion in various chapters of numerous interesting and unusual examples of the application of friction science, proving that tribologists and tribological problems are truly indispensable and multidisciplinary. A few examples covered in the book that highlight the breath of these applications are friction problems in Olympic and other sports, coatings for icebreakers, interparticle friction (toners, pills, powders, etc.), cosmetics, starting a fire caveman style, joint replacement, reducing heat in dental root canal tools, the touch of piano keys, human skin friction, the drag of ships through the water, earthquakes, and the "bounce" in shampoo. This aspect of the book alone makes it an interesting read for both highly technical people as well as those with more than the usual curiosity for how things work.

Dr. Robert M. Gresham
Director of Professional Development
Society of Tribologists and Lubrication Engineers

Preface

It is amazing that friction, a phenomenon that influences so many aspects of our daily lives, is so widely misunderstood. Even after centuries of study by bright and inquiring minds, friction continues to conceal its subtle origins, especially in practical engineering situations where surfaces are exposed to complex and changing environments. With the possible exception of rolling element bearings under thick-film lubrication, the prediction of the friction between materials in machinery is often based more on experience and experiments than on first-principles theory. The richness of friction science is revealed to those with the patience to dig deeper, and requires a willingness to surrender preconceived notions that may oversimplify physical reality.

Although there is a lot of new material in this second edition—particularly as regards engines and brakes—my essential writing philosophy has not changed. I wanted to take the reader on an intellectual journey that begins with common introductions to friction, in which friction coefficients are simply numbers to look up in a table, and travel to a new place, in which we question where those numbers came from, whether they actually apply to specific problems, and why things are not as simple as those watered-down explanations of friction we are taught in high school and introductory college physics might lead us to believe.

When I began to write the first edition more than 10 years ago, the word "tribology" was foreign to many people, even to some in science and engineering. And although the term remains obtuse to the general public, the advent of computer disk drives, microdevices, and nanotechnology has thrust friction science and tribology to the forefront. Designers must now confront the challenges of controlling interacting surfaces in relative motion at sizes far too small for the naked eye to see. Despite the current focus of popular science on nano things (think little and propose big ...), many macroscale challenges remain. These larger-scale challenges should not be ignored, and so they populate the pages of this second edition. I hope that the next generation of tribologists will be motivated to study friction problems across a broad spectrum of sizes, and not lose sight of the forest for the trees.

Almost every day I become aware of new and interesting studies and applications of friction science, and it was difficult to call an end to this project for fear of leaving something out. Yet, any treatise on science or engineering is at best a snapshot of the author's thinking at the time. I have learned a lot since completing the first edition of *Friction Science and Technology* and wish I could change a few things even before this second edition appears in print.

I am indebted to a number of individuals for encouraging and educating me in tribology. First, I would like to thank Dave Rigney, professor emeritus of the Ohio State University, for introducing me to the subject. Next, I want to thank many kind individuals who have expanded my perspective of the subject over the years: Bill Glaeser, Ken Ludema, David Tabor, Olof Vingsbo, Ward Winer, Ernie Rabinowicz, Marshal Peterson, Lew Ives, Bill Ruff, Vern Wedeven, Ray Bayer, Ken Budinski,

Doris Kuhlmann-Wilsdorf, Mike Ashby, Brian Briscoe, Koji Kato, Maurice Godet, Ali Erdemir, and many others.

Finally, many thanks to my wife, Evelyn, for tolerating my long hours of isolation on the iMac, and to Allison Shatkin of Taylor & Francis/CRC Press whose encouragement motivated me to set aside other writing projects and focus on this second edition.

Peter J. Blau
Knoxville, Tennessee

1 Introduction

We, as a group of specialists, are familiar with the fact that the friction coefficient is
just a convenience, describing a friction *system* and *not* a materials property.

Dr. Ing. Geert Salomon,
in the Introduction to *Mechanisms of*
Solid Friction (1964), p. 4

Friction is a remarkable phenomenon. As pervasive as friction is in daily experience, there is still much to learn about its nature, how it changes under different
circumstances, and how it can be predicted and controlled. Its effects on the behavior of machines and materials have been the source of study and contemplation for
hundreds and even thousands of years, reaching back at least as far as Aristotle
(384–322 BC).[1] In fact, it could be argued that the undocumented first use of a log
or rounded rock to move a heavy object was an engineering solution to a prehistoric
friction problem.

Great thinkers like Hero, da Vinci, Hooke, Newton, Euler, and Coulomb, all
considered friction; however, a complete description of its fundamental causes and a
single quantitative model—which is generally applicable to any frictional situation—
remains elusive. The fact that so much learned effort has failed to uniquely discern
the fundamental nature of friction might seem surprising at first, but as the reader
will grow to appreciate, the complexities and interactive variables that influence
frictional systems sometimes defy easy definition. A great deal is now known about
friction in specific circumstances but not in the elusive *general case*, if indeed there
is such a thing.

In recent years, there have been attempts to "bridge the gap" between friction
studies at nanometer scales and the behavior of contacting bodies that operate at mac-
roscales, millions of times larger. Partly as a consequence of those efforts, the definition of a "friction coefficient" has been extended far beyond classical approaches that
concern macroscopic bodies rubbing together into realms that can only be investigated
with electron microscopes or probes that are far too small for an unaided human eye
to see. This book reviews, at various levels of detail, conceptual approaches to understanding, modeling, testing, and applying concepts of solid friction to engineering
systems, both lubricated and nonlubricated. It will be shown that the appropriate size
scale and investigative tools must be selected on a case-by-case basis.

According to *The Oxford English Dictionary* (1989 edition), the word *friction*
derives from the Latin verb *fricare*, which means to rub. Interestingly, the word
tribology, which encompasses not only friction but also lubrication and wear, derives
from the Greek word τριβοσ (tribos), which also means rubbing, but the use of this
term is much more recent. It can be traced back to a suggestion of C. G. Hardie
of Magdalen College, and it emerged around 1965 when H. P. Jost, chairman of a
group of British lubrication engineers, attempted to promulgate its use more widely.

In fact, four national tribology centers were established in England a few years after the Jost report revealed the major impact that friction, lubrication, and wear had on the industry and economy of the United Kingdom. The word friction has a number of less-used relatives including the following:

1. Fricase, *v.*—to subject to friction
2. Fricate, *v.*—to rub (one body on another)
3. Frication, *n.*—the action of chafing or rubbing (the body) with the hands; the action of rubbing the surface on one body against that of another; friction
4. Fricative, *adj.*—sounded by friction, as certain musical instruments (also relates to the sounds produced by the breath as it passes between two of the mouth organs)
5. Fricatory, *adj.*—that rubs or "rubs down" (Latin *fricator*, one who rubs down)
6. Frictile, *adj.*—obtained by friction

Interestingly, the word *fricatrice*, which was used in the 1600s and derives from the same Latin origin, is defined as *a lewd woman*.

Frictional phenomena exact a high cost on society. It has lifesaving positive benefits, such as braking moving vehicles to avoid property damage, injury, or death. But it also has powerful negative effects, such as robbing machines of energy that could otherwise produce useful work. Studies[2,3] have estimated that millions of barrels of oil or their equivalent could be saved by lowering the friction in engines. The precise cost is very difficult to determine, but in 1985, Rabinowicz[4] estimated the annual cost of resources wasted at interfaces in the United States. Table 1.1 indicates that tens of billions of dollars are expended each year due to both friction and wear. Considering the vast number of additional situations not listed in Table 1.1, it is clear that the understanding and control of friction has great economic consequences. Although frictional losses have been estimated to account for about 6% of the U.S. gross national product, there have unfortunately been no comprehensive updates of the decades-old studies concerning the costs of friction.

TABLE 1.1
Resources Wasted at Interfaces (ca. 1985)

Interface	Dollars Dissipated/Year
Piston ring/cylinder—internal combustion engines	$20 billion
Human body—seat in clothing	$20 billion
Tires on road surfaces	$10 billion
Tool/workpiece in metal cutting	$10 billion
Drill/hole in oil drilling	$10 billion
Head/medium in magnetic recording	$10 billion

Source: Adapted from Rabinowicz, E. in *Tribology and Mechanics of Magnetic Storage Systems*, ASME, New York, 1986, 1–23.

1.1 WORLD OF FRICTIONAL PHENOMENA: GREAT AND SMALL

There are many manifestations of friction. Gemant's book[5] describes a host of phenomena, all related to friction. He stated, with remarkable foresight, more than 50 years ago:

> Indeed, it is hard to imagine any process, whether in nature or in industry, that is entirely free of friction. It appears that only processes of the largest and the smallest dimensions, namely astronomical and interatomic motions, can be described without the involvement of friction. However, even this situation might change with a better understanding of the universe on the one hand and of the elementary particles in the atom on the other.

Gemant's book discusses sound waves, viscosity of solutions, viscosity of structures, flow of fluids, lubrication, plastic flow in solids, internal friction in solids, material damping capacity, friction between solids, and other phenomena. Internal friction in metals and alloys has been used to deduce the fundamental processes of diffusion, time-dependent viscoelastic behavior, creep, and vibration damping capacity. The friction of tiny whiskers within a surrounding matrix has been strongly linked to establishing the mechanical properties of advanced ceramic composite materials[6] (see Figure 1.1). Friction occurs in other forms as well: rolling friction, frictional fluid drag in pipes, friction within powder and soil layers, friction in geological formations and glaciers, and aerodynamic friction. Astrophysicists have even used the term *tidal friction* to describe the torque generated between the convective core and the radiative envelope in early stars.[7]

Introductions to friction come early in life; for example, children are taught the frictional benefits of rubbing one's hands together to stay warm. Primitive tribesmen and wilderness campers learned how to create a fire by rubbing wood together. According to Dudley Winn Smith,[8] who claims to hold the world record for starting a fire with a "fire bow," with the proper technique and sufficient practice it is possible to start a fire by this method in under a minute. In Smith's own words, when describing his winning performance in a fire starting competition in Kansas City:

> ... When the starter said "Go" I drew my bow back and forth with long complete strokes. In about three seconds a little pile of smoking black charcoal issued from the pit. Then I stopped rubbing, picked up both the board and the tinder and blew directly onto the smoking pile, which immediately turned into a red ember. In 7-1/5 seconds after I drew the first stroke the tinder burst into flame. Lucky for me, the three timers all agreed ...

Smith recommended using a 29 in. long bow with an octagonally shaped fire drill, approximately 9 in. long and $\frac{9}{16}$ in. in major diameter. His upper pivot, hard but not prone to produce excessive friction, was made from the glass knob of a coffee percolator embedded in a wood block. The $\frac{3}{8}$ in. thick fireboard contains a "fire pit," a hole where the tip of the drill rests, and that is crossed by a U-shaped notch surrounded by tinder. Supposedly, the best woods for the drill and fireboard were said to be yucca and American elm, and red cedar shavings are best for tinder.

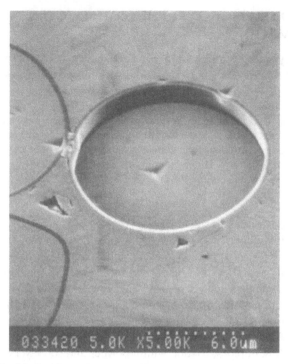

FIGURE 1.1   The friction forces between a fiber and the matrix material in a ceramic composite are estimated from experiments that push the fiber into the matrix with a nanoindentation device. In the center of the ceramic fiber is the impression left by the tip of the three-sided pyramidal indenter used for pushing. (Scanning electron micrograph, courtesy of L. Riester, Oak Ridge National Laboratory.)

As subsequent chapters will discuss, frictional phenomena occur on and within the human body. For example, unpublished studies funded by shampoo and conditioner manufacturers have addressed the friction of hair on hair. The bamboo-type structure of human hair results in directional sliding properties. Friction of hair sliding over hair "against the grain" is much higher than "with the grain." The kinetic friction of hair under various humidity levels affects the "bounce" in styled hair. As will be further discussed in Chapter 9, the friction of skin lubricated by soaps, colorants, and lotions has significant economic implications for cosmetics manufacturers who are expanding their product lines to target specific ethnic groups.

The development of acceptable replacement materials for ivory piano keys is partly affected by the friction of skin on the key material. Studies by Dinc et al.,[9] partly funded by Steinway, Inc., used an apparatus that simulated a piano keyboard to study the friction of skin on polymethylmethacrylate, nylon 66, polytetrafluoroethylene, polycarbonate, and phenolic. It was not only friction, but also the feel of the material that determined its desirability for the application. Sometimes the friction was relatively low, but the tactile sensation was unpleasant to the subject. Increasing humidity and increasing perspiration tended to raise the friction coefficient and make the keys feel more uncomfortable. Slight hydrodynamic effects reduced friction when sliding speed increased, and providing more surface roughness on the keys reduced friction somewhat.

In addition to cosmetics and piano keys, the friction of skin is a key factor in sports. In 2006, there was a controversy over the frictional characteristics of newly introduced composite surfaces of basketballs for professional players in the U.S. National Basketball Association (NBA). Players complained that the balls were too sticky when dry and too slippery when wet with perspiration. The microfibers in the surface were also said to cause small cuts on the hands and fingers and bounce unpredictably when striking hardwood, backboards, or the basket rim. Only 3 months after approving the new microfiber composite basketballs, NBA officials retracted their approval and returned to using the previous leather-covered balls,[10] a decision that greatly disappointed an American special interest group called People for the Ethical Treatment of Animals (PETA), who went so far as to offer free hand cream to players to encourage their use of the new synthetic balls.

Studies of the frictional properties of tooth restorative materials have been conducted in simulations of chewing and in a simple sliding apparatus to assess the effects of the mouth environment on material performance.[11] In hip joint (acetabular caps) and knee replacements, the friction coefficient is typically of the order of 0.02 and is not normally a concern, but if friction becomes too high it will eventually cause loosening of the implants to the point where their function is impaired.[12] Additional examples of friction within the human body and surgical equipment are given in Chapter 9.

Pedestrian slipping accidents are a leading cause of direct and morbidity accident costs in the United States. These safety-related issues have created a considerable interest in measuring the friction of flooring and pavement materials against shoe materials, under a variety of circumstances. The former American Society for Testing and Materials, currently called ASTM, has certain standards (ASTM Standard D-2047 and ASTM Standard D-2534) and Chemical Specialties Manufacturers Association has tests (document BUL 211.1) for the friction of flooring and floor wax products. Additional references may be found in a special ASTM publication.[13] Articulating strut testers have been invented to simulate walking friction.[14] They provide more accurate and realistic frictional information for walking simulation than do simple drag tests of weighted sleds. A further discussion of the friction of footwear on various surfaces, including roofing, is given in Chapters 3 and 9.

Friction in explosives has been of interest for at least 40 years, because it is possible to initiate explosions by friction in sensitive materials. Amuzu et al.,[15] for example, studied the friction of five different explosive compounds: silver azide, α-lead azide, cyclotrimethylene trinitramine (RDX), cyclotetramethylene tetranitramine (HMX), and pentaerythritol tetranitrate (PETN). This unique work established the applicability of using classical concepts of modeling friction as a linear function of the pressure-dependent interfacial shear strength to understand the possibilities of initiating explosions from frictional heating.

The presence of friction in test fixtures used for the mechanical testing of materials can cause significant errors and scatter in test data. One common method for testing the flexure strength of ceramics involves the four-point bend test. In one

TABLE 1.2

Recommended Supply Temperatures for Various Activities on Ice

Activity	Temperature Range (°C)[a]
Speed skating	−6.7 to −5.6
Recreational skating	−7.8 to −6.7
Figure skating	−8.9 to −7.8
Ice hockey	−10.0 to −8.9

[a] Temperatures may vary with the type of cooling system used in the rink.

Source: Adapted from Montebell, G.M., *ASTM Stand. News*, June, 54, 1992.

study, Quinn[16] estimated the error associated with friction in the pins on which the specimens rest may introduce a 4–7% error in calculating the strengths of ceramics. Friction between the ends of right-cylindrical specimens and the horizontal plattens is also an important concern in compression testing.[17]

In addition to the previous example concerning the tactile friction of basketball surfaces, frictional phenomena are important in a wide variety of sports activities. In fact, nearly every sport is in some way affected by or dependent on friction. Some of the most obvious examples include shuffleboard, curling, downhill and cross-country skiing, luge, bobsled, and track and field (traction). The friction of blades on ice is a critical concern in both competition and recreational ice-skating situations. The temperatures of the supply systems vary with the activity, as shown in Table 1.2.[18] As further elaborated in Section 5.6, the frictional behavior of moving skates on ice has a great deal to do with frictional heating, and frictional heating is linearly related to sliding velocity. More recent studies have also focused on the properties and structure of ultrathin films that naturally occur on the surfaces of ice. Nevertheless, Salomon[19] once observed that "the plastic properties of ice are well-known, but we could never have predicted the low friction coefficient experienced in transportation on skis or skates …. Incidentally, even now, we cannot think [of] a suitable material for skating [outdoors] in the tropics!"

While summer non-ice skating with steel blades remains problematical, technology has produced a variety of synthetic skiing and snowboarding surfaces for recreational and athletic practice venues. These multilayer polymer composites are designed to offer not only low friction, but also cushioning, resilience, and moisture control.[20]

Examples of frictional effects in everyday life are endless. The foregoing examples are intended merely to heighten the reader's awareness and illustrate their remarkable diversity. This book focuses on just one group of frictional phenomena: static and kinetic friction between solid materials, both with and without lubrication. The economic and technical implications of this group of frictional problems are both far ranging and important. Aerodynamic friction and fluid friction, such as the resistance of fluid flows through pipes and constrictions, are not treated here.

Numerous mathematical treatments have been developed to describe the influences of friction on machine behavior, the energy efficiency of vehicles, and in metalworking processes. Its influence on a range of practical problems, like those already described, has attracted investigators from many disciplines—solid-state physics, chemistry, materials science, fluid dynamics, mechanical design, and solid mechanics. With such an interdisciplinary history, mathematical models for friction have reflected the diverse backgrounds of the investigators, and there is disagreement about which friction models apply in given situations. To make matters worse, terminology also varies between disciplines. The history of friction studies reveals the interplay between macroscopic concepts and the development of scientific instruments that have fundamentally changed our understanding of surface structure.

1.2   HISTORICAL BACKGROUND

Frictional behavior has been the subject of systematic, documented studies and measurements for over half a millennium. Lubrication has been applied to solve friction problems far longer. One of the most cited examples of this is a drawing discovered in a grotto at El Bersheh, Egypt, and dated at about 1880 BC, which shows a large colossus being pulled by numerous rows of slaves. At the front of the wooden sledge on which the statue rested, a small figure was depicted pouring a liquid, presumably animal fat (since the shape of the vessel was not typical of those used for water at the time), on the large wooden rollers used to transport the great sledge. Davison[21] estimated the number of slaves needed to pull the sledge by assuming that it weighed 60 tons, that each slave could pull with an average of 120 pounds force, and that the friction coefficient (to be formally defined in another section or chapter) between the wood rollers and the wood base of the sledge was 0.16. He calculated that 179 slaves would be needed. In fact, there were 172 slaves in the drawing. In another article, Halling[22] performed a slightly different calculation. He assumed that the slaves, being stout lads, could each pull with a horizontal force of 800 N (about 180 pounds force) and that the weight of the alabaster statue on the sledge was equivalent to a normal force of 600,000 N. Assuming that 172 slaves pulled at once, Halling calculated the coefficient of friction to be 0.23, somewhat higher than Davison's value, but not an unreasonable number.

Leonardo da Vinci's pencil sketches, as presented by Dowson,[23] include several types of apparatus that he designed to study sliding friction, yet in all of da Vinci's voluminous works he never explicitly mentioned the term *friction force*.

The first two classical laws of friction, usually attributed to the Frenchman Guillaume Amontons (1699), are as follows:

1. The force of friction is directly proportional to the applied load.
2. The force of friction is independent of the apparent area of contact.

Interestingly, Amontons developed his concepts about friction not in a research establishment but rather in a shop where glass lenses were being polished. Despite Amontons's association with these two fundamental "laws," the concepts attributed to him are paralleled in the detailed explanations in Leonardo da Vinci's earlier studies (1452–1519). As is discussed later, the so-called "laws of friction" are not always obeyed, especially when sliding occurs in extreme environments such as at high speeds or over a wide range of normal loads. The simple laws of friction have been quite valuable as a basis for understanding the behavior of machines. Still, the well-informed engineer will learn to use these concepts with due caution because there are a number of cases in which these simple laws do not hold.

Robert Hooke considered the nature of rolling friction and plain bearings in the mid to late 1600s.[24] In analyzing the movement of coaches, he identified two components of rolling friction: (a) yielding of the floor during rolling and (b) sticking and adhering of parts. In the beginning of the 1700s, the German Gottfried Wilhelm von Leibnitz[25] published a contribution to the study of friction in which he distinguished between sliding and rolling friction.

Leonhard Euler was one of the most productive scientists and mathematicians of all time. He is credited with over 750 original contributions to scientific knowledge.[26] One of his most important contributions to the understanding of friction is in clarifying the distinctions between static and kinetic friction. In considering a block resting on an inclined plane, he discussed the measurement of static friction in which the plane is slowly tilted until the block begins to move. Pointing out that a very small increase of the tilt beyond the critical point produced a rapid change in the sliding velocity, instead of a very small incremental change, he concluded that the value of the kinetic friction coefficient must be much smaller than that of the static friction coefficient. In later studies, Euler considered the friction of shafts and of ropes wrapped around shafts. In fact, the use of the Greek symbol mu (μ) for the friction coefficient is credited to Euler.

Charles Augustin Coulomb was a French military engineer whose interest in friction was piqued by a prize offered by the Academy of Sciences in Paris in 1777 for "the solution of friction of sliding and rolling surfaces, the resistance to bending in cords, and the application of these solutions to simple machines used in the navy." Coulomb began his work on friction in 1779, after no one had won the 1777 competition and the prize had been doubled. Coulomb's award-winning paper was published in 1781; however, his major work on friction did not appear in print until 4 years later. In that lengthy memoir, Coulomb discussed first the sliding of plane surfaces, then the stiffness of ropes, and finally the friction of rotating parts. He investigated the effects of the nature of the contacting materials, the extent of the surface area, the normal pressure (load), and the length of time that the surfaces remained in contact (the "time of repose"). The effects of these variables are still being studied today in connection with the development of advanced metallic alloys and ceramics for friction-critical applications, such as bearings, seals, brakes, and piston rings. Coulomb's conclusions about the nature of friction dominated thinking in the field for over a century and a half, and many of his concepts remain in use. In fact, the term "Coulombic friction" is still found in recent publications.

The Rev. Samuel Vince, a fellow of the Royal Society, developed a vision of the nature of friction independently from Coulomb, and in 1784 he presented a paper in London titled, "On the motion of bodies affected by friction." That paper was subsequently published in 1785.[27] In it, Vince attributed the nature of static friction to cohesion and adhesion. Later, John Leslie, a professor of physics at the University of Edinburgh, wrote extensively on the friction of solids, calling into question earlier concepts of friction's relationship to energy. He understood that frictional energy could not be adequately explained by the continuous rising of asperities up slopes on opposing surfaces, because the potential energy of that type of system would be recovered when the asperities slid down the other side. He further questioned the role of adhesion in friction, arguing instead the time-dependent nature of asperity deformation (flattening). These conclusions were based on experiments in which bodies were placed in contact and then allowed to rest for various periods of time before sliding was attempted. The same type of problem is significant today in designing spacecraft whose antenna bearings and other moving parts must remain in contact for month after month, then move smoothly, without undue torque, when small motors are eventually activated by a radio signal from the ground.

At about the same time that Vince was working on cohesion and adhesion, important work was being conducted by Sir Benjamin Thompson of North Woburn, Massachusetts. Under his more well-known title, Count Rumford, Thompson set out to explore the nature of frictional heating in 1784. Applying his work to turning cannon bores, he was the first to equate horsepower (mechanical energy) to heat.[28] The dissipation of energy by friction remains important in understanding how frictional heating can alter the properties of the materials in the interface and, in so doing, influence not only wear, but also the nature of subsequent variations of the friction force itself.

Two major industrial problems existed in the early 1800s: the construction of bridges and arches and the launching of ships on slipways. In constructing arches, it was found that using higher-friction mortar materials permitted the use of lower angles between the stones comprising the arch. Friction problems in launching ships spurred a great deal of experimentation. Imagine how embarrassing it might have been for shipyards' engineers to construct a ship and then, with great ceremony and in the presence of high officials, be unable to slide it down the slipway into the water. George Rennie[29] conducted a variety of experiments on solid friction during the early to mid 1800s. His basic apparatus was a horizontal, weighted sled attached by a cable over a pulley to a tray of weights. Using this type of device, he conducted studies of the friction of cloth, wood, and metals. Rennie addressed the ship launching problem by noting that the hardness of woods affects the friction, and further, that using soft soap on the slipways reduced the friction to one twenty-sixth of its former value.

During the industrial revolution, many other practical friction problems emerged: the friction in bearings for grain mills, the friction in windmills and waterwheel parts, friction in belting, and friction in brakes. In the 1830s, Arthur Jules Morin, a French artillery captain, conducted a long and extensive series of rolling and sliding friction studies at the Engineering School of Metz. He continued his work as a professor in Paris and later rose to the rank of general in the French army. A 1860 translation of Morin's book contains a 60-page chapter on "friction," describing its measurement and application to common machine elements such as slides, journals, belts, and pulleys.[30] Remarkably, friction coefficient data for wood-on-wood, found in some handbooks published today, can be traced back to that original work.

During Morin's time, railroads were emerging as an important transportation technology. The same kinds of friction and lubrication problems that existed in early railways must still be addressed today, even though there has been considerable progress in reaching solutions for them. In 1846, Bourne[31] published a history of the Great Western Railway and in it described the factors that affected rolling and sliding friction. Additional effort was devoted to the design and lubrication of bearings of railway cars. In fact, as Dowson[1] pointed out, there is a strong parallelism between the history of tribology and the history of transportation. This parallelism continues as we continue to seek low friction materials and designs for improved efficiency engines and drive trains and controlled friction for more reliable, noiseless brakes and clutches.

In the late 1800s, work on the nature of sliding and rolling friction continued to flourish, enhanced by the development of a number of analytical treatments of solid contact, most notably the works of Heinrich Hertz[32,33] who developed the foundation of present-day contact stress calculations for elastic bodies. In 1886, Goodman[34] developed a series of friction models based on the concept of ratcheting sawteeth, noting that the friction of similar metals was usually higher than for dissimilar metals. Eight-five years later, Rabinowicz's more recent discussions of compatibility[35] echoed these observations, but they were not interpreted in the same manner. Significant progress was also made during the late 1800s in the theory and application of lubricants, such as the seminal papers of Osborn Reynolds (see the discussion in Ref. 1).

In 1898, Richard Stribeck was appointed one of the directors of the newly established Centralstelle für Wissenschaftlich-technische Untersuchen in Berlin. During the next 4 years, he published important papers in basic tribology, particularly in regard to the relationship between friction and the state of liquid lubrication.[36] The "Stribeck curve" is a basic concept taught to all students of lubrication engineering and bearing design. A discussion of this important relationship is given in Chapter 6.

Friction studies in the 1900s benefited from new instruments to study and characterize the structure and microgeometry of real surfaces. Scientific approaches to understanding solid friction in the 1900s returned to considering the role of adhesion, first suggested by John Theophilus Desaguliers in 1734. The work of Tomlinson[37] and that of Deryagin[38] considered friction from a molecular interaction and energy dissipation standpoint. The electrical contacts studies of Holm[39] on true versus apparent area of contact between surfaces laid the groundwork for the famous Archard wear law[40] that was to follow. Holm proposed the existence of "a-spots," regions within asperity contacts in which the electrical current passed between surfaces. The "constriction resistance" produced high current densities in small contact areas, leading to points of high ohmic heating and accelerated wear.

From the 1940s to the 1970s, F. Phillip Bowden and David Tabor of Cambridge University's Cavendish Laboratory made significant experimental and conceptual contributions to the understanding of solid friction. Their two books concerning the *Friction and Lubrication of Solids*[41,42] continue to be considered classical texts on these subjects, and students and surface physicists from that laboratory continue to spread their teachings throughout the world.

Igor Viktorovich Kragelskii, a major force in tribology in the former Soviet Union, made major contributions to understanding and calculating the effects of friction. His fundamental work on friction, lubrication, and wear parallels the efforts of Bowden and Tabor in England. In fact, Bowden and Tabor cowrote the foreword to the English translation of Kragelskii's 1965 text.[43] Kragelskii's scholarly and detailed books provide an excellent basis for understanding not only the historical development of tribology in Russia and elsewhere, but also the manner in which a wide range of external factors influence the friction and lubrication of materials. One of his books focuses on the calculation of friction and wear quantities.[44]

Contributions to understanding friction, both from engineering and from theoretical standpoints, burgeoned in the second half of the twentieth century. Although a number of workers in a variety of disciplines produced insightful work, it is worth noting the important contributions of Donald F. Buckley and his group on understanding the role of adhesion and surface chemistry in friction. While working at the National Aeronautics and Space Agency (NASA), Lewis Research Center, in Cleveland, Ohio, he and his colleagues conducted an extensive series of fundamental friction and adhesion studies during the 1960s and 1970s. The effects of crystal orientation, electronic structure of surfaces, and surface segregation of impurity atoms were investigated. The important results and conclusions of this prolific work are compiled in Buckley's 1981 book.[45]

Figure 1.2 summarizes the history of friction research described here. As relatively new experimental techniques like atomic force microscopy and surface force microscopy emerge, changing perceptions of the structure of solid surfaces and interfacial media between them will continue to prompt rethinking the basic concepts of solid and lubricated friction. The advent of such fine-scale techniques is leading researchers to consider more carefully the fundamental definition of friction. For example, are the tangentially resolved components of forces between atoms on opposing surfaces really "friction forces," or are they something else? A recent book on macroscale and microscale aspects of friction fuels that debate.[46] Persson[47] invokes friction to explain the behavior of flux-line systems and charge density waves. Is that truly friction in the classical sense, or is it another phenomenon that has certain attributes that are analogous to friction? At the other extreme, there has been a great deal of interest in the subject of plate tectonics over the past decade. Can one ascribe the term "friction" to the process associated with the massive movement of the continental plates over one another? What is the definition of friction? Perhaps, at the extreme ends of the phenomenological size scale, the term friction is used more as a descriptive convenience than as a logical extension of the earlier work on resistance to motion between macroscopic bodies.

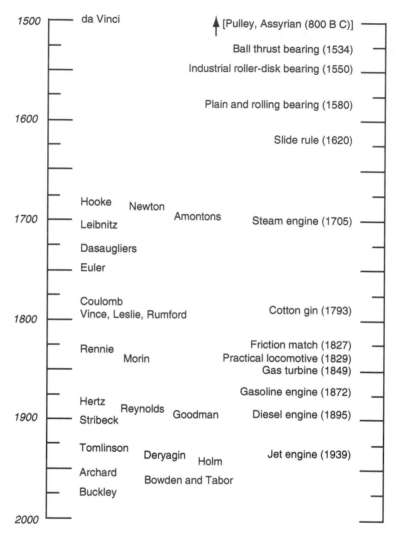

FIGURE 1.2 Timelines showing the correspondence between early work in friction research and the technology of the time.

1.3   TRADITIONAL INTRODUCTIONS TO SOLID FRICTION

Traditionally, students are introduced to the study of friction from a solid mechanics point of view. That is, they are presented with two contacting bodies that are acted upon by a system of forces, which, in turn, results in motion or impending motion parallel to a contacting surface or surfaces. Chapter 2 provides a number of examples that illustrate the traditional approaches to solving such problems, adapted from several introductory texts on physics and mechanics. Regrettably, in the majority of cases, an engineering or science student's exposure to friction ends with that sort of treatment. The classical mechanics approach usually assumes that the friction coefficient will take on either one of two characteristic values: a static value or a somewhat lower kinetic value. This is an approximation at best.

Situations in the real world of friction, lubrication, and wear (i.e., tribology) are not usually so simple. Surfaces are not perfectly clean, materials are not perfectly uniform, velocities and relative sliding motions vary in complex ways, and there are exceptions to the notion that the starting friction is always higher than the sliding friction. The friction force may not remain steady, even when the sliding velocity of the system remains constant. It can be affected markedly by the temperature of the bodies, or the stiffness of the fixtures in the system, or even the relative humidity of the air in some cases. Friction–vibration interactions are important in systems such as bearings, brakes, and seals. Sometimes the friction between surfaces changes unexpectedly after a period of relatively steady behavior. Such complex behavior cannot be explained or predicted with simple friction models. Therefore, more complex, specialized models for the friction of specific situations are required.

1.4   APPROACH OF THIS BOOK

This book is intended to address mechanics, materials, and applications-oriented aspects of friction and friction technology. Chapter 2 emphasizes mechanics-based treatments of friction, including typical problems and equations for estimating the effects of friction in simple machines. Chapter 3 describes a wide range of devices crafted to measure the magnitude of friction. Some are relatively simple but others were developed to simulate the behavior of engineering tribosystems. Chapters 4 and 5 delve into the concepts involved in modeling static and kinetic friction and describe the response of different types of material combinations to frictional contact. Chapter 6 describes basic lubrication concepts. It describes the functions of not only lubricants but also other fluids, such as fuels whose primary function is not lubrication. Chapter 7 addresses the effects of tribosystem variables such as load, speed, temperature, surface texture, and vibration on frictional behavior. It shows how the same materials can exhibit much different frictional behavior when the contact conditions are changed. Chapter 8 focuses on a special subtopic of friction, time-dependent transitions in frictional behavior. It is intended for those interested in understanding the details of such phenomena as running-in and catastrophic transitions in friction when components fail. Chapter 9 hints at the breadth of practical applications of friction science to technology, ranging from friction in machine components and engines to the friction in manufacturing. It addresses the friction of skin and human body parts as well as particle agglomerates, cables, and micromachines.

The practicing engineer will often be introduced to the intricacies of friction with some urgency, for it is usually a pressing problem in friction, adhesion, or lubrication that forces him or her to delve into the subject. Unfortunately, relatively few have the opportunity to be classically trained in the subject of friction or tribology, and often they must forsake the "luxury" of academic study under the press of business. Production-critical machines standing idle for want of friction solutions place a high premium on time. Hopefully, engineering schools of the future will provide better undergraduate instruction in applied tribology, a subject meriting more time and respect than it is sometimes accorded.

An engineer delving into friction science for the first time in the hope of finding "quick answers" more often than not is faced with confusion, frustration, and even a sense of hopelessness. Sometimes employing consultants can even compound the confusion, since each brings his or her biases to the problem, and a second opinion might sometimes be different than the first. Some effective solutions to friction problems may actually begin with an educated guess, but before one can even hazard such a guess, it helps to develop a firm grounding in the subject.

Effectively dealing with friction problems and its many ramifications requires a broad perspective comprising the following elements:

1. The nature of macro- and microgeometric contact
2. The role of dynamic materials properties in friction
3. The mechanics of the surrounding structures and how their interaction with friction forces can cause contact conditions to change
4. The functions of lubricants, contaminants, and interfacial particles
5. The influence and nature of frictional heating
6. Recognition that friction forces may, under some circumstances, change with time, due to externally imposed or self-induced changes in the interface
7. Practical experience

Friction undeniably has a significant daily impact on us, and sometimes the definition and solutions of engineering friction problems are unknown. Yet throughout history, engineers have often been quite clever and successful in solving friction problems even though the fundamental causes for such behavior are elusive. It is the goal of this book both to provide a balanced view of the mechanics and materials aspects of friction and to describe a number of approaches that have been employed successfully for solving important friction problems, even though *why* they worked may not yet be fully understood.

REFERENCES

1. D. Dowson (1979). *The History of Tribology*, Longman, London, p. 48.
2. *Strategy for Energy Conservation Through Tribology*, 2nd ed., American Society for Mechanical Engineers, New York (1981).
3. *A Strategy for Tribology in Canada*, National Research Council of Canada, Publication 26556 (1986).
4. E. Rabinowicz (1986). The tribology of magnetic recording systems—An overview. In *Tribology and Mechanics of Magnetic Storage Systems*, Vol. III, B. Bhushan and N. S. Eiss (eds.), ASME, New York, pp. 1–23.
5. A. Gemant (1950). *Frictional Phenomena*, Chemical Pub. Co., Brooklyn, NY, p. 4.
6. L. N. McCartner (1989). New theoretical model of stress transfer between fibre and matrix in a uniaxially fibre-reinforced composite, *Proc. Royal Soc. London*, A425, 215.
7. P. Goldreich and P. D. Nicholson (1989). Tidal friction in early-type stars, *Astrophys. J.*, 342(Part 1), 1079–1084.

8. D. W. Smith (1937), described in C. F. Smith, *Games and Recreational Methods for Clubs, Camps, and Scouts*, Dodd Mead & Company, New York.

9. O. S. Dinc, C. M. Ettles, S. J. Calabrese, and H. J. Scarton (1990). *Some Parameters Affecting Tactile Friction*, ASME Preprint 90-Trib-28, American Society of Mechanical Engineers, New York, p. 6.

10. M. Stein (2006). *Leather Ball with Return on Jan. 1*, Internet, Retrieved December 12, 2006, at http://sports.espn.go.com/nba/news/story?id=2694335.

11. J. M. Powers and S. C. Bayne (1992). Friction and wear of dental materials. In *ASM Handbook, Vol. 18, Friction, Lubrication, and Wear Technology*, 10th ed., ASM International, Materials Park, OH, pp. 665–681.

12. D. Dowson (1992). Friction and wear of implants and prosthetic devices. In *ASM Handbook, Vol. 18, Friction, Lubrication, and Wear Technology*, 10th ed., ASM International, Materials Park, OH, pp. 656–664.

13. C. Anderson and J. E. Senne, eds. (1978). *Walkway Surfaces: Measurement of Slip Resistance*, ASTM Spec. Tech. Pub. 649, ASTM, Philadelphia, PA.

14. M. I. Marpet and R. J. Baumgartner (1992). Walkway friction: Experiment and analysis, presented at National Educators' Workshop—Standard Experiments in Engineering Materials Science and Technology, Oak Ridge, TN., November 11–13.

15. J. K. A. Amuzu, B. J. Briscoe, and M. M. Chaudhri (1976). Frictional properties of explosives, *J. Phys. D, Appl. Phys.*, 9, 133–143.

16. G. D. Quinn (1992). Twisting and friction errors in flexure testing, *Ceram. Eng. Sci. Proc.*, July–August, pp. 319–330.

17. N. H. Polakowski and E. J. Ripling (1966). *Strength and Structure of Engineering Materials*, Prentice-Hall, Englewood Cliffs, NJ, pp. 302–314.

18. G. M. Montebell (1992). Ice skating surfaces, *ASTM Stand. News*, June, pp. 54–59.

19. G. Salomon (1964). Introduction. In *Mechanisms of Solid Friction*, P. J. Bryant and M. Lavik (eds.), Elsevier, Amsterdam, pp. 3–6.

20. All-Season Extreme, Lake Geneva, WI, Internet http://www.snowmaker.com/snowflex.html.

21. C. St. C. Davison (1957/1958). Wear prevention in early history, *Wear*, 1, 157.

22. J. Halling (1976). *Introduction to Tribology*, Springer-Verlag, New York, p. 4.

23. Ref. 1, p. 98.

24. Ref. 1, p. 145.

25. G. W. Leibnitz (1706). Tentamen de natura et remedlie resistenziarum in machines, *Miscellanea Berolininensia. Class. mathem.* 1710 (Jean Budot, Paris) 1, 307 pp.

26. Ref. 1, p. 164.

27. S. Vince (1785). On the motion of solid bodies affected by friction, *Phil. Trans. Royal Soc. London*, 75(Part I), 165–189.

28. B. Thompson (Count Rumford) (1798). An experimental Inquiry concerning the source of the heat which is excited by friction, *Phil. Trans.*, LXXXVIII, 80–102.

29. G. Rennie (1829). Experiments on the friction and abrasion of the surfaces of solids, *Proc. Royal Soc. London*, 34(Part I), 143–170.

30. A. J. Morin (1860). *Fundamental Ideas of Mechanics and Experimental Data*, D. Appleton, New York, 442 pp. Revised, translated and reduced to English units of measure by J. Bennett (scanned version available on line via http://books.google.com).

31. J. C. Bourne (1846). *The History and Description of the Great Western Railway*, Dave Bogue, London.

32. H. Hertz (1881). On the contact of elastic solids, *J. Reine Angew. Math.*, 92, 156–171.

33. H. Hertz (1882). On the contact of rigid elastic solids and on hardness, *Verh. Ver. Berford. Gew Fleiss*, November.

34. J. Goodman (1886). Recent researches on friction, *Proc. Inst. Civ. Engr.*, ixxxv, Session 1885–1886, Part III, pp. 1–19.

35. E. Rabinowicz (1971). The determination of the compatibility of metals through static friction tests, *ASLE Trans.*, 14, 198–205.
36. R. Stribeck (1902). Die Wesentlichen Eigneschaften der Gleit- und Rollenlager, *Z. Verein. Deut. Ing.*, 46(38), 1341–1348, 1432–1438; (39) 1463–1470.
37. G. A. Tomlinson (1929). A molecular theory of friction, *Phil. Mag.*, 7, 905–939.
38. B. V. Deryagin (1934). *Zh. Fiz. Khim.*, 5(9).
39. R. Holm (1938). The friction force over the real area of contact, *Wiss. Veroff. Siemens-Werk*, 17(4), 38–42.
40. J. F. Archard (1953). Contact and rubbing of flat surfaces, *J. Appl. Phys.*, 24(8), 981–988.
41. F. P. Bowden and D. Tabor (1950). *The Friction and Lubrication of Solids—Part I*, Oxford University Press, Oxford, England.
42. F. P. Bowden and D. Tabor (1964). *The Friction and Lubrication of Solids—Part II*, Oxford University Press, Oxford, England.
43. I. V. Kragelskii (1965). *Friction and Wear*, Butterworth, Washington, DC.
44. I. V. Kragelsky, M. N. Dobychin, and V. S. Kombalov (1982). *Friction and Wear Calculation Methods*, Pergamon Press, New York.
45. D. F. Buckley (1981). *Surface Effects in Adhesion, Friction, Wear, and Lubrication*, Elsevier, New York.
46. I. L. Singer and H. M. Pollock, eds. (1992). *Fundamentals of Friction: Macroscopic and Microscopic Processes*, NATO ASI Series, Series E: Applied Sciences, Vol. 220, Kluwer, Dordrecht, The Netherlands, 621 pp.
47. B. N. J. Persson (1998). *Sliding Friction: Physical Principles and Applications*, Springer-Verlag, Berlin, pp. 399–407.

2 Introductory Mechanics Approaches to Solid Friction

Most of the introductory textbooks on physics or mechanics contain a section on friction. Usually, fewer than one or two lectures are devoted to explaining how to treat such problems in high school science class or undergraduate college courses, and a few homework questions may be assigned. In such cursory treatments, it is customary to define the static and kinetic friction coefficients, to show how free-body diagrams and force polygons (sometimes called string polygons) can be constructed to account for friction forces, and to show how the student may approach macroscopic friction problems when the friction coefficient either is extracted from tables or can be "back-calculated" from the conditions of the problem. Sometimes simple explanations of surface roughness-based origins for friction are given. Usually little or nothing is said about the metallurgical aspects of friction, the role or nature of lubricating films, or the possibility of time-dependent frictional transitions. Although such introductory approaches are useful, they can be misleading since they do not prepare students for the complexities of frictional behavior in real-world, practical situations.

This chapter begins by presenting common approaches to the treatment of friction problems in high school and undergraduate college courses. Subsequent chapters show how the details of frictional interactions can complicate the solutions to practical friction problems, especially in dynamic, interfacially contaminated environments. But it is important to start from a common frame of reference, and to that end this chapter is presented.

2.1 BASIC DEFINITIONS OF FRICTION QUANTITIES

When two solid bodies are placed together under a nonzero normal force and acted upon by another force that has a component parallel to the contact surface (a tangential force), sliding or slipping may or may not occur, depending on whether the applied force can overcome the friction force opposing it. In some cases, the normal force may be due only to the weight of the upper body resting on the lower, whereas in other cases, it may be due to applied forces other than that due to gravity. The problem in determining whether relative motion will or will not occur is one of balancing the forces involved. The following definitions, from ASTM Standard G-40-93 on Standard Terminology Relating to Erosion and Wear, will serve our present purposes:

> *Friction force*—the resisting force tangential to the interface between two bodies when, under the action of an external force, one body moves or tends to move relative to the other.

Coefficient of friction—the ratio of the force resisting tangential motion between two bodies to the normal force pressing those bodies together.

Any force field for which the work done in moving an object from one point to another is independent of the path taken is considered to be a conservative force field. Gravitational and electrostatic force fields are examples of conservative force fields. Therefore, gravitational and electrostatic forces are called *conservative forces*. However, friction forces acting on a body moving from one place to another are *nonconservative forces*. Consider moving a book from one corner of a square, horizontal table to the opposite corner. The direct, diagonal path would expend less energy than one that paralleled the edges of the table.

In this book, we use the term *friction coefficient* interchangeably with the ASTM term *coefficient of friction*, since they are equivalent and in common use. Using **F** to represent the friction force and **N** the normal force, we define the friction coefficient μ as follows:

$$\mu = \frac{\mathbf{F}}{\mathbf{N}} \tag{2.1}$$

More specifically, the force that is just sufficient to resist the onset of relative motion or slip (**F$_S$**) allows us to define the static friction coefficient (μ_s)

$$\mu_s = \frac{\mathbf{F}_S}{\mathbf{N}} \tag{2.2}$$

and the force that resists relative motion after sliding is under way (**F$_k$**) gives rise to the definition for the kinetic friction coefficient (μ_k)

$$\mu_k = \frac{\mathbf{F}_k}{\mathbf{N}} \tag{2.3}$$

For the purposes of this book, if no subscript is used for μ, it will be assumed that the kinetic or sliding friction (μ_k) is being used.

It is important to recognize that Equations 2.1 through 2.3 are definitions and not laws or models of friction. They simply define a proportionality between two forces. In many practical sliding systems, especially in the absence of effective lubrication, the friction coefficient varies with sliding time, and is not necessarily independent of normal force. Examples and reasons for such behavior are discussed in other chapters in this book. For the following illustrations, however, we assume that the static and kinetic friction coefficients can be represented by single-valued constants in the traditional manner.

2.2 TIPPING AND ONSET OF SLIP

In some statics problems involving the potential for relative motion to occur, it is necessary to determine whether or not slip occurs. One example involves tipping versus slipping. Consider a rectangular solid box of weight **W** and length *l* resting on a flat plane (Figure 2.1). The normal force is **N** and it just equals and opposes **W**. When acted upon by a force **P** at height *h*, above and parallel to the plane of rest,

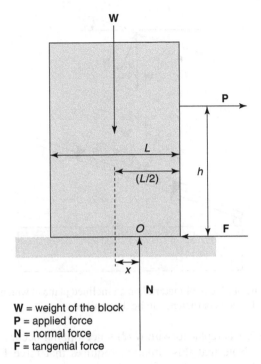

FIGURE 2.1   System of forces on a rectangular solid that may either tip or slip.

the first body will either remain in place, begin to slip (slide), or tip over. To decide which of these possibilities will occur, we set up a system of forces and create a free-body diagram. To prevent tipping over and to maintain static equilibrium, the normal force **N** moves closer to the right edge of the box, off-center by a distance x. We establish an origin **O** at the position of **N** and sum moments about it:

$$\Sigma_M = \mathbf{W}x - \mathbf{P}h \qquad (2.4)$$

Thus,

$$x = \frac{\mathbf{P}h}{\mathbf{W}} \qquad (2.5)$$

and

$$\mathbf{P} = \frac{\mathbf{W}x}{h} \qquad (2.6)$$

The box will tip over at the lower right corner if no slip occurs along the plane of rest, providing that a critical tipping force $\mathbf{P}_t$ is applied:

$$\mathbf{P}_t \geq \frac{\mathbf{W}}{h}\frac{L}{2} \qquad (2.7)$$

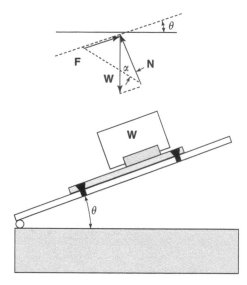

FIGURE 2.2   Sliding of a block of material on an inclined plane in which the angle of repose and hence the static friction coefficient can be determined.

In Equation 2.7, x is replaced with $(L/2)$ since tipping occurs at the lower right corner of the box. Note that this situation requires that force **F**, which balances applied force **P**, is less than the force needed to overcome the static friction force,

$$\mathbf{F} \le \mu_s \mathbf{N} \tag{2.8}$$

otherwise slip would occur before the box tips over. Also, if slip occurs, $0 \le x \le (L/2)$. If the force **P** and the force **F** were just equal, both slipping and tipping might occur.

One traditional method of measuring the macroscale static friction coefficient is to place a block of one material on a plane composed of a chosen counterface material and to slowly tilt the plane until relative motion just begins. Figure 2.2 illustrates this sort of arrangement. A lever is hinged at the left end of a rigid beam to which a flat strip of one material of interest has been affixed. Resting on the strip is a mass to which a coupon of the same or another material of interest is affixed. The total weight of the resting body, that is, the specimen plus the weight of the holder, is **W**. The normal force that opposes **W** and results in static equilibrium perpendicular to the slip surface is **N** when the tilt angle is 0. If the plane is tilted by angle $\theta > 0$, a system of forces is developed. For θ other than 0, $\mathbf{N} < \mathbf{W}$ and

$$\mathbf{N} = \mathbf{W} \cos \theta \tag{2.9}$$

The component of **W** along the contact surface and directed "downhill" is

$$\mathbf{P} = \mathbf{W} \sin \theta \tag{2.10}$$

Before relative motion, the static friction force $\mathbf{F}_s$ $(=\mu_s \mathbf{N}) \ge \mathbf{P}$. Then α is the angle between the normal force and the resultant between the normal force and static friction force, as shown at the top of Figure 2.2. When there is no impending motion,

TABLE 2.1
Commonly Reported Static Friction Coefficients
(Dry or Ambient Air Conditions)

Material Combination	Typical Range in μ_s
Clean, well-adhering metals	1.0–2.0
Typical, sliding metals (ambient air conditions)	0.3–0.8
Lightly oxidized or tarnished metals	0.1–0.35
Clean glass on glass	0.90–0.95
Graphite on graphite	0.10–0.12
Teflon on itself	0.04–0.08
Ice on ice at the melting point	0.05–0.15
Wood on wood	0.25–0.50

$\alpha > \theta$. When $\mathbf{F_S} = \mathbf{P}$, motion is said to be impending and the definition of the static friction coefficient obtains:

$$\mu_s = \frac{\mathbf{F_S}}{\mathbf{N}} \tag{2.11}$$

$$\mu_s = \frac{\mathbf{W} \sin \theta_s}{\mathbf{W} \cos \theta_s} \tag{2.12}$$

or simply,

$$\mu_s = \tan \theta_s \tag{2.13}$$

Under these conditions, $\alpha = \theta_s$, the *friction angle* or the *angle of repose*. Should there be high friction between the two contacting materials, or if the slider is relatively tall in comparison to its length, tipping could occur. In that case, a free-body diagram such as that described in the previous section can be used. In the construction of such free-body diagrams, the directions and senses of the forces due to the weights and applied forces are first determined. Then the directions and senses of the friction forces opposing these forces can be determined.

Equations 2.12 and 2.13 imply that the static friction coefficient is independent of the weight of the slider; however, as discussed in subsequent chapters, this is not always true. Static friction coefficients have been measured for hundreds of years, with most of the investigators using test bodies of convenient size for the time (say, a few centimeters to a few decimeters in size). The friction coefficients measured by such means are useful in treating many types of statics problems but may not be accurate when precise values are required for mechanical design or in special situations, which differ from the conditions of the early experiments. Some commonly reported ranges for static friction coefficients are given in Table 2.1.

2.3 INTRODUCTORY FRICTION PROBLEMS

The following examples illustrate how friction problems are presented in typical statics and dynamics textbooks. These involve the concepts of force balances and free-body diagrams in which friction forces are included.

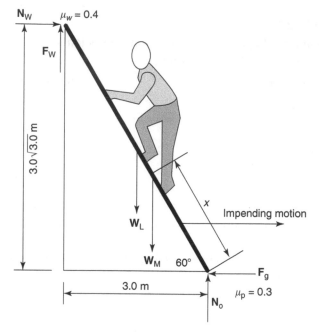

FIGURE 2.3   Friction force analysis for a ladder propped against a wall.

2.3.1   CASE 1. LADDER AGAINST A WALL

A ladder 6 m long is resting against a wall with the bottom end 3 m from the wall. How far up the ladder can an 80 kg man climb without the ladder slipping? Assume that the static friction coefficient of the ladder material against the wall (μ_w) is 0.4, and against the pavement (μ_p) is 0.3.

The angle that the 6 m ladder makes with the ground is $\cos^{-1}(3/6)$, or 60°. Draw the free-body diagram (Figure 2.3) and balance the forces in both the vertical and horizontal directions. Then ensure that the moments around the foot of the ladder at equilibrium sum to zero.

N_O = normal force at the origin of the system (foot of the ladder)
N_W = normal force at the ladder against the wall
W_L = weight of the ladder = (20 kg) (9.81 kg m/s²) = 196.2 N
W_M = weight of man = (80 kg) (9.81 kg m/s²) = 784.8 N
F_W = static friction force of the ladder against the wall
F_g = static friction force of the ladder on the pavement
x = the distance up the ladder that the man can climb before the ladder slips

First, balance forces parallel to the ground:

$$F_g = \mu_p N_O \qquad\qquad (2.14)$$

Similarly, for the direction parallel to the wall:

$$W_L + W_M = F_W + N_O = \mu_w N_W + N_O \qquad\qquad (2.15)$$

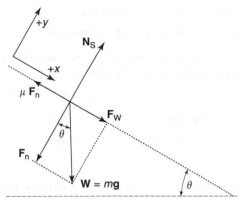

FIGURE 2.4 Friction and gravitational forces acting on a skier.

Now, balance the moments about the origin at the foot of the ladder:

$$\mathbf{W}_M(x \cos 60) + \mathbf{W}_L(3 \cos 60) = \mathbf{F}_W(6 \cos 60) + \mathbf{N}_W(6 \sin 60) \quad (2.16)$$

Solve the vertical and horizontal force balances simultaneously, giving $\mathbf{N}_O = 875.9$ N, and $\mathbf{N}_W = 262.8$ N. Rearranging in terms of x,

$$x = \frac{\mathbf{F}_W(6\cos 60) + \mathbf{N}_W(6\sin 60) - \mathbf{W}_L(3\cos 60)}{\mathbf{W}_M(x\cos 60)} \quad (2.17)$$

Substituting and solving,

$$x = \frac{\mu_W \mathbf{N}_W(6\cos 60) + \mu_p \mathbf{N}_O(6\sin 60) - \mathbf{W}_L(3\cos 60)}{\mathbf{W}_M(x\cos 60)} \quad (2.18)$$

$$x = \frac{0.4(262.8)(3) + 0.3(875.9)(5.2) - 196.2(3)}{(784.8)(0.5)}$$

$$x = 2.79 \, \text{m}$$

2.3.2 CASE 2. SPEED OF A SKIER

A skier starts from rest and proceeds down a mountain. How fast is he/she traveling after 8 s if the slope at the top of the mountain is 30°, and the kinetic friction coefficient of waxed skis on snow (μ) is about 0.12?

Construct the free-body diagram, as shown in Figure 2.4. Since motion has started, use the kinetic friction coefficient, μ. Resolve the components normal to the slope $\mathbf{N}_S$ and along the slope $\mathbf{F}_W$ as follows:

$$\mathbf{N}_S = \mathbf{W} \cos \theta \quad \text{and} \quad \mathbf{F}_W = \mathbf{W} \sin \theta \quad (2.19)$$

In this case, it is more convenient to establish a coordinate system in which the x direction is positive downhill and the y direction is positive normal to the slope of the hill. Therefore, there is no motion in the y direction, and the relationship between the skier's weight ($\mathbf{W}$) and the normal force ($\mathbf{F}_n$) is simply:

$$\mathbf{F}_n = \mathbf{W} \cos \theta \quad (2.20)$$

For the x direction, there is an acceleration, a_x:

$$\mathbf{F}_x = ma_x = \mathbf{W} \sin \theta - \mu \mathbf{F}_n = \mathbf{W}(\sin \theta - \mu \cos \theta) \tag{2.21}$$

But $\mathbf{W} = \mathbf{m}g$, so we can divide both sides by m and solve for a_x simply in terms of the angle of the slope and the acceleration due to gravity:

$$a_x = \mathbf{g}(\sin \theta - \mu \cos \theta) \tag{2.22}$$

From the basic laws of motion, $v = v_o + at$, so starting from rest ($v_o = 0$),

$$v = \mathbf{g}t(\sin \theta - \mu \cos \theta) \tag{2.23}$$

Then after 8.0 s,

$$v = (9.81)(8.0)[0.50 - (0.12)(0.866)] = 31.1 \text{ m/s (69.5 mph)} \tag{2.24}$$

Had the snow been wet and μ increased to 0.25, the skier would be traveling at about 49.7 mph.

2.3.3   CASE 3. MOTORCYCLE ACCIDENT

A motorcyclist lost control of his vehicle, resulting in a skid 46 ft long on the highway surface. How fast was the motorcycle traveling when it fell and began to slide to a stop? Assume that the effective friction coefficient for the motorcycle on asphalt is 0.8, and that all the kinetic energy of the man on his motorcycle at the time he lost control was entirely dissipated by friction over the length of the skid.

We equate the kinetic energy of the motorcycle with the energy dissipated by moving a friction force $\mathbf{F}$ over a distance d as follows:

$$\frac{1}{2}mv^2 = \mathbf{F}d = \mu(mg)d \tag{2.25}$$

Solving for v,

$$v = (2\mu d\mathbf{g})^{-1/2} = [2(0.8)(46)(32)]^{-1/2}$$
$$= 48.5 \text{ ft/s (71 mph)} \tag{2.26}$$

Note that the masses of the vehicle and the rider are not required for this calculation. Vehicle accident reconstruction is an important area of litigation in the United States, and commercial software has been developed specifically for this application. The issue of roadway friction against tires will be considered in more detail in Chapter 9.

2.3.4   CASE 4. ANGLE OF BANK TO PREVENT SLIDING OF AN AUTOMOBILE
ON A CURVE UNDER WET OR DRY CONDITIONS

With what angle should a road be banked such that there would be no friction force perpendicular to the direction of motion of an automobile on the curve? Assume the situation shown in Figure 2.5.

The force normal to the plane of the road is $\mathbf{F}_n$. This force can be resolved into a vertical component, $\mathbf{F}_n \cos \theta$, and a horizontal component, $\mathbf{F}_n \sin \theta$. The vertical component

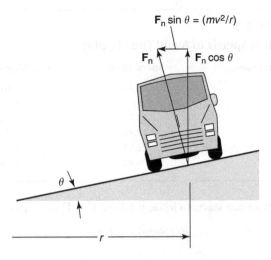

FIGURE 2.5 Forces on a vehicle traversing a banked curve in the road.

will just equal the weight of the vehicle (mg). Solve for $\mathbf{F}_n$ in terms of the angle of the road's slope: $\mathbf{F}_n = (mg/\cos\theta)$. The horizontal component must balance the centripetal force (mv^2/r). Let r be the radius of the curve and v the velocity of the car. Then

$$\mathbf{F}_n \sin\theta = \frac{mv^2}{r} \qquad (2.27)$$

Since $(\sin\theta/\cos\theta) = \tan\theta$,

$$\tan\theta = \frac{v^2}{rg} \qquad (2.28)$$

Note that the bank angle is independent of the weight of the vehicle and its selection depends on the maximum anticipated speed on the curve.

2.3.5 CASE 5. FRICTION COEFFICIENT REQUIRED TO AVOID SLIDING ON AN UNBANKED CURVE IN THE ROAD

What friction coefficient will just keep a 1600 kg motor vehicle from sliding off an unbanked curve at a given velocity? The radius of the curve is 75 m.

To keep the vehicle moving in a circle,

$$\mathbf{F} = \frac{mv^2}{r} = \frac{1600v^2}{75} = 21.33v^2 \qquad (2.29)$$

and the normal force on the road is

$$\mathbf{F}_n = \mathbf{W} = mg = (1600)(9.81) = 15.696\ \text{N} \qquad (2.30)$$

The friction on the tires is

$$\mathbf{F}_{\text{friction}} = \mu_s \mathbf{W} \qquad (2.31)$$

TABLE 2.2
Critical Speeds to Avoid Tire Slipping

Speed (mph)	Speed (m/s)	μ (Required)
10	4.47	0.03
20	8.94	0.11
30	13.41	0.24
40	17.88	0.43
50	22.35	0.68
60	26.82	0.98

so at the point where one starts to loose traction, $\mathbf{F} = \mathbf{F}_{\text{friction}}$, and

$$\mu_s(15696) = 21.33v^2 \tag{2.32}$$

or

$$\mu_s = 0.001359v^2$$

Using the preceding values, one can make a table of velocity versus required friction coefficient to avoid slipping (Table 2.2).

Reported values for the friction of rubber on dry pavement are in the range $\mu \sim 0.6$–0.7 and for wet pavement 0.15–0.20. Thus, one should keep the speed below or about 25 mph on wet days and below 40 mph at any time.

The foregoing illustrations are typical of those given in introductory physics or mechanics treatments of friction. Here, the friction coefficient was assumed to have a single, constant value. Section 2.4 extends this treatment to simple machine elements. In another section of in the chapter, the complications of rolling will be introduced.

2.4   FRICTION IN SIMPLE MACHINE COMPONENTS

In this section, several common types of machine components involving static friction and sliding contact are examined:

1. Wedge-based mechanisms
2. Pivots, collars, and disks
3. Belts and ropes
4. Screws
5. Journal bearings and pulleys

Treatments of other types of components may be found in the references provided at the end of the chapter.

2.4.1   WEDGE-BASED MECHANISMS

Many types of machines contain wedge-like moving parts. A basic wedge system is illustrated in Figure 2.6. A weight **W** is being raised against a flat wall by the application of a force **P** to a wedge of angle θ. We want to know the minimum force

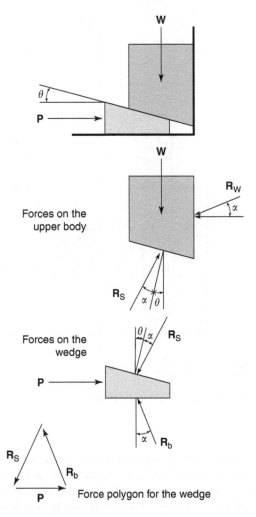

Forces on the upper body

Forces on the wedge

Force polygon for the wedge

FIGURE 2.6 Forces on a wedge system.

required to raise the upper body. We shall assume that the weight of the wedge can be neglected in this case (it is straightforward to include the weight of the wedge, if necessary), and that the static friction coefficients against the base and against the wall are both equal to μ_s. As shown at the bottom of the figure, two free-body diagrams can be drawn. From the previous section, the tangent of the friction angle α equals μ_s. Recognizing that the downward friction force, opposing the motion of the upper body up the wall, is $\mathbf{R_W} \sin \alpha$ and stating the balance of forces in the y direction for the upper body gives

$$\mathbf{W} + \mathbf{R_W} \sin \alpha = \mathbf{R_S} \sin(\alpha + \theta) \tag{2.33}$$

Similarly, balancing forces in the x direction,

$$\mathbf{R_S} \sin(\alpha + \theta) = \mathbf{R_W} \cos \alpha \tag{2.34}$$

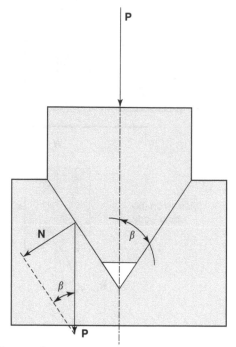

$F_{\text{V-groove}}$ is normal to
the plane of the figure

FIGURE 2.7   Normal forces on a V-guide. Friction forces act perpendicular to the plane of
the figure.

If we know the wedge angle, weight of the upper body, and friction angle, we can
solve the two equations in two unknowns to determine $\mathbf{R}_S$ and $\mathbf{R}_W$. Knowing the value
of $\mathbf{R}_S$ and using the law of sines permits solution of the force polygon for the wedge
(Figure 2.6, at the lower left). The angle between $\mathbf{R}_S$ and $\mathbf{R}_b$ is $[\alpha + (\alpha + \theta)]$ or $(2\alpha + \theta)$,
and the angle between $\mathbf{R}_b$ and $\mathbf{P}$ is $(90 - \alpha)$. Therefore, from the law of sines,

$$\frac{P}{\sin(2\alpha + \theta)} = \frac{R_S}{\sin(90 - \alpha)} \qquad (2.35)$$

and

$$P = R_S \frac{\sin(2\alpha + \theta)}{\sin(90 - \alpha)} \qquad (2.36)$$

For example, given a 200 kg mass ($\mathbf{W} = 1962$ N), $\mu_s = 0.45$ (i.e., $\alpha = 24.2°$), and
$\theta = 5°$, we find $\mathbf{R}_S = 2922.6$ N, and $\mathbf{P} = 2922.6$ [sin(53.4)/sin(65.8)] = 2573 N. Note
that the force required to raise the mass using the wedge is about 31% greater than
that required to lift the mass straight up.

One machine component based on the wedge is the V-guide (Figure 2.7). For a
symmetric slot, the force normal to the wedge faces is $\mathbf{N} = \mathbf{P} \sin \beta$. Thus, the friction

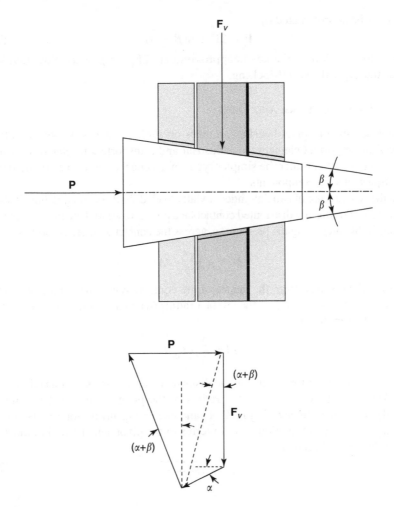

FIGURE 2.8 Forces on a taper key.

force opposing sliding along the axis of the wedge, assuming no elastic seating of the wedge into the guide groove, is

$$\mathbf{F}_{\text{V-groove}} = \mu_s \mathbf{P} \sin \beta \tag{2.37}$$

Another adaptation of the wedge is the *taper key* (Figure 2.8). Its analysis is slightly more complicated. One begins with the analysis similar to that for the simple wedge, replacing the weight with the downward force of the inner slide. Solution of the force polygon (bottom of Figure 2.8) permits calculating the force **P**, in terms of the vertical force $\mathbf{F}_v$, taper angle β, and the friction angle $\alpha = \tan^{-1} \mu_s$:

$$\mathbf{P} = 2\mathbf{F}_v \frac{\cos\alpha \sin(\beta + \alpha)}{\cos(\beta + 2\alpha)} \tag{2.38}$$

which can be approximated by

$$P = 2F_v \tan(\beta + \alpha) \tag{2.39}$$

The force to loosen the key is approximately $2F_v \tan(\beta - \alpha)$. Note that when $\beta < \alpha$, the key will be self-locking.

2.4.2 PIVOTS, COLLARS, AND DISKS

Components such as pivot bearings, flanges on shafts, clutches, brakes, and thrust washers are important elements of industrial and transportation-related machinery. In this section, we consider the simpler types of approaches used to estimate friction and torques on such components.

In the general case of two flat-ended, axially loaded shafts of equal radius, R, placed together under uniformly distributed contact face pressure, we find the moment of thrust friction, M, by integrating the pressure, p, across the contact area, A, as follows:

$$M = \mu_s p \int_0^R r\,dA \tag{2.40}$$

where r is the distance from the center of the contact. When the contact is circular, we integrate from $r = 0$ to $r =$ the contact radius (R), and the moment (M) required for impending motion is

$$M = \frac{2}{3}\mu_s F_{th} R \tag{2.41}$$

where F_{th} is the thrust force holding the disks together. In the case of a right-circular pivot bearing (Figure 2.9a), R is the radius of the pivot. If the pivot bearing is a tapered cone of apex angle β (Figure 2.9b) and neglecting any minor effects of the tip sharpness, the area in Equation 2.40, in terms of the pivot height h and radius of the base of the cone r becomes

$$A = \pi\sqrt{h^2 + r^2} \tag{2.42}$$

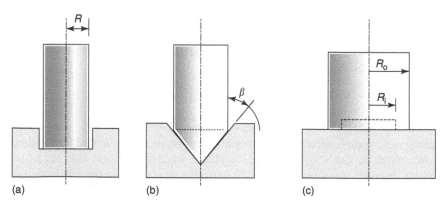

FIGURE 2.9 Types of pivots (a) flat-ended, (b) taper-ended, and (c) annular. A small delivery tube leading to the confined area inside the annulus can be used to admit pressurized oil to improve the pivot bearing's performance.

and integrating for r going from 0 to R,

$$M = \frac{2}{3}\mu_s PR \sin \beta \tag{2.43}$$

Pivot bearings may also be constructed as shown in Figure 2.9c. The contact surface is an annular ring with inner radius R_i and outer radius R_o. Equation 2.43 then becomes

$$M = \frac{2}{3}\mu_s \mathbf{F}_{th} \frac{R_o^3 - R_i^3}{R_o^2 - R_i^2} \tag{2.44}$$

Note that Equation 2.44 can also be used to estimate the torque on a rotating annular pivot bearing by replacing the static friction coefficient with the kinetic friction coefficient.

The friction coefficient in such bearings can be significantly lowered if an oil film is present. Pressurization of the oil film either by external means or by special design of the bearing reduces the friction still further. A Kingsbury thrust bearing uses a special design to create a pressurized oil film during rotation. If the annular pivot bearing is pressurized by an external oil supply, the moment can be calculated in terms of the oil film thickness in the ring h (inches), the lubricant viscosity Z (centipoises), and the rotational speed in revolutions per minute (rpm) as follows:

$$M = (2.388 \times 10^{-8})Z(R_o^4 - R_i^4) \tag{2.45}$$

where the radii are given in inches.

The effective kinetic friction coefficient in pressurized pivot bearings can range from about 0.15 to 0.001 depending on the state of lubrication and the speed of rotation.

2.4.3 BELTS AND ROPES

Consider a belt or rope rubbing against a fixed, circular cylinder, as shown in Figure 2.10a. The relationship between the tension on the belt on both right and left sides is determined by the fact that the tractions vary from a minimum at the ends of the arc of contact to a maximum at the center. If a force, equal to the tension on the left side, is applied, and the weight hanging on the right just tends to move (impending motion), then the ratio of the tension on the left $\mathbf{T}_1$ to that on the right $\mathbf{T}_2$ is

$$\frac{\mathbf{T}_1}{\mathbf{T}_2} = e^{\mu\beta/57.3} \tag{2.46}$$

where e is the base of natural logarithms, μ is the friction coefficient between the belt and the cylinder surface, and β is the angle of arc (in degrees). The same equation can be written with β in radians by omitting the division by 57.3. Note that $\mathbf{T}_1$ is always greater than $\mathbf{T}_2$ when friction is finite because the tension on the left must overcome not only gravity but the friction of the belt as well.

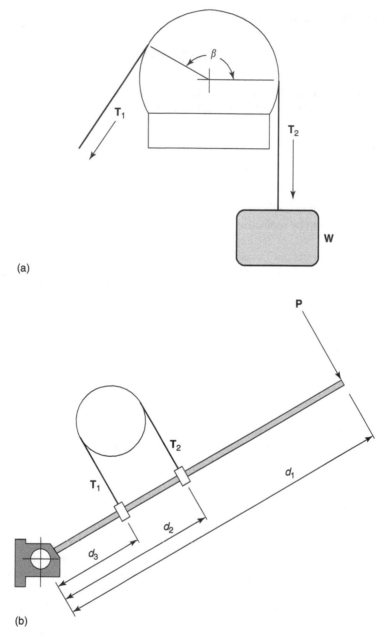

FIGURE 2.10   Friction of belting against cylindrical shafts: (a) general case and (b) simple band brake.

Common applications of belt friction are belt drives that transmit power to a shaft or belt brakes (band brakes) that serve to slow the rotation of a shaft due to the frictional drag. An example of the latter is shown in Figure 2.10b. Analysis of this system is relatively straightforward: (a) sum moments about the hinge and

TABLE 2.3
Friction of Belting Materials and Ropes

Material Combination	μ
Leather belt on a cast iron pulley	
Moist	0.38
Slightly greasy	0.28
Very greasy	0.12
Leather belt on a wood pulley	
Slightly greasy	0.47
Hemp ropes	
On a cast iron drum	0.25
On a wood drum, polished	0.33
On a wood drum, rough	0.50

Source: Adapted from Kragelsky, I.V., Dobychin, M.N., and Kombalov, V.S. in *Friction and Wear Calculation Methods*, Pergamon Press, 1982, 219.

set equal to zero, (b) determine the ratio of the tensions on each side of the shaft, and (c) solve simultaneously. Thus,

$$Pd_1 = \mathbf{T}_2 d_2 + \mathbf{T}_1 d_3 \tag{2.47}$$

so we can use Equation 2.47 to solve for the maximum tension on the belting ($\mathbf{T}_1$) in terms of the known or measurable quantities in the system:

$$\mathbf{T}_1 = \frac{Pd_2}{d_3 + d_2 / e^{\mu\beta/57.3}} \tag{2.48}$$

Typical values for the friction of ropes and belting are given in Table 2.3.

The friction between the belt or rope and the shaft is only one aspect of such problems. There is also friction internally in ropes, belts, and chains. These considerations are very important in the design and efficiency of hoists, pulleys, and drives.

2.4.4 SCREWS

Friction in screws can be analyzed more easily when it is recognized that a screw thread is an inclined plane wrapped around a cylinder such that for each 360° of rotation, the nut advances by a distance h. Letting r be the mean radius of the thread from the center of the cylinder, then the length of the incline per 360° rotation is related to the rise, h, by $\tan \theta = (h/2\pi r)$. Here, h is sometimes called the *pitch* of the screw, and consequently, θ is defined as the *pitch angle* of the threads. The two common kinds of screws are screws with square threads—used primarily in power drives (*power screws*), and screws with V-threads—used primarily for fastening.

The simplest analysis of square threads used for driving the shaft (e.g., in a jack screw) assumes that any moment, due to torque on the screw, acts as a horizontal force $\mathbf{F}_h$ at a distance r, defined earlier and the moment is just $M = \mathbf{F}_h r$. To effect

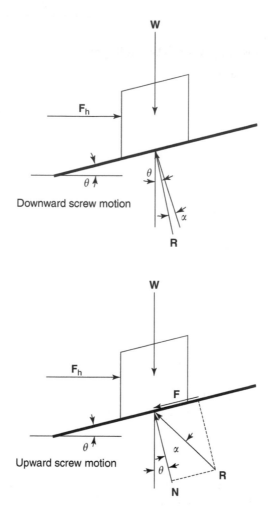

FIGURE 2.11　Force diagram for a screw with friction in its threads.

a force along the axis of the screw, that is, to raise a weight or perform similar work, we can assume that the reactive forces to the axial force and the torque are distributed along the incline of the thread for the entire length so that the thread passes through the supporting nut. Figure 2.11 illustrates the force diagrams for upward screw motion, as if to raise a weight, and downward motion, as if to lower a weight. $\mathbf{F}_h$ acts horizontally on a slope of pitch and angle θ, and $\mathbf{W}$ is the weight acting perpendicular to it. This force situation is similar in some ways to the analysis of wedges given previously in this chapter. Friction force $\mathbf{F}$ opposes the motion of the screw by acting down the slope. The resultant $\mathbf{R}$ is composed of force vectors acting normal to and along the slope. In the instance of impending motion, a balance of forces in horizontal and vertical directions can be written as follows:

$$\text{Horizontal balance of forces: } \mathbf{F}_h = \mathbf{R} \sin(\theta + \alpha) \qquad (2.49)$$

$$\text{Vertical balance of forces: } \mathbf{W} = \mathbf{R} \cos(\theta + \alpha) \qquad (2.50)$$

From $M = \mathbf{F}_h r$, and Equations 2.49 and 2.50,

$$\text{Upward motion: } M = \mathbf{W}r \tan(\theta + \alpha) \qquad (2.51)$$

When the screw is turned in the opposite direction, say to lower the weight, the direction of the moment M is reversed, and the friction angle α lies on the other side of the normal. Solving as done earlier,

$$\text{Downward motion: } M = \mathbf{W}r \tan(\theta - \alpha) \qquad (2.52)$$

Note that if $\theta < \alpha$, then the screw will be self-locking. If not, a screw lock of some kind will be needed to prevent the screw from rotating downward. This situation can be a problem in vibrating machinery in which θ is about equal to α and the vibration supplies just enough extra horizontal force to rotate the screw.

For a V-grooved screw and nut, the moments (torque) required to tighten or loosen a nut under an axial force $\mathbf{A}$, as given in *Marks' Standard Handbook for Mechanical Engineers*,[1] are

$$\text{To tighten: } M = \mathbf{A}r[\tan(\delta' + \theta) + 1.5\mu_{ns}] \qquad (2.53)$$

$$\text{To loosen: } M = \mathbf{A}r[\tan(\delta' - \theta) + 1.5\mu_{ns}] \qquad (2.54)$$

where μ_{ns} is the friction coefficient between the nut and thread (typical range 0.03–0.25, depending on the state of lubrication), θ is the pitch angle as defined previously r is the mean radius of the threads, and δ' is defined in terms of δ, the angle between the plane tangent to the thread face and the (horizontal) plane perpendicular to the screw axis at the same location. That is,

$$\delta' = \tan^{-1}(\mu_{ns} \sec \delta) \qquad (2.55)$$

2.4.5 SHAFTS AND JOURNAL BEARINGS

Journal bearings containing shafts are among the most common types of rotary motion bearings. A simple analysis considers the friction acting at a point on the circumference of the shaft turning within the bearing (Figure 2.12). The moment required to maintain motion in a horizontal shaft of radius r and exert a downward load $\mathbf{W}$ is found simply by summing moments about the central axis of the shaft,

$$M = \mathbf{R}r \sin \theta \qquad (2.56)$$

where $\mathbf{R}$ is the reactive force acting at the contact point of the shaft in the journal bore. A *friction circle* can be defined as the *minor circle* of radius r_f to which the reactive force $\mathbf{R}$ is always tangent. The friction force $\mathbf{F}$ is tangent to the shaft outside diameter. For well-lubricated bearings, $M = \mathbf{R}r\mu$, because at small angles, $\sin \alpha \approx \tan \alpha \approx \alpha$, the kinetic friction angle.

Pulleys on shafts are one case where the previously mentioned analysis can be applied. A pulley with good belt traction and of diameter D is shown in Figure 2.13. The free-body diagram shows the tension $\mathbf{T}$ pulling downward at the right to raise a load $\mathbf{L}$ at the left. As $\mathbf{T}$ increases and motion impends, the point of contact of the

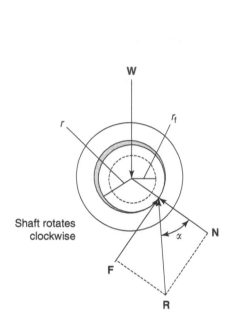

FIGURE 2.12   Forces on a simple journal.

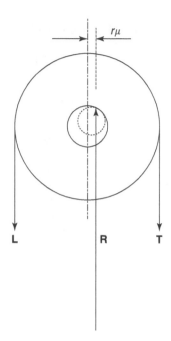

FIGURE 2.13   Forces on a pulley.

pulley internal diameter with the shaft rotates counterclockwise by an angle ϕ. The radius of the friction circle when motion impends is $r \sin \phi$. But at small angles $\sin \phi \approx \phi \approx \mu$. Therefore, the radius of the friction circle in terms of the radius of the shaft, r is $r\mu$, and the origin of the system about which to sum moments is displaced to the left of the vertical line that passes through the shaft center by a distance $r\mu$. Therefore, the minimum tension in the belt to overcome the load **L** at constant velocity is found from

$$\mathbf{T}(D + r\mu) = \mathbf{L}(D - r\mu) \tag{2.57}$$

and

$$\mathbf{T} = \mathbf{L}\frac{D + r\mu}{D - r\mu} \tag{2.58}$$

2.5   ROLLING FRICTION

Friction during rolling is important for applications such as the performance of rolling element bearings and the rolling of sheet products. If a cylinder of radius R rolls along a stationary base such that when it rolls through an angle ψ the axis of the cylinder is displaced relative to the base by an amount equal to $R\psi$, then pure rolling is said to exist. In pure rolling, the point of contact of the cylinder on the base plane remains stationary (Figure 2.14), and the axis passing through that point and perpendicular to the base plane is called the instantaneous axis of rotation. If the

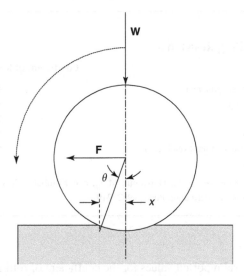

FIGURE 2.14 Simplified analysis of rolling friction forces.

cylinder and plane were both perfectly rigid and there was no friction from the air, the cylinder would continue to roll indefinitely. However, as shown in Figure 2.14, there is not a single point of contact but rather a distribution of the contact pressure along a slight ridge that is developed at a location toward the direction of travel (toward the left in the present case).

A driving force **F** must be applied to maintain motion. The pressure distribution on the buckled section of the surface can be represented by a resultant, normal force **N**, acting at a point offset a distance x from the perpendicular through the center of the cylinder. If we assume force equilibrium during continuous rolling motion, we can sum moments about that point and find that

$$\mathbf{W}x = \mathbf{F}(R \cos \theta) \tag{2.59}$$

However, when $R \gg x$, Equation 2.59 can be approximated, $\mathbf{W}x \sim \mathbf{F}R$. The distance x is often defined as the *coefficient of rolling resistance*:

$$x = \frac{\mathbf{F}}{\mathbf{W}} R \cos \theta \approx \frac{\mathbf{F}R}{\mathbf{W}} \tag{2.60}$$

Note that this coefficient is not dimensionless but is given in units of length. Review of the literature suggests that there is no universally agreed upon symbol for the coefficient of rolling resistance. Therefore, the symbol used for the coefficient of rolling resistance must be determined from the context of the work.

Consider a cylinder of known weight lying at rest on a flat board. If the board is lifted at one end and at an angle of θ degrees the cylinder begins to roll down the incline, then Equation 2.60 can be used to compute the coefficient of rolling resistance.

TABLE 2.4
Coefficients of Rolling Resistance

Material Combination	Coefficient of Rolling Resistance (mm)
Hard, polished steel on hard, polished steel	0.008
Well-oiled surfaces (average)	0.038
Steel on steel (typical)	0.051
Rusty surfaces (average)	0.190
Properly inflated rubber tires on hard road surface	0.203
Hardwood on hardwood	0.508

Source: Adapted from Kragelsky, I.V., Dobychin, M.N., and Kombalov, V.S. in *Friction and Wear Calculation Methods*, Pergamon Press, 1982, 219.

Table 2.4 shows several reported values for the coefficient of rolling resistance. Additional data are available from Ref. 2.

The type of rolling that occurs in most rolling element bearings is not pure. In fact, it is usually accompanied by a degree of slip. A more detailed analysis of the latter case is also provided by Kragelsky et al.,[2] and the following example will illustrate that situation.

When a ball rolls down a conforming, rectilinear, circular groove (Figure 2.15), it rotates around the central axis such that the various points on the curved contact area are at different distances from this axis. Since the contact area moves along the groove while rotation occurs, some slip must occur within that area. It has been shown that there are only two points, symmetrically placed with regard to the centerline of the groove, at which pure rolling occurs. Between these points—slip occurs opposite to the direction of ball motion and exterior to these points—slip occurs in the opposite direction. Figure 2.15 indicates this. The resistance to rolling due to slip is a function of the relative size of the radii of the ball R_b and the groove R_g. Let d be the groove width, N the normal force, and f the rolling resistance, then the moment M_r about the axis of rotation is

$$M_r = 0.324 \frac{fNd^2}{8\rho} \tag{2.61}$$

where

$$\rho^{-1} = 0.5\left[\frac{1}{R_b} + \frac{1}{R_g}\right] \tag{2.62}$$

It has been established that when the radii of curvature of the ball and groove differ by less than about 20%, the rolling resistance due to slip predominates over hysteresis losses.

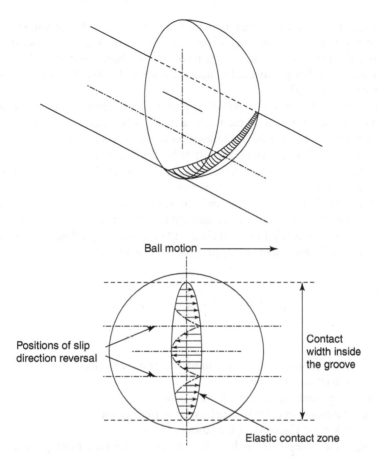

FIGURE 2.15   Spherical rolling element in a groove.

2.6   FRICTION IN GEARS

Spur gears operate with a combination of slip and rolling contact, depending on the design of the gears and the specific location on the gear teeth during their engagement. Friction between surfaces is one of the main sources of power loss in gears, so gear designers consider the frictional implications as they compute tooth profiles. Pure rolling is said to occur only at a certain orientation between the meshing gear teeth, called the *pitch point*. On either side of this point, the slip increases. In general, the percentage of slip (S) between surfaces having surface velocities of v_1 and v_2 can be calculated from the following:

$$S = 100\frac{2|v_1 - v_2|}{|v_1 + v_2|} \tag{2.63}$$

Without the factor of 100, this expression is also known as the *slide-to-roll ratio*. Experiments in rolling contact are commonly conducted using stacked disks whose spindle speeds can be independently adjusted to produce different values of S. For example, the kinetic friction coefficient using two hardened steel disks rolling on one another[3] remained relatively constant ($\mu \sim 0.04$) above a slide-to-roll ratio of 0.2 (or less than −0.2) but plummeted to near zero when the slide-to-roll ratios were in the range 0.03 to −0.03.

The design and operating conditions of gears also determine the regime of lubrication between the surfaces of the teeth, a subject discussed in Chapter 6. A more detailed discussion of gear types, nomenclature, wear modes, and regimes of lubrication can be found in the handbook article by Dudley.[4] Nevertheless, it is useful to consider the factors that enter into rolling plus sliding analyses by considering one expression that has been used to calculate the friction coefficient between gear teeth. It involves not only the dimensions and geometry of the gears but also their surface finish and the characteristics of the lubricant. Hohn et al.[5] developed the following expression for the friction coefficient (subscripted M to acknowledge coauthor Michaelis), which was used to analyze carburized steel gears:

$$\mu_M = 0.048 \left(\frac{F_{bt}/b}{v_{\Sigma c}\rho_c} \right)^{0.2} \eta^{-0.05} R_a^{0.25} X_L X_{SC} \qquad (2.64)$$

where

$\quad F_{bt}$ = the normal force on the tooth in a transverse section
$\quad b$ = width of the gear
$\quad v_{\Sigma c}$ = the sum velocity at the pitch point
$\quad \rho_c$ = equivalent curvature radius at the pitch point
$\quad \eta$ = kinetic viscosity of the lubricating oil
$\quad R_a$ = the composite surface roughness of the two mating surfaces (see Chapter 6)
$\quad X_L$ = a correction factor for the additives in the oil (X_L = 1 for mineral oils)
$\quad X_C$ = a correction factor to account for the presence of a coating on the gear(s) (X_C = 1 for noncoated gears)

Unlike the previous relationships presented in this chapter that were derived primarily from mechanics, Equation 2.64 reflects a more empirical approach to account for the role of surface finish, lubricant viscosity, and the possible use of coatings. In general, the more rigorous the attempt to model the friction of a practical mechanical system, the larger the number of variables that must be included. Unfortunately, the adjustable "correction factors" that are so prevalent in engineering can be difficult if not impossible to determine from first principles and, therefore, must be measured or inferred to enable the model to fit the data. Wear-induced changes in the contact surfaces also affect friction, a subject further discussed in Chapter 4.

In this chapter, we considered traditional mechanics treatments of the following friction problems: simple sliding friction, tipping versus slipping, developing force diagrams, simple machines, and rolling friction. Experimental methods for measuring friction coefficients in the laboratory are discussed in Chapter 3.

Such information, if properly used, can facilitate the design and analysis of machines, and measurements often show that the friction coefficient in practical devices is not necessarily constant over time.

FURTHER READING

E. A. Avallone and T. Baumeister III, eds. (1987). *Marks' Standard Handbook for Mechanical Engineers*, 9th ed., McGraw-Hill, New York, NY (Section 3.2 by D. Fuller).

D. C. Giancoli (1991). *Physics—Principles with Applications*, 3rd ed., Prentice-Hall, Englewood Cliffs, NJ (Sections 4–9).

R. C. Hibbeler (1983). *Engineering Mechanics—Statics and Dynamics,* 3rd ed., Macmillan, New York, NY (Chapter 8).

M. M. Khonsari and E. R. Booser (2001). *Applied Tribology: Bearing Design and Lubrication*, Wiley, Oxford, UK.

I. J. Levinson (1971). *Statics and Strength of Materials*, Prentice-Hall, Englewood Cliffs, NJ (Chapter 6).

REFERENCES

1. E. A. Avallone and T. Baumeister III, eds. (1987). *Marks' Standard Handbook for Mechanical Engineers*, 9th ed., McGraw-Hill, New York, NY.
2. I. V. Kragelsky, M. N. Dobychin, and V. S. Kombalov (1982). *Friction and Wear Calculation Methods,* Pergamon Press, 219 pp.
3. J. Kleemola and A. Lehtovaara (2007). Experimental evaluation of friction between contacting discs for the simulation of gear contact, *Tribotest*, 13, pp. 13–20.
4. D. W. Dudley (1980). Gear Wear, in *Wear Control Handbook*, eds. W. O. Winer and M. B. Peterson, ASME, New York, NY, pp. 755–830.
5. B.-R. Hohn, K. Michaelis, and T. Vollmer (1966). *Thermal Rating of Gear Drives: Balance Between Power Loss and Heat Dissipation*, AGMA Technical paper.

3 Measuring Friction in the Laboratory

In principle, determining friction coefficients seems straightforward: one simply measures the normal force or weight of an object and the force resisting relative tangential motion between it and a counterface. Investigators have measured frictional quantities for centuries by using ramps, pulleys, spring scales, and similar devices. Yet, basic researchers who envision friction as the net result of nanoscale interaction, and engineers who attempt to simulate the frictional environment within specific device employ more sophisticated and elaborate techniques. Some employ strain gauges, deflecting beams, and piezoelectric force sensors, and others have devised ways to infer the magnitude of friction indirectly.

Direct friction measurements on mating parts buried deep inside machines can be impractical or expensive; therefore, friction data are obtained in the laboratory using instruments called tribometers. A variety of laboratory friction measurement methods are described in this chapter. Some techniques use small coupons, others use prototypical parts, and still others use full-sized components. Numerous standards for friction testing have been developed; however, customized nonstandard methods are also useful. Both are described in this chapter.

Selecting a friction test method begins with the following considerations in mind:

1. Understand the purpose for which the data are needed.
2. Be aware of the strengths and limitations of any potential test methods.
3. Recognize that friction is a characteristic of the tribosystem and not a fundamental material property. Therefore, a given set of materials can rank in a different order when friction is tested in different kinds of tribometers.

3.1 CLASSIFICATION OF TRIBOMETERS

The 1989 edition of the *Oxford English Dictionary* defined a tribometer as an *instrument for estimating sliding friction*. It is interesting that the verb *estimating* rather than *determining* or *measuring* is used, and that the term *friction* is used in the qualitative sense (cf. friction coefficient or friction force). The earliest reference provided by that dictionary is to the 1774 writings of Goldsmith, who used the word *tribometre* to mean a "measurer of friction." In 1877, Knight defined it as *an apparatus resembling a sled, used in estimating the friction of rubbing surfaces.*

Laboratory techniques for measuring friction are developed and used for a variety of purposes, including the following:

- To simulate the tribocontact situation in a particular machine
- To evaluate candidate-bearing materials for a friction-critical application

- To evaluate lubricants for a particular application
- To qualify lubricants for use on the basis of established criteria
- To monitor surface contamination on a product
- To acquire nontribosystem-specific (generic) friction data as a means to compare and develop new materials, coatings, or lubricants
- To investigate the fundamental nature of friction of solids or lubricated solids

The purpose for testing guides the selection of equipment and procedures. Unfortunately, and to the chagrin of new entrants to the field, a host of variables can affect friction. These include the normal force, velocity and acceleration characteristics, direction of motion relative to surface features, system stiffness, surface cleanliness and roughness, contact temperature (both ambient and that produced by frictional heating), relative humidity, lubricant properties, presence of loose particles, and more.

As will be apparent from the discussions in Chapters 4 and 7, the relative importance of each variable on test results depends on the materials, the test configuration, and the applied conditions. Voluntary and government-sponsored organizations the world over have developed Guides, standards, and recommended practices to help engineers and scientists systematize the measurement of friction-related data. Standards typically identify which variables should be controlled and which are allowed to vary (by exception), how data are to be collected, and how they are reported. However, simply because a certain variable is not explicitly mentioned in a standard does not mean that it cannot affect friction under another set of circumstances.

Since no laboratory friction test can stimulate all possible tribosystems, handbook compilations of friction coefficients should be used with care. As the eminent physicist Richard Feynman pointed out in his lectures on physics:[1] "The tables that list purported values of the friction coefficient of 'steel on steel' or 'copper on copper' and the like, are all false, because they ignore the factors which really determine μ." One of the factors that "really determine μ" is the means by which it is measured.

The German DIN Standard 50 322[2] defines size scales and degrees of complexity for wear tests, but the underlying concept can apply equally well to friction tests. The DIN categories for wear testing are as follows:

   I. Full-size field trials (e.g., a flat-bed truck)
  II. Tests of the actual component but on a test stand (e.g., a truck with its wheels on a dynamometer stand)
 III. Tests of the subassembly on a test stand (e.g., removal and testing of a motored transmission)
  IV. Tests of a smaller-scale version of the assembly (e.g., testing a downsized version of the transmission)
   V. Tests of triboelements (e.g., a gear-testing machine that uses actual gears)
  VI. Tests involving the basic geometry of contact (e.g., ball-on-flat, roller-on-roller.)

According to this structure, "laboratory tests" could fall under any category except the first. Hogmark and Jacobson[3] followed Czichos' scheme for tribotesting and discussed systematic methods for selecting the proper type of test, cautions in regard to accelerated testing, and ways of using a scanning electron microscope to analyze the effects of surface contact.

In 1976, the Subcommittee on Wear of the American Society of Lubrication Engineers (now called the Society of Tribologists and Lubrication Engineers [STLE]) published the results of an extensive survey of friction and wear devices.[4] The publication contained 234 total responses that were grouped into categories based on the macrogeometry of contact, namely:

1. Multiple sphere
2. Crossed cylinders
3. Pin-on-flat (reciprocating or linear motion)
4. Flat-on-flat (reciprocating or linear motion)
5. Rotating pins-on-disk (face loaded)
6. Pin-on-rotating disk (face loaded)
7. Cylinder-on-cylinder (face loaded)
8. Cylinder or pin-on-rotating cylinder (edge loaded)
9. Rectangular flat-on-rotating cylinder (edge loaded)
10. Disk-on-disk (edge loaded)
11. Multiple specimens
12. Miscellaneous

Many of the devices identified in the ASLE study were designed to mimic the contact conditions in specific types of bearings, gears, drives, bearings, bushings, and mechanical seals, but most of them fell into categories V or VI of DIN Standard 50 322. Each entry in the compilation describes the range of operating parameters for each of the testing devices, typical materials tested, and several references that illustrate how the machine was used in specific cases.

Despite the widespread availability of commercial tribometers, investigators continue to design and build their own tribometers. Reasons for this include the need to provide certain testing conditions, specific component simulations, or specimen dimensions that are not readily available. Some researchers find the cost of commercial equipment prohibitive, do not care for the design features of commercial units, or may have a need to obtain only a few data. It also might be more cost-effective to contract testing services than acquire a tribometer. In that case, it is important to select appropriate test methods. A commercially motivated vendor may promote his or her equipment even though it does not suit the need for specific kinds of data as well as it should.

Another way to classify tribometers is based on the macrogeometric conformity of the contact. *Conformal surfaces* fit together such that the apparent area of contact does not appreciably change as wear occurs. Examples include a flat block sliding on a flat plate and a curved brake shoe sliding against the internal surface of a brake drum. By contrast, *nonconformal surfaces* have been defined[5] as surfaces whose centers of curvature are on opposite sides of the interface, as in rolling element bearings or spur gear teeth. Other examples include test geometries such as the flat-faced block-on-rotating ring, hemispherical pin-on-disk, and the Shell four-ball test to characterize lubricants.

Three advantages of conformal tests are as follows: (1) the nominal area of contact (hence, the nominal pressure under constant load) does not change as a result of wear, (2) the pressure distribution tends to be more uniform than in nonconformal tests, and (3) the state of lubrication can be better controlled because the film thickness tends to be constant over a wider area. One of the main disadvantages of conformal contact testing is that it can be difficult to align the specimen surfaces in some cases. This is especially true for flat-ended pin-on-disk tests. Even small misalignments between sliding surfaces (much less than a degree of parallelism) can result in the digging-in of the leading edge of the slider, uneven distribution of contact pressures, and uneven trapping of wear debris or lubricants. Uneven wear can alter a conformal contact's alignment sufficiently to create problems. The edges of flat specimens can sometimes be chamfered, rounded, or beveled to prevent digging-in.

A major advantage of nonconformal tests, such as the rounded-end pin-on-disk test, is that the contact is a point of tangency such that precise parallelism and coplanar alignment of the fixtures or prepared surfaces are not required. However, contact stresses in nonconformal tests tend to decrease nonlinearly in the region of contact as wear enlarges the nominal area of contact. Figure 3.1 shows how the nominal contact pressure (normal force per unit apparent contact area) changes as a function of volumetric wear loss from two nonconformal specimen geometries: a 5.0 mm wide flat block sliding on a 50 mm diameter ring at 10.0 N normal force, and a 10.0 mm diameter sphere sliding on a flat disk at the same normal force, assuming that there is negligible wear of the disk and ring specimens. The very high initial apparent contact pressure falls off faster for the pin-on-disk geometry than for the block-on-ring

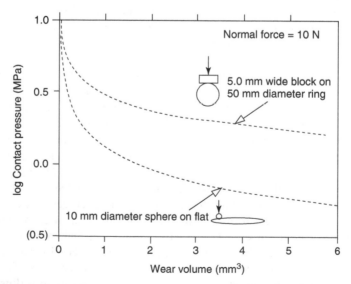

FIGURE 3.1 A rapid decrease in the nominal contact pressure occurs as nonconformal contacts experience wear.

because it starts with a "point contact" rather than a "line contact," respectively. However, as a consequence of the geometry, at comparable specimen wear volumes, the nominal contact pressure is higher on the block specimen. Note that the rate of change of contact pressure with increasing wear becomes similar for the two specimen geometries at higher wear volumes. Running-in transients in friction and wear behavior are therefore a consideration in nonconformal tests.

Some kinds of laboratory-scale friction tests begin with nonconformal contact and end with conformal contact due to wear. For example, an initially flat-ended pin pressed against a rotating ring may soon wear to conform to the curvature of the ring. C. S. Yust, at Oak Ridge National Laboratory (ORNL), favored running-in a spherically ended pin (in pin-on-disk tests) to reduce the initially high contact pressure by wear. Then he measured the contact diameter of the pin and adjusted the load to obtain the desired nominal contact pressure. Friction and wear were then determined for increments of sliding, the normal force being adjusted at the end of each interval based on the measured pin wear scar diameter.

Voitik[6] developed a four-digit numerical coding scheme to systematize tribocontact geometry. Matching the tribological aspect number (TAN) of an application to that of the testing machine would, he suggested, result in better laboratory simulations. Table 3.1 summarizes the categories that comprise TAN codes, and Table 3.2 provides a few examples to compare TAN codes for engineering components with common laboratory test configurations.

Although TAN codes are a convenient way to categorize contact geometry, they do not include any information about materials, temperatures, or the tribochemical environment. Therefore, additional information beyond the TAN code is needed to characterize a tribosystem.

TABLE 3.1
Tribological Aspect Numbers

First Digit (Contact Velocity)	Second Digit (Contact Area)	Third Digit (Contact Pressure)	Fourth Digit (Entry Angle[a] [Degrees])
1) Unidirectional	1) Point-to-point	1) Unidirectional	0) 0 (infinite planes)
2) Cyclic	2) Line-to-line	2) High frequency	1) 90 → >75
3) Roll/Slip	3) Point-to-area	3) Cyclic loading	2) <75 → >65
4) Fretting	4) Line-to-area		3) <65 → >55
	5) Area-to-area		4) <55 → >40
	6) Area-to-larger area		5) <40 → >25
	7) Open, fixed area		6) <25 → >15
	8) Open, variable area		7) <15 → >10
			8) <10 → >2
			9) <2 → 0 (ball-on-flat)

[a] Angle at the leading edge of the tribocontact, beginning with the direction of the normal force and tilting toward the direction of motion.

Source: Adapted from Voitik, R. in *Wear Test Selection for Design and Application*, ASTM, Philadelphia, PA, 1993.

TABLE 3.2
TAN Codes for Engineering Parts and Laboratory Tests

Application	Configuration	TAN Code
Internal engine components	Tappet on cam shaft	3219
	Valve guide	2511
	Piston ring	2521
	Thrust washer	1510
Laboratory tests	Ball (pin)-on-disk	1318
	Block-on-ring	1418
	Reciprocating flat pin-on-disk	2519
	Shell four-ball test	1317

Source: Voitik, R. in *Wear Test Selection for Design and Application*, ASTM, Philadelphia, PA, 1993.

3.2 SPECIMEN PREPARATION AND CLEANING

There are three elements involved in surface preparation of friction test specimens: (a) forming the proper contact shape, (b) providing the proper surface finish, including roughness and lay (directionality of features), and (c) cleaning the specimens. If the surface is intended to simulate a specific machine part, it should have the same surface finish, lay, and heat treatment as that of the part. Machining processes like grinding and turning can produce subsurface residual stresses and work hardening in metals, thereby altering the response of the near-surface layers to surface contact. In simulative work, if the actual parts are usually cleaned in a certain way in service, so should the test specimens. If the surface is meant for more fundamental studies of friction, it should be prepared carefully to remove the effects of prior machining. A mirror-like finish on a surface is no guarantee that residual machining damage does not lie a few micrometers below it. In other words, different preparation processes can result in the same final surface roughness parameters, but the condition of the near-surface material may not be the same.

Barwell[7] illustrated the effects of surface preparation of stainless steel specimens on the sliding wear damage produced in castor oil–lubricated tests. One surface was produced by buffing, leaving prior machining artifacts intact. Another surface was produced by anodic polishing (electropolishing). Microindentation hardness tests were performed transverse to the sliding direction. The results are reproduced schematically in Figure 3.2. The buffed surface was harder due to residual machining damage. Its initial hardness, coupled with its smoothness, appeared to have prevented further work hardening during frictional contact. The other surface was softer, and the friction-induced shears resulted in considerable work hardening.

In some fundamental studies of friction, surfaces are polished, etched, solvent cleaned, dried, and then finally exposed to shortwave ultraviolet or a vacuum oven to desorb contaminants. Vacuum annealing and ion sputtering have also been used in basic friction and adhesion studies. In most cases, dust and extraneous particles

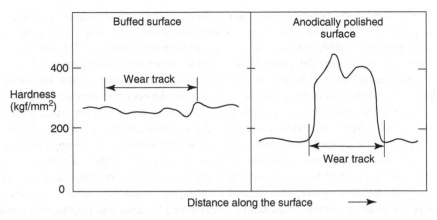

FIGURE 3.2 Effects of surface preparation on the microindentation hardness of sliding wear tracks on castor oil–lubricated surfaces. (Adapted from Barwell, F.T., *Metallurg. Rev.*, 4, 141, 1959.)

should be removed just before testing with a puff of air from a rubber bulb or with an oil-free compressed air source, not with the breath.

There is no "best way" to prepare surfaces for friction testing. The method selected must fit the objective of the test. Acetone may be used to clean metals for friction testing, but it will attack certain polymers, making them sticky. Surface films, such as tarnishes or adsorbed moisture, can have significant effects on friction, especially when the contact pressure is low and when the surfaces are very smooth. However, when the surfaces experience abrasive sliding or if the applied contact pressure is high enough to cause shear to take place deep within the subsurface, thin films present at the onset of the experiment may have little or no noticeable effects on friction. Even in ASTM standard test methods, specimen preparation procedures differ. In some wear-testing procedures, cleaning instructions are vague. For example, the ASTM G 76 test for solid particle erosion states (in Section 11.2), "Clean the specimen surface carefully." The procedure for the block-on-ring test, ASTM G 77 (Section 9.1), is more detailed:

> Clean the block and ring using a procedure that will remove any scale, oil film, or residue without damaging the surface. For metal, the following procedure is recommended: clean the block and ring in trichloroethane, ultrasonically if possible; a methanol rinse may be used to remove any traces of trichloroethane residue. Allow the blocks to dry completely.

The pin-on-disk standard, ASTM G 99 (Section 8.1), recommends a slightly different approach:

> Immediately prior to testing, and prior to measuring or weighing, clean and dry the specimens. Take care to remove all dirt and foreign matter from the specimens. Use non-chlorinated, non-film forming cleaning agents and solvents. Dry materials with open grains to remove all traces of the cleaning fluids that may be entrapped in the material.

In recent years, concerns about atmospheric pollution and human exposure to carcinogens have caused changes in recommended solvent-based cleaning procedures. Trichloroethane, a known carcinogen and toxic substance, and Freon™ 113, an ozone layer–destroying chlorinated fluorocarbon, have fallen out of favor as cleaning solvents. Test procedures need to be updated not only to incorporate new knowledge and improved measurement techniques, but also to address environmental and health concerns. Certain solvents, such as ethanol and acetone, tend to absorb water and can leave stains or films. Methanol is a suitable alternative to ethanol. Lubricated friction testing, especially using used engine oils, can present significant cleaning problems. Gasoline is a good alternative to acetone in such cases. The surface of polymers can be cleaned with mild lens-cleaning solutions. Sometimes scrubbing samples with detergent and warm water, rinsing, and then drying is suitable. Whichever way is chosen to clean specimens, it should be documented because that information could help one to sort out differences in friction data.

Recontamination and oxidation after cleaning can affect the friction of metal surfaces, especially when testing at low contact pressures. If a metal surface is abraded just before testing, the exposure of fresh metal to the air may stimulate rapid oxidation. Consequently, the specimen may be in the process of reforming an oxide when the test begins. Depending on the chemistry of the oxide, the rate of surface film thickening may be parabolic or logarithmic, so that the rate of change in the new oxide layer thickness is highest just after substrate cleaning. In such cases, waiting for several hours, until the oxide has reached a relatively constant thickness might produce more repeatable data than testing immediately, during the time when thickness is changing most rapidly.

Porous or polymeric materials adsorb surface contaminants readily, and some polymers may dissolve or become tacky when exposed to common cleaning solvents such as acetone. It is therefore important to check the compatibility of testing materials with the proposed cleaning procedure. In some instances, it may not be feasible to use liquid cleaners; instead, use a jet of air or insert gas to remove particles before testing. Again, document the chosen cleaning procedure.

Handling test specimens with bare hands may introduce errors into friction tests. Sweat is a clear liquid produced by glands in the body to remove certain wastes and to effect cooling. An average person has about 1,000,000 sweat glands, and there are two types: apocrine glands (armpits) and ecrine sweat glands (the majority of the skin surface). Sweat is approximately 98% water and 2% other substances such as sodium chloride, urea, sulfates, albumin, and fatty acids. Some of these constituents, such as fatty acids, are lubricative.

The effects of handling specimens with sweaty fingers were studied using reciprocating ball-on-flat sliding tests on two material combinations. The slider ball specimens were 9.525 mm in diameter. Test conditions are given in Table 3.3. Three sets of replicate tests were performed: both specimens cleaned in acetone and methanol, the ball specimen rubbed once with a sweaty thumb, and the flat specimen rubbed once with a sweaty thumb from the same individual.

The average friction force per stroke was computed for each test, and the value for each stroke number was averaged for two tests to produce the plots shown in Figures 3.3a and 3.3b. For both steel-on-steel and ceramic-on-ceramic pairs, the

TABLE 3.3
Test Conditions Used for Studying the Effects of Sweat Films on Friction Data

Material combinations	M-50 tool steel ball on 01 tool steel flat specimen, silicon nitride ball on silicon nitride flat specimen
Applied normal force	1.02 N
Stroke length	10 mm
Duration of the test	20 reciprocating strokes
Rate, strokes per minute	44 (average speed 14.7 mm/s)
Relative humidity	57 ± 3%

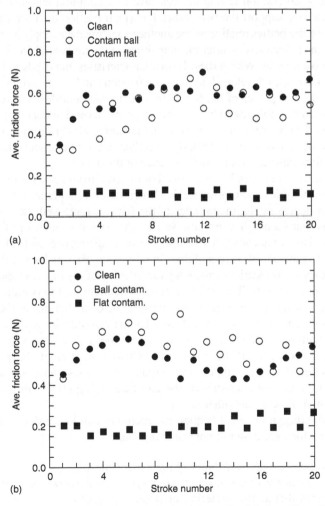

FIGURE 3.3 Effects of perspiration contamination on the friction of steel-on-steel and silicon nitride-on-silicon nitride. (a) M-50 on 01 tool steel (two tests per datum), (b) NBD 200 on SRBSN (two tests per datum).

effect of handling the flat specimen was much more pronounced than for the ball specimen. For steel-on-steel, the friction was reduced by a factor of about 6 at the end of the test, and in the silicon nitride case, the friction was halved due to contamination of the flat. The presence of sweat on the flat silicon nitride specimen also eliminated the rise-drop-rise trend in friction exhibited by the clean couple. These effects are quite understandable: first, the ball is in constant contact so it can be wiped clean more quickly during sliding, and second, the amount of sweat on the ball is much less than that transferred to the flat specimen, whose larger deposit can replenish the lubricant during sliding.

3.3 DESIGN AND SELECTION OF FRICTION-TESTING METHODS

Fundamentally, all friction-testing devices have four common features: (1) a means to fix or otherwise support the two bodies for which friction data are desired, (2) a means to move the bodies relative to one another, (3) a means to apply a normal force, and (4) a means to measure or infer the magnitude of the tangential friction force that opposes relative motion. Within these broad characteristics, hundreds of friction-testing schemes have been developed. Macroscale friction can be measured with a simple protractor and a few pieces of flat board, or with a table, a pulley, a wire, and a pan balance. However, sophisticated high-precision probes moved under computer control have been used to measure friction on the molecular scale. Friction-testing devices might appear deceptively simple to design, yet there are subtle aspects that can affect the repeatability, reproducibility, and relevance of the data.

Friction coefficients can be determined or inferred from accurate measurements of just a few quantities such as the angle of inclination of a plane to the horizontal when motion begins or the magnitudes of the friction force and the normal force. As simple as that seems, it is not always straightforward to measure friction with sufficient precision or accuracy to discriminate small differences in friction between materials or the effects of trace additives in lubricants. Except for the fundamental research studies or applications involving microdevices, only macroscopic forces are measured in friction tests. That is, the external force required to overcome the frictional sliding resistance is measured, not the separate contributions of the individual asperities (high spots) on contact surfaces. The tribosystem in essence becomes a "black box" with inputs like the applied force and imposed velocity and outputs like the tangential force. As will be discussed in Chapter 4, attempts have been made to model friction strictly as "inputs and outputs" of a tribosystem, but such models ignore the physical characteristics of the interface, aging effects in the materials, lubricants, environment, and other factors.

Those who design or select friction test methods should consider a host of factors, some of which are easier to control than others:

- *The form, size, and availability of specimen materials.*
- *Whether or not to clean or otherwise prepare the surfaces:* Realistic simulations may dictate that materials be tested "as-received" without cleaning.
- If cleaning is used, *what method is best*? Is the time between cleaning and testing critical because of recontamination or oxidation?
- What *surface roughness* and *lay* are required?

- *The test direction relative to the surface texture* should match that for the application.
- *Is sliding motion continuous or intermittent*, and is stick-slip an issue?
- Should *more than one load, contact pressure, and velocity* be used?
- Are the materials frictionally sensitive to *the environment*? Should one attempt to *control humidity* or gaseous environment, or just *monitor and report it*?
- When doing a simulation, do *the stiffness and natural frequency* of the tribometer match those of the application?
- *What kind of force sensing system* is adequate? *Are the data to be recorded*? What *sampling rate or interval between samples* is required?
- *How many repetitive tests* are needed to obtain a representative value?
- If lubricated, *how should the lubricant be applied*? How often should it be replaced? Should the lubricant be preconditioned to better simulate its use characteristics?
- *Are third bodies (particles or wear debris) a factor*? If so, should they be artificially removed or allowed to accumulate on the contact surface?
- What kinds of *descriptors* are needed to fully characterize *test conditions, lubricants*, and *materials*? Such information is helpful when interpreting anomalous results. The need for such detailed testing information is not reserved solely for basic research, but should be part of any friction measurement plan.

The foregoing considerations underscore the challenges of developing friction test methods. Since frictional behavior is tribosystem-specific, there is no known combination of sliding materials that produces the same friction coefficient irrespective of the manner of sliding (i.e., there is no such thing as the intrinsic friction of a given pair of materials). Consequently, the development of standard reference materials to calibrate tribometers is not possible. Instead, the devices to measure normal and tangential forces of the apparatus should be calibrated accurately across the expected range of operation. Sometimes the use of friction-testing standards is appropriate, but at other times the procedure must be customized to achieve the desired results.

Five factors that influence the selection of friction test methods and apparatus are as follows:

1. Macrocontact geometry, stiffness, and vibration damping
2. Type and magnitude of the relative motion
3. Magnitude of the normal force applied to the contact
4. Requirement of environmental or temperature control
5. Conditioning and preparation of the materials

Macrogeometry affects the contact stress distribution and magnitude. It also affects the flow of lubricants into and out of the contact. Stiffness affects the uniformity of sliding motion and the potential for mechanical vibrations to be generated and transmitted. The type of friction force sensing system selected can affect the stiffness of the system. The type of motion (e.g., unidirectional, reciprocating) affects the evolution of surface texture, debris entrapment, lubricant flow, and dissipation of frictional heat. It also affects the exposure or protection of surfaces from the surrounding environment. Temperature and environmental controls, if required, can make friction force measurement more difficult because force sensors can be affected by heat or corrosive environments, requiring more sophisticated isolation or calibration schemes. Chapter 7 describes the effects of specific test variables on friction data.

Size effects in friction testing are important—so important in fact that entire conferences and symposia have been devoted to the subject.[8] For example, the size of the nominal contact area is affected by macrogeometry and applied force. The higher the force, the more the contacting surface features are deformed so that the load can be supported. If a very small nominal contact area is used, such as in miniature ball-on-flat tests at low loads, then the frictional behavior will be much more sensitive to local surface features and microtopography. In other words, the scale of macrocontact will be closer to the scale of asperity sizes and microstructural feature sizes. If a relatively large contact area is selected, then the effects of individual asperities will tend to be averaged out. This effect has been discussed in other sections in more detail.[9]

In 1957, J. F. Archard[10] analyzed seven different cases of the dependence of normal load P on the area of frictional contact A. For purely elastic Hertzian contact, like a smooth sphere-on-smooth flat plane, A is proportional to $P^{2/3}$, but as the contact becomes more plastic, A becomes proportional to P. A further consideration is the slider shape. Intuitively, if a very hard needle is placed on a surface under a certain load, it will penetrate more deeply than if a blunter cone or sphere is placed on the same surface under the same load. The tangential force required to pull a sharp needle through a soft surface is intuitively higher than that required to pull a gently curved sphere along the same surface. As will be discussed in the next chapter, the selection of the slider tip geometry affects the relative contributions of sliding, plowing, and chip-forming (cutting) to the total friction force (see Figure 3.4). Research on scratching has shown that tip sharpness affects the sensitivity of the data to crystallographic anisotropy (i.e., the crystal orientation at the surface relative to sliding direction of the tip).

In recognition of the effects of indenter sharpness on friction force, ASTM scratch hardness test method G 171 defines a *stylus drag coefficient* as *the ratio of the normal force to the tangential force on the stylus*. Despite the similarity of this definition to that for μ_k, it is not called the friction coefficient in the standard. If a flat or gently curved diamond surface is slid on metallic surfaces in humid air, the friction coefficient tends to be quite low (say 0.03–0.12). However, if the diamond is in the form of a sharp indenter, the friction can be quite high (>1.0). One could therefore argue that the latter results are not so much a measure of frictional properties of a certain pair of materials as they are of the deformation of a surface by a hard indenter of the given geometry and orientation relative to the direction of travel.

The distinction between a friction coefficient, as it is conceptualized on a molecular scale, and the relative resistance to abrasion is further complicated in the case of multiasperity contacts and larger nominal contact areas (i.e., greater than a few square millimeters). For example, it could be argued that the tangential forces

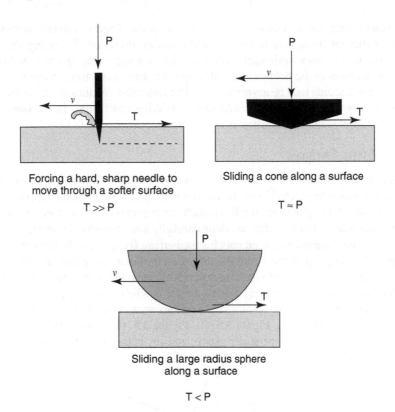

Forcing a hard, sharp needle to move through a softer surface

$T \gg P$

Sliding a cone along a surface

$T \approx P$

Sliding a large radius sphere along a surface

$T < P$

FIGURE 3.4 The sharpness or bluntness of the slider affects the measured friction force because the adhesion, plowing, and cutting contributions differ.

measured during wear tests of metals on abrasive papers reflect a kind of scratching resistance and not the friction coefficient. Likewise, instrumenting three-body abrasive wear tests, such as the ASTM G 65 dry-sand/rubber-wheel abrasion test, produces neither a scratching coefficient nor a friction coefficient in the strictest sense because particle rolling, sliding, and scratching may all be involved in the interface. This is, again, akin to a "black box" scenario in which testing inputs such as normal force are transformed into outputs such as tangential force, but the minute details of that transformation are not precisely known. Obviously, one could report the ratio of tangential force to normal force as the "effective friction coefficient" without worrying about the mechanism(s) that gave rise to it. In that case, it becomes even more important to document the details of the test method.

Not all tribometers are designed correctly. Many are built by investigators with little or no training or experience in mechanical design. As a result, some machines are too elastically compliant and produce unwanted vibrations or slick-slip response. Other machines may be mechanically robust, but the means of inserting and removing test specimens is so awkward that the contact surfaces can be marred or damaged. Some tribometers have excessive friction in their fixtures and bearings, producing problems when attempting to measure friction forces at low normal force. Reciprocating tribometers with force sensors not colinear with the friction contact sometimes require different force calibrations for each sliding direction. Orienting the mid-line of the force sensor collinearly with the friction force in the interface helps, but assembly stresses in the holders may still require bidirectional force calibrations and friction-time records may be asymmetrical. The following sections describe a variety of approaches to measuring static and kinetic friction coefficients in the laboratory. Some are simple and robust, whereas others are more sophisticated.

3.3.1 STATIC FRICTION

In some respects, the simple friction tests are among the most elegant. It is conceptually easy to visualize the mechanics of the experiment and to comprehend the parameters being measured. Despite their simplicity, such experiments can be just as illuminating as very sophisticated ones if they are done carefully and interpreted properly.

Two of the simplest arrangements for measuring friction coefficients are shown in Figure 3.5. The top of the figure illustrates an *inclined-plane device*. This simple device, sketched by Leonardo da Vinci over 500 years ago, forms the basis for several current-day formal standards. Such a device can be constructed relatively easily and inexpensively; however, commercial versions with built-in electronic inclinometers are also available (see Figure 3.6). In Figure 3.5, m_1 is the mass of the added weight, m_2 is the mass of the slider block, and m_3 is the mass hung on the pulley needed to produce sliding.

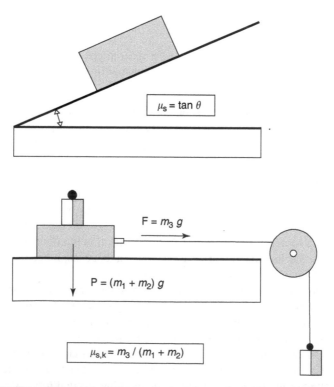

$$\mu_s = \tan\theta$$

$$F = m_3 g$$

$$P = (m_1 + m_2) g$$

$$\mu_{s,k} = m_3 / (m_1 + m_2)$$

FIGURE 3.5   Two of the simplest tribometers: (top) the inclined plane and (bottom) the sled.

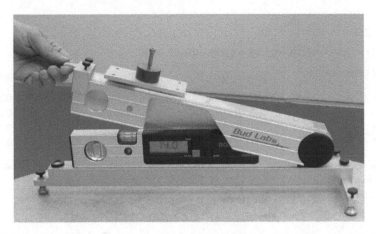

FIGURE 3.6   Commercial inclined-plane friction apparatus. (Courtesy of Bud Labs, Inc., Rochester, NY.)

A simple mechanics analysis implies that it is unnecessary to know the mass of the object resting on the tilting plane to measure static friction coefficients: the plane is slowly tilted until slip occurs. The angle of inclination at that instant, θ, is measured using a protractor or similar device, and the static friction coefficient μ_s is equal to its tangent. Owing to the nature of the tangent function at low angles and the design of tilting plane devices, it can be difficult to accurately differentiate between materials that display low values of static friction, and such tilting plane methods lose efficacy at low μ_s.

It is an accepted practice to report static or kinetic friction of unlubricated solids to a precision of two decimal places. This convention has been widely adopted for the following reasons: (1) the static and kinetic friction coefficients of the majority of solids sliding on solids under unlubricated ambient conditions generally fall between 0.10 and 1.00, and (2) the friction force tends to vary during sliding, and that variation is typically greater than 0.01. In contrast, friction coefficients for well-lubricated tribosystems can measure well below 0.10, and higher precision is required.

Several standard friction test methods are based on the inclined-plane concept. These include ASTM D 3248 (static friction of corrugated and solid fiberboard) and ASTM D 3334 (fabrics woven from monofilaments). Others were prompted by the need to determine the residual lubricants on photographic films: ANSI/ISO 5769-1984 and ANSI/NAPM IT9.4-1992 (for photography processed films) and ASTM G 164 (for lubrication of flexible webs).

As noted previously the inclined-plane method does not require that the experimenter know the weight of the slider, but one should not assume that μ_s is generally independent of the normal force. For example, static friction data obtained by Rabinowicz[11] for a range of high-purity, noble metals clearly indicate that the weight of the sliding specimen can significantly affect the measured static friction coefficient. Table 3.4 highlights these findings, obtained over more than five orders of magnitude in load. At low loads there seems to be a greater tendency for seizure, a disturbing factor for those engaged in the design and study of micromechanical devices

TABLE 3.4

Static Friction Coefficient as a Function of Load

Material Combination	Load = 0.01 gf	Load = 1.0 gf	Load = 1000 gf
Au on Pd	0.98	0.60	0.38
Au on Rh	0.71	0.34	0.25
Ag on Ag	1.38	0.48	0.32
Ag on Rh	1.30	0.27	0.22

Note: Adapted from Rabinowicz, E., *Wear*, 159, 89, 1992, inclined-plane method in laboratory air, selected data.

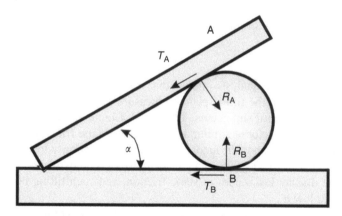

FIGURE 3.7   Adapted from Sphere-between-two-plates testing configuration described by Studman and Field.

involving sliding contact (computer hard drives, precision positioning mechanisms, micro-electromechanical systems [MEMS], etc.). Rabinowicz noted that his data were not entirely consistent with more recent findings using atomic-scale measurements of lateral (friction) forces, and he suggested that static friction coefficients might go through a maximum and subsequently decline at extremely low loads. Furthermore, one might argue that for very small specimens, static electrical charges could delay the onset of sliding and thus raise the friction force.

An inverse adaptation of the inclined plane was devised by Studman and Field[12] to measure the static friction coefficient of materials, such as glass, that can be formed into spheres or cylinders and flat coupons. Figure 3.7 shows one of their friction testing geometries in which a sphere (ball) is trapped between two plates: a flat lower plate and an inclined upper plate. Either a hinge or a groove in the lower plate could be used as a pivot point for the upper plate. In this unique device, the procedure was to use a rod to push the ball under the tilted plate until slip occurred, then slowly withdraw the rod until the ball returned to the point where static equilibrium occurs and there is no further slip. The static friction coefficient is computed with the largest angle α for which no slip occurs can be used. The values of static friction coefficient at each location were shown to be

$$\mu_B = \frac{\sin \alpha}{1 + \cos \alpha} \tag{3.1}$$

$$\mu_A = \frac{\sin \alpha}{1 + \cos \alpha} \left(\frac{R_B}{R_B + Mg} \right) \tag{3.2}$$

where R_B = the normal force at point B, M = mass of the sphere, and g = acceleration due to gravity. Taking moments about the origin of the incline, note that $R_B = R_A - Mg$. If needed, the value of R_B can be determined from the location of the center of gravity of the inclined plate. From the geometry of the arrangement, when using identical plate materials, slip should always occur at point B before point A; therefore, the value of μ_B was used as the basis for the measurement. Using a 5 mm diameter steel sphere and glass plates treated with different surface preparation and cleaning methods, μ_B ranged between 0.12 and 0.27. The repeatability of 10 measurements on each condition was $\mu_B \pm 0.01$.

A horizontal sled configuration, envisioned by da Vinci, is shown at the bottom of Figure 3.5. Here, the normal force is determined from the total mass of the moving specimen plus any added weights. The tangential force is applied horizontally. As with the inclined-plane device, commercial versions of sled-type device also exist, some of which have data acquisition systems (see Figure 3.8). Such devices are routinely used in measuring the friction of polymers and textiles. The horizontal force is raised until the onset of relative motion. Thus, the "breakaway" or starting

FIGURE 3.8 Commercial sled-type friction device. (Courtesy of Testing Machines, Inc., Ronkonkoma, New York.)

friction can be determined. The means of applying the tangential force can be as simple as placing increasing amounts of weight on the pan hung from a pulley, pulling with a "fish scale" (or one of its more sophisticated relatives), or pulling the block specimen with a motor drive while monitoring the friction force. The latter requires a very smoothly operating, low-speed motor.

Budinski[13] points out that using nylon fishing line to pull the specimen along the inclined plane may produce more stick-slip behavior and higher static friction coefficients than using a stiffer chain or wire. Thus, he recommends that the choice of the pulling wire material should be based on the nature of the simulation required. The kinetic friction coefficient can also be obtained using such a "friction sled" device by measuring the friction force during steady sliding, that is, after starting friction transients have subsided and the sled is moving along at a constant speed.

A rather unusual type of tribometer, designed for fundamental studies of stick-slip, was constructed by D. Kuhlmann-Wilsdorf's group at the University of Virginia and subsequently described in an article by Brendall et al.[14] The so-called "hoop apparatus" consisted of a narrow ring of metal tubing whose inner diameter serves as the sliding surface (see Figure 3.9). A conformal slider rides up to the inside of the hoop within a shallow groove designed to prevent its wandering laterally as the hoop is slowly rotated about its axis. Fixed to the slider is a polarizing filter, which rotates as the slider climbs the inner diameter of the hoop. Light transmitted through an externally mounted polarizing filter and through the filter on the slider is changed in intensity as the latter is rotated. Thus, the light intensity can be measured to permit determination of the friction force at "slip."

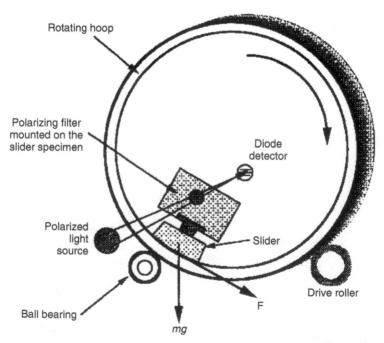

FIGURE 3.9 The "hoop" friction apparatus developed at the University of Virginia.

Static friction is of considerable interest in micro- and nanotribological systems, as is a related phenomenon known as *stiction*, which is further discussed in the next chapter. Specialized tests have been developed by the computer hard drive industry to investigate this phenomenon because it applies to their products. These generally involve a slider-on-disk arrangement.

3.3.2 SLIDING FRICTION

Unlike static friction tribometers that characterize resistance to impending motion, sliding (kinetic) friction test devices comprise a wide range of operating conditions, ranging from constant-speed unidirectional sliding to articulating machines in which a complex series of accelerations, decelerations, and changes in direction are imposed. Each has its own place.

Eight common types of tribometers are shown in Figure 3.10. Excepting for the sled test and thrust-washer arrangement, these methods use nonconformal contact.

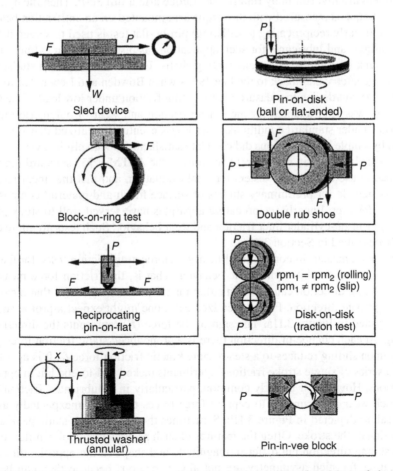

FIGURE 3.10 Eight common types of friction-testing geometries. Most of these are commercially available.

The four geometries shown at the top of the figure are the most common and can be purchased from commercial apparatus suppliers. The block-on-ring geometry is the basis for ASTM G 77 for dry sliding wear, but another variation is used for lubricant testing (ASTM D 2714). The pin-on-disk test forms the basis for ASTM G 99 and was the result of an international effort involving at least 30 participants.[15,16] One version of a double rub-shoe test is performed using the so-called Hohman A-6 tester. A paper describing its use in screening solid lubricants was published in 1963.[17] Either nonconformal, flat rub-shoes or conformal rub-shoes have been used in such tests.

Thrust-washer type, flat-on-flat tests are often used for evaluating the friction of face-seal materials or materials for metalworking processes where relatively high face loadings occur. The basic thrust-washer geometry is described in Section 2.4.2. ASTM D 3702 prescribes an annular-contact thrust-washer geometry to determine the wear rates of self-lubricating materials, including polymeric composites.[18,19] An upper specimen is deadweight loaded vertically against a lower specimen, which is mounted on a rotating stage that can be moved by a torque arm extending from it. Specimens are first run in by rotating the torque arm a full 360°. Then the dynamic friction torque is measured with the torque arm mounted on an antifriction bearing.

The linearly reciprocating (oscillating) pin-on-flat test is used to screen materials, coatings, and lubricants for such applications as piston rings on cylinder lines, slider-crank assemblies, and reciprocating shafts. Fundamental friction studies that use this geometry date back to the late 1930s when Bowden and Leben[20] developed a rather elegant device to measure reciprocating friction under low loads. The same geometry was later adapted for measuring friction and wear of polymeric implant materials under standard conditions,[21] but it was eventually realized that wear produced by simple linear motions did correlate satisfactorily with clinical wear results. Qualifying language in more recent versions of the ASTM F 732 standard indicates the need to use articulating testers for final evaluation, and that the linear reciprocator is suitable for preliminary studies of surface finish and material composition effects. However, linearly reciprocating apparatus have served well to study piston ring segment-on-cylinder liner friction and wear behavior. Such tests are more completely described in Section 3.4.

It is not unusual, in conducting linearly reciprocating friction tests, for the tangential force to exhibit asymmetric behavior. That is, the friction force measured in one sliding direction can differ in either magnitude or shape from that measured in the opposite direction. For an idealized case, one might expect a profile such as that depicted in Figure 3.11a. The sign of the force (F) represents the direction of sliding. At each change of direction there is a pronounced static friction force peak f_s that upon sliding reduces to a steady-state kinetic friction force f_k. It is as if there were a series of single stroke friction experiments tacked end-to-end but in opposite directions. However, it is fairly common, particularly in unlubricated friction tests in which wear is occurring, to capture force traces that look unexpectedly asymmetrical, as depicted in Figure 3.11b. Sometimes there is no clear static peak at the beginning of the stroke. Often the features of such profiles may look similar for the same sliding direction but differ in magnitude and shape in the opposite direction. The causes for such asymmetry are not always obvious because they can be due to apparatus characteristics, errors in calibration of the friction sensor, the actual

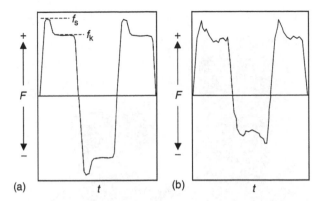

f_s
f_k

(a) t (b) t

FIGURE 3.11 Asymmetrical frictional behavior is common when testing in the reciprocating mode: (a) Idealized behavior and (b) commonly observed behavior.

material's behavior, or a combination of the three. If the axis of the friction force sensor (a load cell) is not colinear with the friction force at the sliding contact, different moment systems can develop as the direction of sliding changes and the component of tangential force resolved in the primary sensing direction changes when the sliding direction is reversed. If the elastic response of the sliding system is not the same in each direction due to assembly stresses or if the sensor does not respond similarly in tension and compression, then it is imperative to calibrate the tangential force for each direction of sliding. Contamination or wear debris can lubricate the surface, reducing the static friction spike at the beginning of the stroke, or conversely, debris can pile up or transfer to the slider near the end of each stroke, raising the friction force.

It is useful to visualize the mechanics of the turnaround point in a reciprocating tester that involves a strain-gauge-type of load cell or a proving ring that requires elastic displacement to sense the force. Initially, there is no relative motion in the contact, but force builds up until the static friction is exceeded and the contact springs forward to catch up with the specimen drive system. Speed decreases to match the imposed velocity and then decreases to zero at the end of the stroke. If there is a pileup or a strong adhesive interaction between materials, there may be residual tangential force to be overcome when the sliding direction changes. Factors affecting this scenario include the stiffness of the machine, the type of reciprocating drive system (e.g., slider crank vs. a stepper motor), and the type of sensor used (piezoelectric, strain gauge, load cell, etc.), the speed of reciprocation, and inertial forces that must be overcome to change direction. In lubricated testing, the regime of lubrication may change along the length of the stroke (see Chapter 6). In summary, the phenomenon of tangential force asymmetry requires a case-by-case analysis.

The pin-in-vee block test, shown at the lower right of Figure 3.10, is not used so much for friction as for screening the antiseizure properties of lubricants. ASTM standard D 2670 for evaluating wear in oils uses that method. Ives and Boyer[22] evaluated the usefulness of the pin-in-vee block test for screening the friction and wear of materials in recycled engine oils with various additives and found that it had problems when applied for that purpose. One problem concerned the load calibration

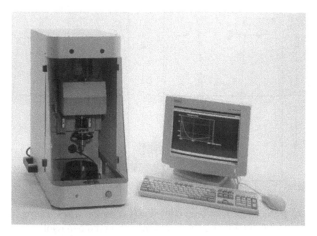

FIGURE 3.12 Computer-controlled, multifunctional tribometer. (Courtesy of N. Gitis, Center for Tribology, California.)

procedure and was associated with the need to contend with four contact surfaces (two faces on each vee block).

In contrast to single-purpose tribometers that are built to impart specific motions and loads or which simulate a specific application, some testing machines are designed to be more flexible, enabling the investigator to add modules that provide a range of testing conditions. For example, Figure 3.12 shows a commercially made, computer-controlled multifunctional tribotesting platform that can be equipped with various sensor options, programming modes, and mechanical drives.

In addition to the aforementioned macrocontact tests for sliding friction, recent years have witnessed the evolution of micro- and nanotribometers. This interest was prompted not only by the need to characterize the performance of miniaturized moving parts, like MEMS and nano-electromechanical systems (NEMS), but also by a desire to study the smallest details of interaction between atomic arrangements or molecular layers on surfaces. Several of these specialized devices are discussed in Section 3.4.

3.3.3 ROLLING FRICTION

Rolling friction measurement comprises a special category that differs from the approaches to measuring sliding friction. As mentioned in Section 2.5, rolling friction usually involves some degree of slip; therefore, rolling friction devices can range from simple disk-on-disk machines to more complex multiaxis machines that can vary the degree and directionality of slip as well as the shape of the elastic contact patch.

Interestingly, rolling element–bearing catalogs contain a wealth of technical data on the load, speed, and life ratings for a bewildering variety of bearings, but they omit any information on starting friction. Yet the starting friction of rolling bearings is important for engineers who must select motors for industrial machinery that either starts and stops frequently or must remain under load without moving and then be called upon to function. Consequently, in 2006, ASTM Committee

G2 approved a "Standard Test Method for Determination of the Breakaway Friction Characteristics of Rolling Element Bearings" (ASTM G 182). In this case, a small bearing is placed on a tilting plane under load, and the plane is slowly tilted until the bearing just begins to roll. Actually, the term "friction" is used as a global characteristic of the component because friction occurs at many contact points inside a roller bearing, and it is the sum of these reactions that gives rise to the externally observed resisting torque.

Part of the reason why manufacturers refrain from providing starting torque (friction) data for their products may be that it is difficult to control how the bearings are stored or used. Some bearings are shipped without lubricants (or are minimally lubricated), or they may be stored for varying times before being placed in use. Therefore, the starting friction is difficult to specify with any degree of accuracy. It is especially important to know the starting friction for bearings in such applications as high-precision telescopes or antenna pointing devices and motors for mechanical assemblies in spacecraft that need to conserve power.

Various testing machines for rolling contact were described by Blau.[23] Most involve lubricated testing. In some instances, the lubricant is fed to the contact by immersing the lower roller in a tray of oil. Others resort to drip-feeding. The disk-on-disk test is a traction test in which rotating rollers (or one fixed and one rotating roller) contact each other. One common version of the disk-on-disk test is called the Amsler machine, described by O'Donaghue and Cameron.[24] The machine can be used to study contact friction, wear, and various states of lubrication under rolling and sliding conditions. One can use a crowned roller mated to a flat roller to ensure better alignment or to produce a more concentrated contact. To produce slip, some machines use gearing between the parallel roller drive shafts and others use separate motors for each drive shaft.

In rolling contact measurements, the term traction is preferred to friction because both rolling and slip are involved. Analogous to the friction coefficient, the traction coefficient is the ratio of the tangential force across a tribocontact to the imposed normal load. Traction under elastic (Hertzian) contact conditions is generally associated with the limiting shear strength of the entrapped lubricant and is therefore a function of the pressure and temperature. Wedeven[25] patented a device in which a motor-driven ball in an inclined holder is rolled on a rotating lubricated disk. A schematic diagram of the system is shown in Figure 3.13. By adjusting the angle of tilt and the relative velocities of the ball and disk, it is possible to achieve varying degrees of slip, simulating a variety of rolling conditions. This kind of arrangement can generate what is commonly called a traction curve in which the traction coefficient is plotted as a function of percent slip in the contact.

3.3.4 TESTS OF FLEXIBLE SURFACES

Flexible surfaces include fibers, woven bundles, webs, thin sheets, and strips. Materials and products of interest range from human hair to arresting cables on aircraft carrier decks and from metal sheets passing through rolling mills to high-speed papermaking machines where the product moves at speeds exceeding tens of meters per second. Both static (starting friction, stiction) and kinetic friction are of interest in these applications.

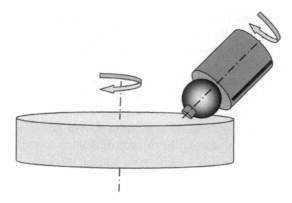

FIGURE 3.13 Inclined ball-on-disk method used by Wedeven to measure traction and rolling friction.

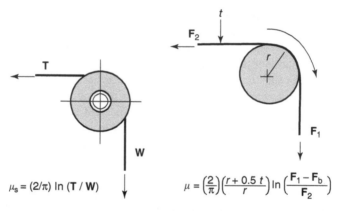

$$\mu_s = (2/\pi) \ln (T/W)$$

$$\mu = \left(\frac{2}{\pi}\right)\left(\frac{r+0.5\,t}{r}\right)\ln\left(\frac{F_1-F_b}{F_2}\right)$$

FIGURE 3.14 Two variants of the wrapped-cylinder geometry: capstan (top) and sheet bending (bottom).

Briscoe et al.[26] measured the friction and stick-slip behavior of monofilaments of polyethylene terephthalate sliding on glass fibers and human hair in a crossed-fiber configuration. A cantilevered fiber was rested against a taut, crosswise fiber clamped at both the ends. The deflection of the cantilevered fiber was sensed by a photodiode array illuminated from a focused light beam reflected back through a microscope from the fiber intersection. The friction of various fibers on hair exhibited marked stick-slip behavior, and the effects of the oriented structure of the human hair on the friction could be observed with this technique. Twelve years later, atomic force microscopy was applied to study the structure and friction of human hair, also in conjunction with the development of hair care products.[27] It was of interest to understand the role of microstructure, humidity, and distribution of conditioning products on friction and adhesion between strands.

One family of tribometers for flexible materials drapes or wraps the specimen over a fixed cylinder (see Section 2.4.3). Figure 3.14 illustrates two variants of this method. In the upper figure, where the angle of arc is 90°, the friction coefficient of the wire sliding on the cylinder is computed from the ratio between the tension on the wire (**T**) and the hanging weight (**W**). Draped-cylinder techniques enable measurement of the friction of fibers, either woven into a belt or as individual strands, against a cylindrical surface that simulates a pulley or guide roller. For example, Gupta[28] used a method to measure the friction of single fibers. The friction coefficient for a strand of 10 μm diameter, polymer-coated glass fiber was seen either to increase or decrease with the number of cycles of load application depending on the substrate used. Data in Table 3.5 indicate the effects of substrate and number of cycles on the friction coefficient of the strand against several counterfaces. Gupta also reported data for wet and dry conditions.

The same principle forms the basis of ASTM G 143, the capstan friction test,[29] which is further described by Budinski.[30] One version of that test involves wrapping a web under a roller and over a stationary cylindrical counterface (see Figure 3.15).

TABLE 3.5
Effects of Cycling on the Sliding Friction Coefficient of a Strand of Glass Fibers against Various Substrates

Substrate	μ (First Cycle)	μ (10 Cycles)	$\Delta\mu$ (1–10 Cycles)
PTFE	0.11	0.09	−0.02
Porcelain	0.33	0.33	0.00
Brass	0.18	0.16	−0.02
Stainless steel	0.28	0.32	0.04

Source: After Gupta, P.K., *J. Am. Ceram. Soc.*, 74, 1692, 1991.

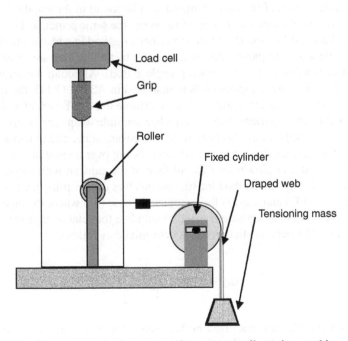

FIGURE 3.15   Variation on the capstan test that uses a tensile testing machine.

TABLE 3.6
Average Static Friction Coefficients for Selected Materials against a Hard-Coated Aluminum Drum Using ASTM G 143

Web Material	Ave μ_s
PTFE	0.05
Cotton cloth	0.16
Kapton™	0.17
Leather	0.55

Source: Adapted from ASTM standard G143, Appendix.

The static friction coefficient (μ_s) is determined using Equation 2.46 of Chapter 2, namely:

$$\mu_s = \frac{\ln(T_2/T_1)}{0.01745 \cdot \theta} \tag{3.3}$$

where T_2 is the force sensed by the load cell, T_1 the tensioning force, and θ is the wrap angle on the cylinder (degrees). Friction coefficients for selected web materials, measured using a capstan test with a 100 mm diameter, hard-coated aluminum drum, were presented in bar chart form in the appendix to ASTM G 143. Four illustrative values read from that chart are given in Table 3.6. Additional data, including the standard deviations of the measurements, can be found in the standard.

Tribometers to evaluate the friction of yarns use the same principle. For example, ASTM Standard Test Method D 3108-89 describes a method in which a length of yarn is run at known speeds (typically 100 m/min) in contact with either single or multiple friction surfaces using a cumulative wrap angle θ (rad). Although the arrangement (see Figure 3.16) is more complicated than that used in ASTM G 143, the measured input and output tensions are again used to determine the coefficient of sliding friction. Note that θ can be greater than 2π rad when multiple wraps are involved.

This draping method has also been used to measure static and dynamic friction coefficients (and the stiction) of magnetic tape sliding over a capstan drive roller.[31] Magnetic tape is drawn slowly over the surface of the capstan with a synchronous motor winder. The capstan method for stiction involves determining the tape tension for the static case (T_s) and tension for the dynamic case (T_d) where the tape is being drawn over the capstan at a constant speed to simulate the relative tape slip speed in the application. The percent stiction %S is determined as follows:

$$\%S = 100\left[\frac{(T_s - T_d)}{T_d}\right] \tag{3.4}$$

A draped-cylinder test was used by Wenzloff et al.[32] to determine the friction of sheet metal against dies in bending operations. The arrangement shown at the

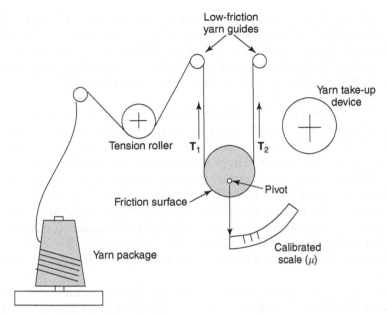

FIGURE 3.16 Device for measuring the friction of yarn on materials. (Based on ASTM Test Method D 3108.)

bottom of Figure 3.14 uses a metal sheet of thickness t, bent over a roller of radius r. First, the strip is drawn over the freely turning roller to determine the force due to the difference between the pulling and backtension forces (F_b). Then the roller is fixed as a second strip is drawn across it to determine F_1 and F_2. The equation in Figure 3.14 is then used to determine μ. The recommended procedure to determine statistically significant values of friction coefficient is to conduct tests over a range of F_2 values, plot them as ($F_1 - F_b$) versus F_2, and calculate μ from the slope of the graph. Childs and Cowburn[33] used the cylinder-wrap geometry to investigate the friction and real area of contact between flat and V-grooved belts and the driving surface. Belt friction was found to be sensitive to the belt weave and construction.

In addition to laboratory-based friction measurements on flexible materials, embedded sensors have also been developed to enable real-time measurements for manufacturing processes. This subject is beyond the scope of this book, but a description of the technology can be found in the paper by Ramasubramanian and Jackson.[34]

3.3.5 STANDARDS

Frequent reference has been made to standards throughout this book for several reasons. First, data obtained under similar, standardized conditions are more easily compared. Second, approval by a standards body implies that a concerted effort was made to understand the test method and its repeatability. Third, the scope statement published with a standard informs the user of its limits of applicability.

Unfortunately, some investigators who claim to have used standards have not applied them correctly or are not in full compliance with requirements. For example, a standard may stipulate a certain test load, but the user may select a lower load when testing a thin film that could be damaged by the specified load. Although such deviations may be justified, it remains the responsibility of the user to report which parts of the standard procedure were not followed. That qualifying step will avoid confusion when comparing results to those obtained when in full compliance with the provisions of the standard.

In 1993, a compilation of international friction and wear test standards was published by the Versailles Project on Advanced Materials and Standards (VAMAS), Technical Working Area 1 on "Wear Test Methods."[35] That compilation presented 251 standards from six countries (France, Germany, England, Japan, Poland, and the United States). Materials ranged from chopped forages to painted surfaces, and from sintered metals to movie film. Many of the standards apply to tires on roads and footwear on flooring materials. A relatively small fraction of the international standards concern friction of metals and ceramics. The growing number of standards for friction and wear testing reinforces the author's assertion that no single friction test can provide information satisfactory for all applications.

Table A1 in the appendix to this chapter lists 17 ISO standards related to friction and Table A2 lists 37 standards published by ASTM International. Even within ASTM, standards are developed by autonomously operating committees who may but need not necessarily choose to coordinate their work with other ASTM committees. Therefore, friction test methods are inconsistent in terms of, for example, test specimen cleaning, preparation, and data reporting format. If friction coefficients for similar material combinations differ when obtained using different standards, the reader should not be surprised.

The time to develop and approve a consensus standard can range from 1 to 10 years or more, depending on a number of factors. Nevertheless, the developmental history of a typical ASTM standard comprises the following activities:

- Recognition that there is a need for a standard and establishment of a task group to look into the matter.
- Agreement by the task group members on a first-round testing protocol. Then, after procuring and preparing test specimen materials, completion of one or more rounds of inter-laboratory testing.
- Discussion of test results and the determination of whether more testing is needed before a draft standard can be prepared.
- Preparation of a draft standard according to ASTM's form and style requirements.
- Discussion of the draft standard to clarify technical points, followed by submission of the draft standard to the Editorial Subcommittee for a preliminary review of clarity, spelling, consistency with format guidelines, etc. (In the case of the ASTM Committee G2 on Wear and Erosion, the editorial function is embodied in subcommittee: G02.93.)
- Balloting occurs at the subcommittee level. If subcommittee approval is not unanimous, all negative votes and comments are resolved before the draft moves to the main committee ballot.

- Main committee ballot and resolution of any negatives at that level.
- Final approval, assignment of a numerical designation, proofreading of the final standard, and publication in the Annual Book of Standards.

ASTM requires reapproval of each standard every five years, but standards can be revised more frequently if needed. Occasionally, an outmoded or obsolete standard is withdrawn or replaced by a new standard with a different designation. For example, friction test standards for sintered metal parts (e.g., ASTM B 460, B 461, B 526) have all been withdrawn. It is therefore advisable to use the most recent version of any standard. The ASTM Internet Web site (http://www.astm.org) maintains a search engine through which one can determine the most recent version of a standard and learn which standards have been withdrawn.

ASTM is one of many U.S. organizations that have standardized friction test methods. For example, the Society of Automotive Engineers (SAE), the Chemical Specialties Manufacturers Association (CSMA), the Packaging Institute, the Technical Association of the Paper and Pulp Industry (TAPPI), and the Society of Motion Picture Technicians and Engineers (SMPTE) also have developed specialized friction test standards and practices. Sometimes, standards are cross-referenced or held jointly between organizations. For example, The American National Standards Institute (ANSI) has an inclined-plane test for the friction of photographic film. Informally called "the paper-clip test" its official designation is ANSI PH1.47, "Methods for Detecting the Degree of Lubrication on Processed Photographic Film by the Paper-Clip Friction Test." According to the method, a "common steel paper clip" mounted so as to protrude from the center of a Y-shaped piece of flat plastic (such as Plexiglas™) that weighs between 60 and 100 g and straddles the incline. A test strip of film is clamped on the inclined beam. Starting with the beam raised to an arbitrary angle, the slider, which rests and balances entirely on the curved end of the paper clip, is given a slight push. If it slides quickly, the operator grabs it and prevents it from sliding off. The angle of the beam is decreased and the procedure repeated to determine the minimum tilt angle that just permits the initial motion to be sustained down the slope. The friction coefficient (tangent of the tilt angle) for both sides of the film is reported. Effective lubricants are said to produce values of $\mu = 0.15$ or lower. The paper-clip test has been adapted by ASTM and given the designation ASTM G 164.

ASTM standard test methods usually provide information on the repeatability of data (within the same laboratory) and the reproducibility of results (between one laboratory and another). In general, the repeatability of friction test standards is better than their reproducibility. However, the sources of observed variations in friction data are not always straightforward to identify. The following list suggests possible causes for test-to-test variations:

- Uncertainties in the alignment of the specimens in the apparatus
- Running-in (wear-in) effects
- Operating at or beyond the optimal operating range of the force sensor
- Instrumental drift or frictional heating of the force sensors
- Inconsistent specimen preparation or cleaning (effects of surface finishing, inconsistency in time between specimen preparation and testing)

- Nonuniform specimen materials
- External influences: external vibrations, cross-contamination from lubricant testers in the vicinity of the tribometer, particulate contamination, differences in relative humidity during the testing campaign, debris accumulation in the interface, development and loss of third-body deposits on test surfaces

One of the most-studied methods for laboratory-scale friction and wear testing is the pin-on-disk test. The VAMAS organization coordinated an extraordinary number of unlubricated pin-on-disk friction and wear tests that involved over two dozen laboratories in seven countries. Two separate rounds were conducted using sliding combinations of bearing steel, alumina ceramic, and silicon nitride ceramics. Spherically tipped pins were slid in air against polished disks at 10 N normal force, 0.1 m/s sliding speed, and 1000 m sliding distance. Results of the friction portion of those tests are given in Table 3.7, and a more detailed description of the program may be found elsewhere.[15,16] In general, the percent coefficient of variation (i.e., 100 × [standard deviation/average]) was 27.0% for self-mated 52100 steel, 23.1% for steel against ceramics, and 25.3% for ceramics against other ceramics. The repeatability of pin-on-disk tests within a given laboratory may be better than these figures if adequate control over the experiments is exercised and the materials themselves are uniform and consistently handled.

ASTM standard E 691, which is available as software from ASTM, describes the calculation of standard statistical measures for conducting interlaboratory tests. A more recent ASTM guideline, ASTM G 117, Guide for Calculating and Reporting Measures of Precision Using Data from Interlaboratory Wear or Erosion Tests, has been aimed specifically at wear testing but may also be used for friction data.

Variations in friction, especially for unlubricated surfaces, can represent a significant percent of the nominal value, and the magnitude of sliding friction can change during the course of an experiment. Say that the friction coefficient in a certain test rises to a maximum, then it monotonically decreases to the end of the test. Reporting such results becomes problematical. Which value should be reported?

TABLE 3.7
Reproducibility of Pin-on-Disk Friction Data

Sliding Combination (Pin/Disk)	Number of Tests	μ_{ave}	Standard Deviation in μ
Data from round 1			
52100 Steel/52100 steel	109	0.60	0.11
Alumina/52100 steel	75	0.76	0.14
52100 Steel/alumina	64	0.60	0.12
Alumina/alumina	76	0.41	0.08
Data from round 2			
Silicon nitride/silicon nitride	83	0.70	0.21
Silicon nitride/52100 steel	83	0.80	0.22
Silicon nitride/alumina	83	0.75	0.20
52100 Steel/silicon nitride	84	0.75	0.20
52100 Steel/52100 steel	83	0.59	0.21

Short of publishing entire friction versus time traces for each experiment, it sometimes becomes necessary to make a choice for the sake of consistency. One might select the starting value, the post-break-in (steady-state) value, the final value, and the value after five min of sliding, or the statistical average and standard deviation for a portion of the test. As is discussed in Chapter 8, the time-dependent characteristics of frictional variations can provide mechanistic clues to understanding interfacial friction processes and wear modes as well as the state of lubrication. From that point of view, it is helpful to examine more friction data than just an average or a nominal coefficient of friction. With the exception of ASTM G115, few standards require this level of detail to be reported, yet it would seem helpful for mechanical designers and those who maintain equipment to know not only the average level of friction to expect, but also what variability to expect above and below that average.

3.4 SPECIALIZED FRICTION TESTS FOR BASIC AND APPLIED RESEARCH

All manner of specialized friction tests have been developed, but they all embody the basic principles of friction testing in one way or another. A representative collection of laboratory test methods is presented in this section.

Contact conditions for specialized tests span a wide size scale. In general, the smaller the contact size, the more directly localized heterogeneities in contact surface can affect the measured friction force. Tests on very clean, flat surfaces with atom-sized probes have clearly demonstrated the periodicity of lateral forces from potential wells. Area-scanning tips, operating at slightly larger scales, have mapped frictional variations. Still larger probes, the size of roller pen balls, have measured the frictional drag of fine particle layers and tractions over 2–20 μm diameter contacts. Although laboratory-scale tests use nominal contact areas of only a few square millimeters, large dynamometers, like those involving stacked plates in large commercial aircraft brakes, have nominal contact areas measured in square meters. Each technique and each method has its place in friction science and technology.

3.4.1 NANOSCALE FRICTION

The static and kinetic friction coefficients are defined as a ratio of the tangential force to the normal force; however, the study of truly nanoscale contact phenomena, as are probed by atomic force microscopes and the like, might be better served by defining a periodically varying *frictomic force* $f(x)$ and the associated nanoscale friction coefficient μ_{nano}:

$$\mu_{nano}(x) = \frac{f(x)}{P}$$ (3.5)

where the frictomic force is periodic in sliding distance, x, and $f(x)$ could be approximated by a periodic function such as

$$f(x) = F_{max}(\cos \omega x - \phi) + c$$ (3.6)

where F_{max} is the maximum lateral force, ω is the frequency of variation, ϕ is the phase correction due to the starting position of the probe tip, and c is the baseline correction due to asymmetry or hysteresis. The feature size for which such a definition could be used might be determined by the limit at which detected fluctuations in tangential force no longer correspond to the periodic spacing between atoms or clusters of atoms on the surface.

McClelland et al.[36] published studies of "friction" on the atomic scale in the late 1980s, and Salmeron[37] employed a similar technique to study the properties of lubricant layers. The apparatus of McClelland et al. was based on an atomic force microscope. The lateral force on a tungsten wire containing a fine point (a few atoms sharp) was measured using a focused laser beam reflected from the side of the wire. Optical interferometry was used to sense the movement of the tip as the specimen, a graphite surface, was traversed below the tip using piezoelectric drivers. Normal loads, on the order of 6–75 µN, were applied, and lateral forces, on the order of 0.5–1.0 µN, were detected. Lateral force variations corresponding to a periodicity of about 0.25 nm were measured, indicating the ability of the technique to detect influences of graphite lattice spacings on lateral force. These investigators estimated a nominal friction coefficient of approximately 0.01, approximately one-tenth the value of the friction coefficients typically reported for graphite sliding on a variety of metals and on itself.[38] Such results fuel the continuing controversy regarding scale effects in friction, a topic that is addressed elsewhere in this book, and it provided yet another example of the tribosystem dependence of friction.

The term *friction force microscopy* was coined to describe the ability of a scanning tip to produce an image of an area whose contrast and brightness are related to the localized variations in tangential force within that area. Overney and Meyer[39] reviewed the subject of friction force microscopy (FFM). Two methods for conducting FFM were discussed: bidirectional measurements with two sensors and bidirectional methods with one sensor. These are schematically shown in Figure 3.17. A thin-cantilevered beam of known spring constant and containing an extremely fine tip applies the normal force to the surface through piezoelectric transducers, which raise or lower it. In the first method, two sensors are used. These can, for example, be capacitive gap sensors or tunneling detectors. In the second method, sometimes called the *torsion method*, a laser beam reflected from the upper side of the beam impinges on a four-quadrant photodiode detector. The relative intensity of the light impinging on each of the four quadrants can be added or subtracted to calculate the

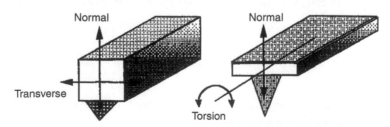

FIGURE 3.17 Two possible instrument configurations for conducting friction force microscopy.

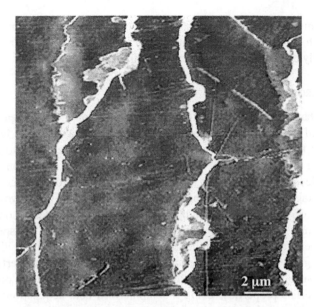

FIGURE 3.18 Friction force microscopy of a human hair. Contact mode: scan field 20 × 20 μm. (Courtesy of JPK Instruments AG, Germany.)

deflection and torsional forces on the cantilever. Commercial FFMs with sophisticated image-processing software are available. An example of an FFM image of a human hair is given in Figure 3.18. It must be remembered that the frictional variations in such images represent the interaction of specific finely tipped probes with specific substrates, and the same variations may not be observed in more macroscopic situations. FFM has also been used to study the mixing of Langmuir–Blodgett films (monolayers) of organic polymers on surfaces to provide insights into the behavior of boundary-lubricated interfaces.[40]

The use of nanocontact probes is relatively new, but studies of frictional phenomena involving microcontacts date back to more than half a century. In the late 1930s, Bowden and Hughes[41] conducted elegant experiments on the role of oxides on friction using a small cylinder sliding on a thin wire inside a sealed glass tube. The wire was degassed by passing electric current through it to heat it. When ready for the measurement, the cylinder was projected along the wire using a spring, released by an electromagnet. The deceleration of the cylinder was measured photographically and thus the friction could be estimated.

Atomic force microscopes (AFMs) can be employed when surfaces are relatively pristine and flat; however, recent attention to the possibilities of fabricating cylindrical bearings from carbon nanotubes has prompted experimentalists to device indirect friction measurement methods. In 2000, Cumings and Zettl published a paper in *Science*[42] that attempted to infer the friction forces for a telescoping arrangement of carbon nanotubes. By not measuring friction directly, but by observing the contraction of the telescoped tubes in a transmission electron microscope, assuming linear spring behavior, and making certain assumptions about the van der Waals forces that cause the contraction of the tubes, they estimated the static friction force at $<2.3 \times 10^{-14}$ N/atom and the dynamic friction force at $<1.5 \times 10^{-14}$ N/atom. The measurement of friction forces on such a fine scale prompts one to question the definition of "friction," which was classically developed for macroscopic sliding situations.

3.4.2 MICROSCALE BALL-ON-FLAT TESTS

Instruments have been developed to measure friction on contact areas larger than those sampled by nanoscale probes like AFMs, but not as large as for traditional macrocontact tests. Small-scale pin-on-flat devices have been built to measure the effects of thin films, powders, or microstructural variations on friction. The friction microprobe developed by Blau[43] in the late 1980s is one such device (see Figure 3.19). A specimen stage constructed of aluminum was suspended between two elastic webs, the deflection of which is measured by a capacitive gap gauge facing one end of the stage. A 1.0 mm diameter spherical slider is lowered onto the flat specimen by a motorized lever arrangement, and the stage is traversed at speeds from 10 to 100 μm/s while the stage deflection (i.e., friction force) is measured. The Hertzian contact diameters in such a device are of the order of 10–20 μm, depending on the normal load and materials properties. Using a computerized data acquisition system, it is possible to take several thousands of data per second, providing the device with a high spatial resolution. This device has been used to measure the friction forces along whisker-reinforced ceramic composites, layers of C_{60} powders (fullerenes),[44] and the frictional behavior of ceramic powder-covered surfaces.[45] Construction of the device to provide sensitivity to small friction forces requires it to be relatively compliant, and that characteristic made it sensitive to the rapid accelerations and decelerations that occur during stick-slip studies. More recently, computer-controlled microfriction

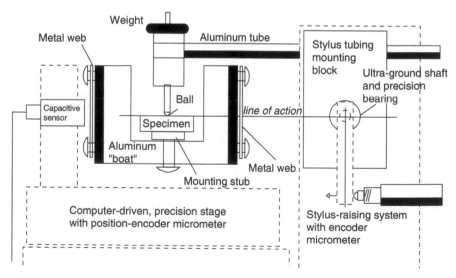

FIGURE 3.19 Friction microprobe built by the author to study phenomena between AFM contact sizes and traditional laboratory tests.

FIGURE 3.20 Close-up of the unique glass spring design of a commercial nanotribometer that allows very small loads to be applied while accurately measuring the tangential (friction) force. In this example, the static contact is a 1 mm diameter glass ball in contact with a glass-coated PCB substrate. (Nano Tribometer image courtesy of CSM Instruments.)

instruments with similar contact sizes to the ORNL friction microprobe have become commercially available. Figure 3.20 shows the slider assembly of one such instrument. It applies normal forces as low as 50 μN and can be operated with linear or rotary motions.

Microtribometers have been used to study the effects of electrical current on metallic frictional contacts. Konchits[46] used such a device in which a hemispherically tipped slider and flat counterface were electrically insulated from the rest of the machine and various current levels were passed through the interface while the friction force was measured. For a 0.5 mm radius steel tip sliding on tin at 2.25 μm/s speed and 0.1 N load, the maximum static friction peaks during stick-slip behavior increased by about 50% as applied current increased from 0 to 1.5 A, suggesting that the presence of electric current increased the adhesive component of the static friction force. In contrast to the results for tin, experiments performed on copper at 0.15 N load exhibited a different kind of behavior. There was no significant stick-slip behavior. Instead, when current was increased, the friction first dropped, and then rose to a value slightly higher than the value obtained at a lower current.

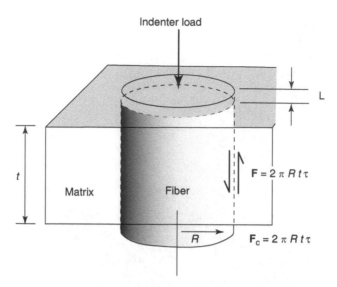

FIGURE 3.21   Diagram of the indentation of a fiber to measure its friction with the matrix. (Adapted from the description in Weihs, T.P. and Nix, W.D., *Acta Metallurg.*, 22, 271, 1988.)

Konchits proposed that when the current density at the contacts exceeded about 10^6 A/cm^2, thermal softening occurred in the contacts, altering their deformational and frictional characteristics.

3.4.3   FRICTION OF A FIBER WITHIN A COMPOSITE

Debonding and frictional sliding of fibers in the matrix behind an advancing crack in ceramic matrix composites control the toughening associated with fiber pullout. Sophisticated friction tests have therefore been developed to determine the friction between reinforcing fibers and matrix materials within composites. For example, Weihs and Nix[47] described experiments in which a computer-controlled, nanoscale indenter pushes on the end of a fiber to generate wall debonding and interfacial sliding. In some cases, relatively thick specimens of material are used and the analysis considers the effects of sliding within the solid. In other experiments very thin specimens are used so that the fiber is pushed partly out the opposite side. Figure 1.1 in Chapter 1 shows a scanning electron micrograph of the results of a push-in experiment, and Figure 3.21 shows it schematically. The force **F** on the indenter results in a sliding length of L of a fiber of radius R, determined by the frictional resistance to slip τ. Thus,

$$L = \frac{\mathbf{F}}{(2\pi R\tau)} \tag{3.7}$$

At some critical length, L_c, the fiber will be pushed out of a specimen of thickness t:

$$L_c = 2\pi R t\tau \tag{3.8}$$

Equations 3.7 and 3.8 simplify the actual situation in which lateral strains during compression force the fibers to expand against the walls of cylindrical boundary. Other complications, such as fiber/matrix interface roughness and transverse fiber cracking, make simple analyses of the experiment approximate at best. Mathematical treatments of the problem are still the subject of debate, because initial assumptions on which such analyses are based may not always be valid in different material systems. Readers may want to consult the analysis of McCartney,[48] Hsueh,[49] and Hutchinson and Jensen[50] to better understand the complexities of such treatments and their applicability to various regimes of material behavior.

3.4.4 MULTIDIRECTIONAL TRIBOMETERS

Pin-on-disk machines are a common means to measure sliding friction and wear in the laboratory. One rather unique variation of this method, described by Briscoe and Stolarski,[51,52] involves rotating a flat-ended pin while the disk is also rotated. These synchronized movements enabled investigators to study frictional energy dissipation in several polymers and highlighted the importance of textural orientation relative to the sliding direction.

To properly simulate the *in vivo* wear response of polymeric bio-implant materials such as ultrahigh molecular weight polyethylene (UHMWPE), cross-path motion is required.[53] Friction can also be important in systems in which there is momentary loss of contact and when impacts are combined with sliding. For example, Fujisawa et al.[54] describe a variant of a pin-on-disk apparatus that can impose macroscale impacts to the surface between periods of sliding.

3.4.5 FRICTION OF IMPACTING SPHERES

A device for measuring the friction of individual or streams of spheres on impact with a plate was described by Kragelskii et al.[55] Recorded by high-speed stroboscopic photography, a spring was used to propel a sphere against a plate at a preset angle (α_o). The angle of rebound (α_k) and the velocities before v_o and after v_k impact are measured. The friction coefficient on impact μ_i was estimated from

$$\mu_i = \frac{v_o \cos\alpha_o - v_k \cos\alpha_k}{v_o \sin\alpha_o + v_k \sin\alpha_k} \tag{3.9}$$

Table 3.8 compares the impact friction of rubber by 2 mm diameter steel balls and polyurethane balls as a function of attack angle α_o.

3.4.6 PENDULUM-BASED DEVICES

Oscillating pendulum devices have been used to measure the frictional characteristics of lubricants used in precision instruments. Rymuza[56] and Barker[57] have used such devices. A schematic diagram of the arrangement is shown in Figure 3.22. A pendulum that is connected to a cylinder in a groove of angle φ is released at some

TABLE 3.8
Impact Friction Coefficients for Steel and Polymeric Balls on Rubber

Angle (Degrees)	μ_i, Steel-on-Rubber, $v_o = 2.0$ m/s	μ_i, Polyurethane-on-Rubber, $v_o = 4.0$ m/s
20	0.50	0.55
30	0.33	0.42
40	0.24	0.30
50	0.15	0.22
60	0.12	0.15

Source: Adapted from Kragelskii, V., Dobychin, M.N., and Kombalov, V.S. in *Friction and Wear Calculation Methods*, Pergamon Press, New York, 1982, 212–215.

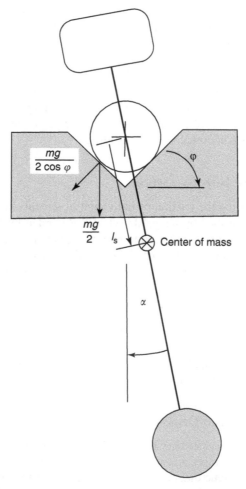

FIGURE 3.22 Oscillating pendulum device for measuring friction in instrument bearings.

angle α from the horizontal (e.g., 25°). The friction coefficient for the system μ is related to the amplitude damping decrement Δ by

$$\mu = C\Delta \tag{3.10}$$

where

$$\Delta = \frac{(A_o - A_n)}{n} \tag{3.11}$$

and

$$C = \frac{(l_s \cos\varphi)}{4r} \tag{3.12}$$

A_o and A_n are the amplitudes of the pendulum on the first swing and after n oscillations, respectively. The radius of the cylinder is r and the distance between the center of the cylinder and the center of the mass of the pendulum is l_s. This technique has been used to characterize lubricants for fine instrument movements and oscillating bearings. A more recently developed, sensitive method to measure additive effects on lubricants is described in Section 3.4.7.

3.4.7   FRICTION MEASUREMENT USING PRECISION CHAINS

Assessing the frictional effects of small changes in lubricant composition can be challenging in light of the relative magnitude of the friction coefficients (μ generally less than 0.10) and the need for high repeatability in the measurements. P. Lacey, working at Southwest Research Institute, developed an innovative approach using fine chains draped over a drum.[58] Subtle changes in lubricant additive effects were detectable using that method.

A subsequent refinement of Lacey's wire-on-capstan method was described by Hargreaves and Tang in 2006.[59] Since the measurement must be sensitive enough to detect differences in the second and third decimal places of friction coefficient, cleaning of the chains before testing was found to be important for achieving repeatable, discriminatory results. The mean Hertzian contact pressure, P (Pa), for a strand of wire on a capstan was approximated as

$$P = 0.625\left(\frac{T_1 E_{red} e^{\mu\alpha}}{2Rr}\right)^{1/2} \tag{3.13}$$

where T_1 = tension on the wire from the applied load (N), μ = friction coefficient, α = wrap angle (radians), R = radius of the capstan (m), r = radius of the wire (m), and E_{red} = reduced elastic modulus (Pa) given by

$$E_{red} = \left(\frac{1 - v_1^2}{E_1}\right) + \left(\frac{1 - v_2^2}{E_2}\right) \tag{3.14}$$

Symbols $v_{1,2}$ are Poisson's ratios and $E_{1,2}$ are elastic moduli for the wire material and capstan materials, respectively. Because μ appears in the exponent in Equation 3.13, lubricants with poor frictional characteristics produce higher contact pressures than those with better lubricity. Friction coefficients for a series of oils, tested between 20 and 100°C, ranged from 0.094 to 0.158 and fell within the ranges expected for so-called boundary lubricants.

3.4.8 PISTON RING AND CYLINDER BORE FRICTION

As much as 40–50% of the total frictional loss within an internal combustion engine can be attributed to the piston ring/cylinder bore interface. Because lowering friction raises engine efficiency and saves fuel, considerable attention has been paid to developing test methods for this application. The relative frictional losses in the various subsystems of engines vary with speed and duty cycle, and it is difficult to extract the frictional contribution of the ring/cylinder interface from other sources using dynamometer measurements of power output. To complicate matters, the friction between the piston ring and the cylinder bore is not constant but varies with crank angle. Intuitively, this complex behavior should not be surprising because the piston rapidly changes speed and direction, and the lubricating film thickness in the interface varies during the stroke (see Chapter 6 on lubrication). Furthermore, the temperature decreases from the combustion end of the stroke to the end nearer the sump. Automotive engineers have struggled with this measurement problem for decades and a number of them have attempted to correlate laboratory, bench-scale friction tests and engine tests.[60–63] Computer programs have also been developed to sort this out, but this will not be described here.

The large number of variables in the environment of the piston ring/cylinder interface presents a challenge to those developing bench tests. Gaydos and Dufrane[64] developed a reciprocating test rig into which hot exhaust from a single-cylinder diesel engine was admitted. Two opposing flat specimens were intended to represent the cylinder liner, and two curved specimens, with a 32 mm radius crown, were used as the "ring" specimens. The testing arrangement is shown in Figure 3.23. The stroke

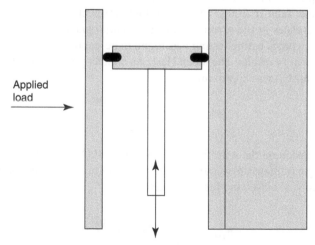

Applied
load

Battelle Columbus Laboratories test
double reciprocating curve-on-flat

FIGURE 3.23 Simulator used for studies of piston ring/cylinder liner friction developed at Battelle Columbus Laboratories.

length in that case was 108 mm and the drive speed was 1500 rpm. Test temperature was maintained at 260°C and the ring loading was 10.3 N/mm, produced by side loading of one of the flat specimens.

Rather than using segments of piston rings and cylinder liners like Gaydos and Dufrane have, some investigators have created what are known as suspended liner (a.k.a. "floating liner") devices. Cater et al.[65] describe a bench-scale device in which a cylinder liner is suspended by sensors while a piston is driven up and down by a crank immediately below it. The suspended liner tribometer was used to model friction force variations during lubricated sliding in which the normal force varies with crank angle in the simulated engine.

Because friction testing embodies mechanical, thermal, material, and tribochemical aspects, most attempts at simulating piston ring/cylinder behavior in laboratory tests have focused more on the first three of these aspects than the fourth. In fact, the condition of the lubricant can play a significant role in ranking of materials, coatings, and surface treatments used in cylinder components. It cannot be assumed that fresh lubricant poured from a can will adequately simulate the condition of engine-conditioned lubricants. Some of the first work in this area is attributed to F. Rounds,[66] who 25 years later published a review.[67]

During the 1990s, M. Naylor[68] at Cummins Engine Company found that friction and wear of advanced coatings and surface treatments for diesel engine piston rings and cylinder liners ranked in a different order of merit when tested in fresh and sooty diesel oil. Sooty oil was not necessarily higher or lower in friction than fresh oil for various combinations of piston and liner materials—it was just different. Therefore, it was not a matter of simply applying a "correction factor" to compensate for oil condition. This finding prompted a subsequent research effort at ORNL to develop more realistic bench tests that used engine-conditioned lubricants in simulative tests,[69,70] and that effort culminated in the development of a new standard, ASTM G 181.[71]

Under ASTM Committee G2 auspices, a task group of automotive engineers and testing machine makers was formed. It was decided to use a commercially made, linearly reciprocating testing machine (Cameron-Plint model TE-77) that was in use in dozens of laboratories, so there was an experience base from which to draw. The ring specimen holder had to be designed to conform to the keystone shape of the piston ring segment, and it was fabricated from a block cut from a production piston from a Caterpillar C-15 diesel engine (see Figure 3.24). Special alignment and running-in procedures were then designed to reduce the variability in friction results from test-to-test. To evaluate the sensitivity of the proposed method to oil condition, a series of chemically analyzed diesel engine drain oils was obtained from Southwest Research Institute. Measuring the high soot content (>5%) accurately in those oils required specialized procedures, beyond those used in routine oil analysis.

An extended series of experiments using production-grade, Cr-plated rings sliding on cast iron test coupons verified that the method could repeatedly differentiate between lubricant conditions. Sample test results are provided in Table 3.9. The average values for μ at each applied load and the coefficient of variation (standard deviation/average, expressed in %) are also given. As shown, friction coefficients varied within a relatively narrow range and were typical of the boundary lubrication

FIGURE 3.24 A block was cut from the top of a piston from a commercial diesel engine to provide the correct shape to hold a ring segment with a tapered cross-section.

TABLE 3.9
Average Friction Coefficients for Test Oils at 100°C and Two Loads

Test Oil	Soot Content (wt%)	Average μ at 20 N Load	Average μ at 200 N Load
Fresh 15W40 diesel engine oil	<0.1	0.125	0.112
M11 HST (high soot test)	6.7	0.133	0.113
M11 EGR test	9.3	0.113	0.113
Mack T8 test	5.2	0.128	0.112
Mack T9 test	2.1	0.092	0.095
Mack T10 test	5.8	0.103	0.106
	Average	0.116	0.109
	CoV	13.8%	6.6%

Note: Selected data from Truhan, J.J., Qu, J., and Blau, P.J., *Wear*, 259, 1048, 2005.

regime (see Chapter 6). Note that a precision of three decimal points was needed in this case to differentiate small differences in frictional behavior.

The M11 high soot test and the M11 EGR test generally produced the highest friction, but the range in friction coefficients among the various test oils decreased at high load for reasons that will become clearer from the discussion of the Stribeck relationship in Chapter 6. Note that μ for fresh oil was higher than the average and that freshness does not guarantee low friction, a finding consistent with Naylor's earlier work.[68]

3.4.9 FRICTION OF BRAKE LININGS

The proper functioning of friction brakes in automobiles, trucks, and aircraft can literally be a matter of life and death, and the friction coefficient is a contributor to that performance. Other factors include the design of the brakes, the area of contact, the brake control system, aerodynamic drag, tire/road grip or slip, and the driver's behavior. Therefore, the effectiveness of a particular braking system is composed of many factors, including, but not limited to, the friction between the contact surfaces.

Automobile and truck brakes come in a wide variety of designs, sizes, and operating systems. The two most common types are drum brakes and disc* brakes. Most automobiles and light trucks in the United States use hydraulic disc brakes. Worldwide, air disc foundation brakes are commonly used in heavier trucks, but historically the U.S. heavy trucking industry has clung to the use of drum-type brakes. However, as described in Chapter 9, this situation is changing.

Similar to other applications, the challenge for friction testing of brake materials is to replicate conditions that are representative of their full spectrum of use. In practical terms, that is truly a tall order. Brakes operate in hot and cold environments, both wet and dry, on steep grades, on and off paved roads, with frequent stops, with all kinds of driver behavior, and with different kinds of control systems. A brake material may perform well under one set of conditions but poorly under another; therefore, the development of acceptable performance criteria for brake materials is challenging, and most commercial brake lining tests comprise a sequence of stages to simulate a range of operating conditions.

Tests of brake lining materials are often conducted on what are called "dynamometers." The kinetic energy needed to simulate the vehicle forces on the brakes can be provided by a series of spinning weights (inertia dynamometers) or by a motor whose power output is controlled to simulate different kinds of braking conditions that are experienced by a selected type of vehicle. Hundreds of computer-controlled, instrumented dynamometers with forced-air cooling systems are in use throughout the world. In just one dynamometer test a great deal of data are generated and the challenge becomes one of interpreting its meaning. The notion that a single value of the kinetic friction coefficient can characterize the braking performance of a given friction material is naive because the same series of brake linings can rank in different orders of merit depending on which frictional characteristics are compared.

---

* When discussing brakes, the author uses the industry's conventional spelling "disc." When referring to pin-on-disk laboratory tests, the alternate spelling is used.

Despite the popularity of brake dynamometers, there is also a role for smaller-scale bench tests in areas like preliminary screening of lining formulations and quality control checking of batches of lining materials. Bench-scale tests are also used to evaluate experimental materials that might otherwise be too costly to fabricate into full-sized brake components during their early stages of development. Laboratory tests for friction brakes typically range in scale from compact, bench-top devices that use 12.7×12.7 mm^2 contact pads to massive 3.7 m (12 ft) diameter inertia dynamometers that are used to evaluate brakes for large aircraft. Therefore, the nominal contact area of friction couples in the laboratory can range from that of a set of full-sized brakes to subscale coupon tests in which the nominal contact area is less than 1/5000th as large.

Past researchers of brake lining formulations built their own tribometers. For example, in the 1970s, Ho and Peterson[72] used an electric motor to spin a 30.5 cm diameter, 0.79 cm thick, steel disk at 1750 rpm to produce a sliding velocity of 22 m/s at the contact position. A high-pressure, compressed air cylinder applied braking loads of up to 10,800 N to a pair of 1.9 cm diameter buttons mounted on a caliper assembly. Ho and Peterson first used a series of tests at constant load but for different sliding times to prescreen materials. Then a second test series, in which the load was increased until failure, was used to down select materials from those that passed the first series. Some investigators[73-75] have modified pin-on-disk testers to conduct brake lining studies, but those tribometers involve small contact areas and usually do not simulate full-sized brake heat dissipation, sliding speeds, or system stiffness. Relative comparisons within the same study may be useful but may not be extrapolated to full-sized brakes.

A series of research projects on advanced materials for friction brakes was conducted at ORNL using a subscale brake materials testing (SSBT) system.[76-78] The apparatus, shown in Figure 3.25, uses a 10 hp, variable-speed motor (producing surface speeds of up to 15.0 m/s), a precision spindle, and a compressed-air-driven force application system. Lining specimens were 12.7 mm square-faced blocks. Data on friction force, normal force, and temperature (via an infrared detector aimed at the track on the disc specimen) were recorded with a data acquisition system at rates of a few hundred data per second. A shroud and water nozzle can be used to simulate the effects on braking friction when contact surfaces get wet.[79]

An attempt to correlate SSBT data with published dynamometer data for the same lining materials was reported elsewhere[76] and is summarized here. The research sponsor was primarily interested in "S-cam" drum brakes common in U.S. heavy trucks. In that design, an S-shaped cam rotates to force a pair of brake shoes outward and against the internal diameter of the drum. The following equation for the braking force on a drum brake (F_b) was presented in a paper by Strawhorn:[80]

$$F_b = \frac{2pAS\mu r_d}{r_c r_t} \tag{3.15}$$

where p = air line pressure, A = cross-sectional area of the piston actuator, S = length of the slack adjuster, r_d = brake drum radius, r_c = radius of the S-cam, r_t = mean static rolling radius of the tire, μ = mean friction coefficient.

FIGURE 3.25 Subscale brake material tester (SSBT) built at Oak Ridge National Laboratory to study the friction of nonconventional brake materials under dry and wet conditions.

At equilibrium, the torque on a tire with a certain radius (r_t) arising from the braking force (F_b) balances that from friction of the lining on the brake drum surface (F_d) with an inner drum radius of r_d. Thus,

$$T = F_b r_t = F_d r_d \tag{3.16}$$

Using Equations 3.15 and 3.16, the friction coefficient corresponding to torque values from dynamometer tests was estimated from

$$\mu = \left(\frac{T}{p}\right)\left(\frac{r_c}{2ASr_d}\right) \tag{3.17}$$

The Performance Review Institute (PRI) is affiliated with the SAE, and it maintains a database of torque values for aftermarket linings intended for use on drum brakes in trucks with gross axle weight ratings in the range of 17,000−23,000 lbs (7,718−10,442 kg). The PRI certifies dynamometer test facilities to participate in the program. The procedure is based on the dynamometer testing portion of the U.S. government's Federal Motor Vehicle Safety Standard (FMVSS) 121,[81] but only the braking torques for 40 psi line pressure results are reported, in accordance with Technology and Maintenance Council (TMC) Recommended Practice TMC RP 628.[82] Figure 3.26 illustrates the distribution of torque values for the aftermarket brake linings for vehicles that were published in the TMC "Fleet Advisor" Internet

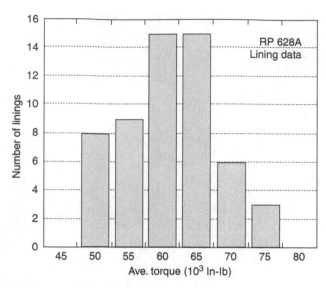

FIGURE 3.26   Published TMC RP 628 torque values for 56 replacement brake linings for heavy trucks (PRI data).

Web site in December of 2003. The figure includes only the data for those 56 linings that used an air chamber size of 30 in.[2]

Table 3.10 compares friction coefficients that were calculated using Equation 3.17, using PRI dynamometer torque data, with friction coefficients measured using the SSBT on small specimens of the same three commercial linings. SSBT tests were performed using gray cast iron as the disc material and several load and speed combinations. Results indicate that some applied test conditions for this block-on-disk apparatus correlated well (the difference was <10%) with friction coefficients calculated from dynamometer testing of S-cam drum brakes, but some results differed by as much as 29%. Two of the commercial lining materials correlated better with SSBT data than the third (AR5 is a high-traction material recommended for dump trucks), suggesting that it is not only the test method, but also the material that influences whether a reasonable correlation will be obtained between subscale and full-scale tests.

The data in Table 3.10 highlight two important implications for correlating bench-scale with full-scale test results, whether related to brakes as in this case, or to other tribosystems:

1. Concluding that a given bench-scale tribotest does or does not correlate with field tests is unjustified unless the full testing parameter space has been explored. Some test conditions may enable correlations better than others.
2. Correlations between tribosystems may be material dependent. In other words, if two tribometers are found to correlate well for a certain friction couple, that finding does not guarantee that results will correlate equally well for other material combinations.

TABLE 3.10

Comparison of Friction Coefficients from Dynamometer Tests with Those from the SSBT for Various Test Conditions

Lining Material	Test Method	Dyno Torque (in lb)	SSBT Contact Pressure (MPa)	SBT Sliding Speed (m/s)	Friction Coefficient[a]	% Difference (Dyno vs. SSBT)
Armada™ AR2	RP 628A	58,581	—	—	0.269	—
	SSBT	—	3.2	1.9	0.312	16.0
	SSBT	—	3.2	6.0	0.283	5.2
	SSBT	—	3.2	10.0	0.318	18.2
	SSBT	—	4.6	2.0	0.269	0.0
	SSBT	—	5.1	6.0	0.297	10.4
	SSBT	—	5.1	10.0	0.327	21.6
Armada AR4	RP 628A	64,716	—	—	0.297	—
	SSBT	—	2.5	1.9	0.303	2.0
	SSBT	—	2.5	6.0	0.278	6.4
	SSBT	—	2.6	10.0	0.306	3.0
	SSBT	—	5.6	2.0	0.251	15.5
	SSBT	—	5.6	6.0	0.262	11.8
	SSBT	—	5.6	10.0	0.343	15.5
Armada AR5	RP 628A	58,581	—	—	0.366	—
	SSBT	—	4.4	1.9	0.416	13.7
	SSBT	—	4.4	6.0	0.432	18.0
	SSBT	—	3.2	10.0	0.472	29.0
	SSBT	—	10.7	2.2	0.395	7.9
	SSBT	—	8.5	6.5	0.419	14.5

[a] Calculated using Equation 3.17 and the 40 psi torque from RP 628A dynamometer tests; or for SSBT tests, calculated directly from average friction force and normal force for a series of constant speed drags.

The following chapter explains how friction derives from a combination of interfacial processes, and effective laboratory simulation depends on duplicating the correct combination of processes.

Sanders et al.[83] from Ford Motor Company, Scientific Research Laboratory, considered how a bench-scale tribometer might be made more simulative of actual braking conditions and provided their analysis and design philosophy for subscale lining tests. Their approach was based on an analysis of the energy dissipation per unit area and heat flow.

Bench-scale tribometers can provide useful information for lining material developers and those studying the mechanistic behavior of friction materials. Lining formulations can contain as many as 40 additives,[84] and running full-scale brake tests of numerous experimental compositions—especially with duplicate runs to establish repeatability—would be prohibitively expensive. For that reason, two laboratory test methods have become *de facto* standards: (1) the friction assessment and screening test (FAST)[85] that was developed by Arnie Anderson of Ford Motor company in the

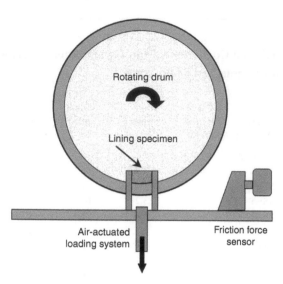

FIGURE 3.27 Chase test geometry used for friction material quality control and the development of lining materials.

TABLE 3.11

Edge Codes Specified for Friction Test Results in SAE Standard J866

Edge Code	Friction Coefficient
C	≤0.15
D	>0.15 and ≤0.25
E	>0.25 and ≤0.35
F	>0.35 and ≤0.45
G	>0.45 and ≤0.55
H	>0.55
Z	Unclassified

1960s and (2) the Chase test (Figure 3.27). The former uses a 1/2-in. (12.7 mm) square pad specimen pressed on the face of a 90 mm diameter cast-iron disc, and the latter uses a 1-in. (25.4 mm) square pad that conforms to the inside diameter (277.4–279.9 mm ID) of a simulated brake drum. The Chase test includes a burnishing procedure, and six subsequent stages (baseline, first fade, first recovery, second fade, second recovery, and baseline rerun) that generate friction data at a series of reported drum temperatures. Data for each application of the pad are presented in both graphical and tabular form.

    Data from SAE J661 (Chase) tests are incorporated into a brake lining marking scheme called "edge codes" that is described in SAE J866. Alphabetical edge codes are assigned to ranges of increasing friction coefficient under what are termed "normal" and "hot" test conditions (see Table 3.11). They derive from laboratory test conditions that do not necessarily simulate driving situations. Therein lies the problem with selecting replacement brake linings based on edge codes.

Although the FAST test and the Chase test continue to be used, direct use of the results for selecting vehicle replacement linings is controversial. Despite their intended use in materials research and development or for quality control, some vehicle fleet purchasing agents use them to select replacement brake linings. Amid protests from brakes engineers and brakes industry working groups, bench-scale data have historically been misapplied. Those who select brake linings should rely more on full-scale brake test methods or road trials to evaluate products for the specific types of vehicles and usages. However, edge codes are published, convenient, and far less costly than vehicle testing programs.

With support from the National Highway Traffic Safety Administration (NHTSA), ORNL and the National Transportation Research Center, Inc., in Knoxville, Tennessee, conducted an extensive study involving SSBT test data, Chase test data, test track results, and on-highway heavy truck braking distances for nine commercial lining materials.[86] Two levels of severity were used in the SSBT tests (low-load/low-speed and high-load/high-speed). Chase tests were conducted at Link Testing Laboratories following the SAE J661 procedure. Results of that work related to the current discussion are as follows:

- Not all friction coefficients obtained in Chase and SSBT tests for the same linings matched; however, six out of nine linings ranked in similar order with respect to the friction coefficient under a given set of test conditions.
- After normalizing for the gross vehicle weight and compensating for the vehicle operators' failure to maintain the desired air line pressure during stopping, the stopping-distance trends for on-highway tests of heavy trucks correlated reasonably well under some Chase and SSBT testing conditions.
- The truck drivers' performance significantly affected the consistency of road test results, and this factor made correlation between laboratory data and truck braking performance more problematic. As is described in Chapter 9, the brake friction is not the only factor capable of decelerating a given vehicle.

Table 3.12 lists some current industry brake lining friction-testing procedures. Differences between these procedures include the kinds of brake system designs, vehicle size and axle weight rating, the scale of testing, and the kinds of braking characteristics being investigated.

A general review of U.S. and European brake testing methods was published by Agudelo and Perro.[87] It discusses procedures from the U.S. Department of Transportation, SAE, the International Standards organization (ISO), and the United Nations Economic Commission in Europe (ECE). These methods continue to evolve.

As of this writing, the U.S. Department of Transportation requirements for brake linings apply only to original equipment products despite the fact that an estimated 70% or more of the commercial trucks on U.S. highways are using aftermarket (replacement) linings. In recent years, there has been a flood of inexpensive imported linings into the United States, and under a privately financed study, a troubling

TABLE 3.12

Standard Tests Used for Braking Systems and Materials

Designation	Title	Comments and Applications
FMVSS 105	Hydraulic and electric brake systems	Full-sized vehicle test for normal and emergency braking conditions on vehicles weighing more than 3500 kg
FMVSS 121	Air brake systems	Contains both a dynamometer portion and a stopping-distance requirement for heavy trucks using original equipment linings
FMVSS 135	Passenger car brake system	Self-certification test for original equipment hydraulic and parking brakes used on vehicles weighing less than 3500 kg
ECE R90	Uniform provisions concerning the approval of replacement linings assemblies and drum brake lining for power-driven vehicles and their trailers	Separate annexes describe a series of test methods for brakes on different kinds of vehicles
PRI-TMC RP 628	Aftermarket brake lining classification	Reports the 40 psi air pressure torque level measured in the FMVSS 121 dynamometer procedure
SAE J661	Brake lining quality control test procedure	A laboratory-scale "Chase" test using a pad-on-disk geometry
SAE J1652	Dynamometer effectiveness characterization procedure for passenger car and light truck caliper disc brake friction materials	Inertia dynamometer-based test for vehicles up to 5954 lb (2700 kg) in gross weight
SAE J1802	Brake block effectiveness rating	Inertia dynamometer test method using a drum-brake configuration
SAE J2430	Dynamometer effectiveness characterization test for passenger cars and light truck brake friction products—brake evaluation procedure	Single-ended dynamometer test for disc-type automotive brakes; also used for the brake effectiveness evaluation procedure (BEEP®), a voluntary certification program
SAE J2681	Inertia dynamometer friction behavior assessment for automotive brake systems	Effects of pressure, temperature, and linear speed on friction material performance; provides stopping distance estimates based on correlated data

number of these imports failed the FMVSS tests prescribed for original equipment linings.

To bolster the ability of law enforcement and safety officials to evaluate brakes at highway testing stations, there has been a recent push toward what are called *performance-based brake testing* (PBBT) technologies.[88] Two main types of PBBTs are a flat plate type that uses force sensors built into a series of parallel tracks in a road surface and roller dynamometers on which the truck drives onto a set of instrumented rollers and performs a specified series of controlled brake applications. As an example of PBBT test criteria, consider the ratio of the braking force (BF) to the weight on the braked wheel (WL). Somewhat analogous to the friction coefficient, minimum (BF/WL) ratio of 0.25 for steer axle brakes and 0.35 for nonsteer axle brakes are recommended. Field deployment of PBBTs is intended to augment highway vehicle inspections, not to replace them.

3.4.10 TIRE/ROAD SURFACE TESTING

A driver's ability to slow or stop a vehicle depends not only on the effectiveness of the friction between the lining and the drum or disc, but also on the traction between the tires and the road surface. As with braking systems, there are a large number of standards and customized test procedures for tire/road traction. Those who manufacturer or select tires tend to use standardized road surfaces and vary the tire composition and design, but those who investigate pavement properties might limit the number of tire compositions and design variants and focus instead on different kinds of road surfaces. Needless to say, the number of variables in tire/road testing becomes quite large, and testing standards have been helpful. Several ASTM methods for testing tires and pavements are included in Table A2 in the appendix to this chapter.

There is a difference between friction coefficients for tire surface materials and the traction of those tires. For the purposes of this book, the term tire friction refers to that between the outer surface of the tire tread material and a particular counterface. Tire traction, however, reflects the functional performance of the tire based on its shape, materials of construction, tread design, internal structure, inflation pressure, temperature, and surface over which it is traveling. Internal friction, within the tread and casing, affects the rolling resistance of the tire, its tendency to heat up during use, and its durability. Lowering the tire rolling resistance can increase the fuel efficiency of a car or truck by several percentage points.[89]

Unlike friction materials development for which laboratory-scale tribometers have been used, tire friction/traction tests have utilized full-scaled tires, either in dynamometer test cells, on instrumented trailers, or on the actual vehicles for which they are intended. So-called "split-mu" tests involve driving along prepared sections of a road surface in which one side of the vehicle is traveling on a wet or icy surface and the other side is traversing a dry surface. Test track facilities, like the U.S. Department of Transportation's Turner-Fairbank Highway Research Center in Virginia and the Transportation Research Center, Inc., a contract testing facility that occupies over 4500 acres in East Liberty, Ohio, specialize in road surface, braking, and on-road tire testing.

Table 3.13 lists the types of road surfaces that were defined in a report from the state of West Virginia.[90] Each type is defined in terms of its constituents, the manner of blending, whether the surface has been graded for water run-off control, the rigidity of the road base, and sometimes, in terms of the thickness of the stabilized layers. In other parts of the world, packed sand surfaces would also be included in such a list.

TABLE 3.13

Types of Road Surfaces Categorized in the State of West Virginia

General Category	Subcategory
Unsurfaced	Primitive
	Unimproved
	Graded and drained
Soil surfaced, gravel, and stone	Soil surface
	Gravel or stone
Paved	Bituminous surface treated
	Mixed bituminous
	Bituminous penetration
	Asphaltic concrete
	Concrete
	Brick

Source: Summarized from http://www.wvdot.com/3_roadways/rp/facts/Chapter% 202/ Road%20Surface% 20Type.pdf.

The roughness profile of a paved road surface, also a factor in tire traction, is remarkably similar in appearance to that for a ground metal surface that was measured using a stylus profiling system; however, instead of having a traverse length in mm and asperity heights of a few micrometers or less, the traverse length is expressed in tens of meters and typical feature heights are in the order of 5–50 mm.[91]

A review of tests for road surfaces and the effects of road condition on traction can be found in a report by Liang.[92] Müller et al.[93] have proposed a means to estimate the maximum friction of tires on highway surfaces. However, another major area of tire friction testing involves aircraft safety during landings in different weather conditions, as described by Norheim et al.[94] These authors attempted to establish a relationship between braking performance and the microstructure of ice and snow on runways. Water, in the form of ice and snow, develops a myriad of structures that can affect its tribological behavior.

3.4.11 WALKWAY FRICTION TESTING

The National Safety Council (Itasca, Illinois) estimates that over 25,000 slip and fall injuries that result in hospital care occur each day in the United States, and that annual workman's compensation claims exceed US$ 200,000,000. Safety-related and economic consequences of slips, trips, and falls have therefore prompted considerable interest in the testing, analysis, and design of walking surfaces.[95–98]

So-called *articulating strut devices* have been developed to measure footwear friction. Consider, for example, the one described by R. J. Brungraber, Bucknell University, and M. I. Marpet, St. John's University,[99] illustrated in Figure 3.28. The specimen surface (e.g., a floor tile) is clamped onto a baseboard and the angle board is set at a specific angle. The weight is released, and it is determined whether the strut assembly "breaks away," as evidenced by the separation of the front of the weight from

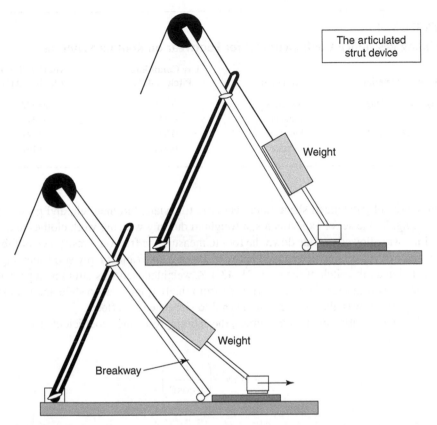

FIGURE 3.28   The articulating strut device used to simulate walking friction of flooring.

the ramp. A full set of angles is used and those angles that result in breakaway are noted. The tangent of the breakaway angle is the critical friction coefficient. Other ASTM tests for footwear on flooring, based on the so-called James machine and on Brungraber's strut devices, are listed in Table A2 in the appendix to this chapter.

In 1996, ASTM Subcommittee F13.30 on Footware first published a standard guide for reporting the details of slips, stumbles, trips, and falls.[100] The production of surfaces with low or nonslip characteristics is an important focus within the flooring industry; however, sidewalks, flooring, and other walking surfaces are only one place where footwear traction is important. Another place is the construction industry.

Nagata[101] reviewed several methods to measure the friction of footwear worn in the Japanese construction industry. He reported that falls from roofs, scaffolds, ladders, and other overhead structures accounted for an estimated 43% of construction industry fatalities in Japan in 1998. A simulated, variable-pitch roof apparatus, 6.1 m long and capable of inclining at 21–35° to simulate the range of slopes on typical Japan roofs, was built at the National Institute of Industrial Safety in Tokyo, Japan. The device was used to measure the sliding resistance of three types of footwear (safety shoes, sneakers, and split-toed, rubber-soled *jikatabi*) on roofing surfaces

TABLE 3.14

Sliding Resistance Coefficient (μ_k) for Footwear on Roofing Materials

Roofing Material	Footwear	Dry Conditions (Pitch = 42°)	Wet Conditions (Pitch = 31°)
Plywood sheeting	Sneakers	0.835	0.322
	Jikatabi	0.84	0.371
Galvanized metal	Sneakers	0.84	0.341
	Jikatabi	0.774	0.161

composed of galvanized sheet metal, ceramic tile, slate, bitumen felt, and plywood sheeting. In one set of experiments, a weighted dummy wearing work clothes and a helmet was allowed to slide down the roof to measure the frictional resistance of fabric and exit velocity, which ranged between about 2 and 7 m/s on plywood sheeting, depending on the pitch of the roof (21–42°). A weighted sled was built and a pair of test shoes was inserted into it to protrude from the bottom so as to slide against the roofing. Water was also sprayed on the roof to simulate rain effects.

For a roof with a pitch of θ degrees, the "sliding resistance coefficient" (μ_k) was defined by

$$\mu_k = \tan\theta - \left(\frac{\alpha}{\mathbf{g}\cos\theta}\right) \tag{3.18}$$

where α = the slide acceleration (m/s) and $\mathbf{g}$ = 9.8 m/s². Acceleration was measured using a rotary reel device with an optical encoder to count revolutions as a tethered cord on the slider unwound. Table 3.14 lists Nagata's μ_k data for sneakers and jikatabi sliding on plywood and sheet metal. The results suggest that sneakers are better than jikatabi for walking on galvanized metal roofing in the rain.

3.4.12 METALWORKING

The need to control friction in metalworking and manufacturing processes has led to the development of specialized laboratory test devices to simulate situations in which friction affects the ability to efficiently change the shape of a work piece or to control its surface quality.[102] For example, Lin et al.[103] described a technique to measure friction during deep drawing of can blanks in which they instrumented the blank holder. Using a split work piece holder and a known hold-down force on the sheet, these investigators measured the sliding friction coefficient as the material was drawn into a die. The effects of different lubricants were evaluated. More recently, the subject of friction-testing methodology for sheet metal forming has been reviewed.[104] One of the most comprehensive reviews of friction and lubrication concepts in metalworking was first published by Schey in 1983,[102] and a more recent handbook, edited by Totten,[105] also provides a comprehensive review of the subject.

3.4.13 FRICTION OF ROCK

The frictional properties of rocks and minerals are of interest in the prediction and explanation of earthquake phenomena.[106] Earthquakes may result from complex, intermittent sliding of multiple rock types within a fault system,[107] and reducing the complexity of the system to the friction of certain types of rock is one way to reduce the tractability of the problem. Common methods for measuring the friction of rocks are described by Scholz.[108] As shown in Figure 3.29, these are (a) triaxial compression, (b) direct shear, (c) biaxial loading, and (d) rotary. The first method is limited by the small amount of slip it can produce but is often used for elevated-temperature friction studies. The second method can accommodate large specimens but suffers from an uneven distribution of normal forces (a moment develops about the loading axis as sliding proceeds). The third method, like the first, produces limited slip, and the fourth permits unlimited slip distance. For example, as Tullis explains,[106] in

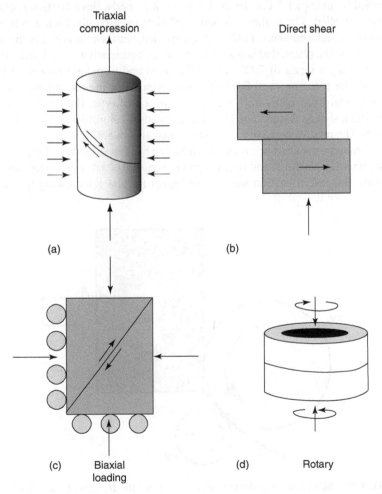

FIGURE 3.29   Configurations used for friction testing of rock.

earthquakes, the shear force builds up until there is a sudden slip with its attendant release of energy. Understanding the time-dependent effects on friction in rock interfaces is a key to modeling instabilities in geological systems; therefore, apparatus used to measure the friction of rocks must be capable of measuring rapid transients in the normal and friction forces.

3.4.14 FRICTION OF CURRENCY

Counting paper currency by automated machinery such as bank currency counters and automated teller machines (ATMs) requires the control of friction between the paper currency and several kinds of contact surfaces. Middleton and Copeland[109] describe the process of counting currency using a friction wheel device. A schematic diagram of the process is shown in Figure 3.30. A stack of bills (B) is forced against a friction wheel (FW). A retarder wheel (RW) of lower friction coefficient holds back the second bill from being drawn into the device by friction against the first bill, thus avoiding a so-called "double-pick." For the device to work properly, three friction coefficients must be controlled. The highest friction coefficient must be bill/friction wheel, next bill/retarder wheel, and lowest bill/bill. Edge friction, between the edges of the stack of bills sliding on the platen that supports it, is also a consideration in such machines.

Completing a series of 7000 tests, Middleton and Copeland measured friction coefficients for U.S. notes and Bank of Scotland currency in new and used condition using a standardized method.[110] It consisted of a sled-type, flat-on-flat Davenport device with a sliding distance of 150 mm, a speed of 2.5 mm/s, and a normal force of 1.96 N. The relative humidity was maintained at about 50%.

The properties of currency paper are affected by their direction of travel through the papermaking machine, and friction experiments involved tests in two orthogonal directions. Bill-to-bill friction was always higher for new bills, owing to the raised

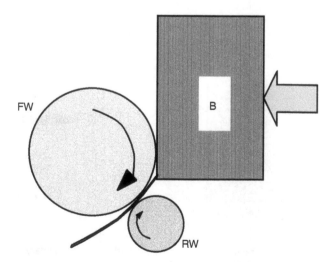

FIGURE 3.30 Schematic of a device used to measure the friction of currency in money counting machines.

TABLE 3.15
Static and Kinetic Friction Coefficients for Currency against Several Counterfaces

Currency	Counterface	μ_{static}	$\mu_{kinetic}$
New bills	New bills	0.467	0.361
Old bills	Old bills	0.378	0.316
New bills	Serrated polyurethane	0.841	0.789
Old bills	Serrated polyurethane	0.649	0.573
New bills	Molded and polished polycarbonate	0.302	0.221
Old bills	Molded and polished polycarbonate	0.365	0.268

Source: Adapted from Middleton, D.E.S. and Copeland, W., *Wear*, 193, 126, 1996.

surface features created by intaglio printing. However, as wear occurs, these features flatten and the bill-to-bill friction tends to decrease. The friction of used bills against possible roller materials, however, is not necessarily less than that for new bills, as shown in Table 3.15 that provides an illustrative selection of test results from Middleton and Copeland's investigation.[109]

This section has exemplified some of the diverse methods that have been employed in studies of friction. Additional examples may be found throughout this book.

3.5 FRICTION SENSING AND RECORDING

When using devices other than those that depend on measuring the angle of inclination of a plane, it is necessary to measure the friction force and normal force to determine static or kinetic friction coefficients. Force transducers of various types have been used in friction measurements. Simple macrotribometers utilize a spring-operated dial indicator mounted in line with the tangential force. The same scale can be used to measure the normal force (weight) of the slider in the manner of a fish scale. Although this is useful in some situations, estimating the friction force by observing a fluctuating needle, or even one that locks at the maximum force, is not the most accurate method to measure friction. Although not as sophisticated as some laboratory tribometers, the use of portable force transducers may be quite satisfactory for rapid friction measurements in the field.

Battery-operated, portable digital force gauges that capture peak forces are commercially available for several ranges of force. Such force gauges are the more sophisticated brothers of the spring-loaded "fish scale" used to measure the weights of a sport fisherman's prize catch. Their portability and simplicity of operation make them popular for in-the-field measurements of the friction of objects sliding on floors, etc. Output can also be linked to a laptop computer to record the data.

Portable, bench-top devices have been used on the checkout counters of automotive parts stores to promote certain brands of lubricants. One of these uses a torque wrench to determine the seizure load, and hence, the limiting friction coefficient for sliding. Typically, a steel pin is loaded against a rotating shaft in the presence of the test lubricant while increasing pressure is applied manually on a torque wrench affixed to the pin holder. At a certain load (applied torque) the shaft will stall or a shear pin will fail. Side-by-side lubricant trays allow customers to compare "Brand X" with some "new and improved" lubricant. As described previously this kind of low-speed, ambient temperature bench test does not faithfully simulate lubricant behavior in a fired engine that has been run for a while.

A larger and more elaborate version of the torque wrench method is aimed at measuring the extreme-pressure characteristics of thick oils or greases. It uses a rotating test pin clamped between two V-shaped blocks (ASTM D 2670, ASTM D 3233) in which the pressure is increased in steps until seizure occurs. The seizure condition is reported in units of "teeth," which refers to the calibrated gear wheel that applies the load.

Various types of galvanometer movements have been used to measure friction in laboratory experiments. Although these are quite sensitive, these are also very delicate and, in the face of new electronic sensors, are now rarely, if ever, used. In most macroscopic testing, friction forces are measured either from the deflection of a beam on which strain gauges are mounted or directly with a load cell. As described in the previous section, the reflection of a laser spot from a tiny cantilever beam is employed in nanoscale friction force measurements. In the design of the ORNL friction microprobe, described in Section 3.4.2, the deflection of an elastic beam was measured not by strain gauges but by a capacitive gap sensor.

As the size of the contact and the magnitude of the friction forces decrease, the force sensors require greater sensitivity and attention to calibration. Sensitivity can be increased by using thinner elastic beams; however, this approach brings additional problems. With a thinner beam, as friction force changes, the sliding contact point changes its position. Thus, as friction force varies during sliding, the contact point on a compliant system may move back and forth, changing both the sliding velocity and the accelerations and decelerations of the contact. Low-load tribometers whose sensors are based on elastic deflections tend to be more prone to stick-slip behavior, and this characteristic should be checked when calibrating the system at various speeds. Thinner beams are also more susceptible to external vibrations that can overwhelm the friction force signal.

Piezoelectric transducers are sometimes employed on microscale tribometers owing to their ability to capture signals at a high data rate; however, the piezoelectric effect can cause output signals to decay even at constant load, and electronic signal processing is needed to account for this characteristic. This also makes piezotransducers more difficult to calibrate directly (e.g., by hanging static weights over a pulley). Strain-gauge type load cells that have been allowed to warm up and stabilize are easier to calibrate using dead loads.

Real-time, *in situ* monitoring of friction forces using sensors embedded within operating machinery or industrial processes is likely to see increased use as wireless telemetry and miniaturization becomes increasingly affordable. For example, a diaphragm-type strain-gauge device that can be embedded into metal workpieces to measure friction during forming has been developed.[111]

One of the more challenging areas of tribometry is measuring friction in harsh environments, such as corrosive fluids, high vacuum, or high temperature. In such cases, it may be necessary to isolate the sensor from the contact area using rigid rods, bellows, high-vacuum mechanical feed-throughs, long lever arms, or various other means to transmit the friction force to the sensor, which must be protected from the environment. It is important in such cases to calibrate the force system carefully and periodically to prevent errors based on such factors as thermal expansion of the fixtures, feed-through friction, friction in lever pivots, and other mechanical elements involved in force transmission.

In designing such systems it is best to keep the friction force collinear with the axis of the sensor system to reduce errors from off-axis force moments. It is relatively common in reciprocating tribometers that the force calibration is different in opposite directions. This problem can arise either from asymmetrical fixture assembly stresses or from the use of designs in which the moment of the friction force/normal resultant about the position of the force sensor is not the same for each direction of sliding. Dual-direction calibrations can be performed and computerized data acquisition systems can be programmed to use the correct calibration factors when the sliding direction changes. Figure 3.11, discussed previously, shows a reciprocating friction force trace in which the friction is asymmetrical about the midline. One might be inclined to simply average the friction force on each side of the line to determine the average friction coefficient, but if the force sensor was only calibrated in one direction the friction force data for the opposite direction may not be correct, thereby biasing the average value toward an inaccurate value. However, if the calibration was correct and the system configuration was properly designed, it is possible that the friction behavior was actually different in opposite directions. Possible causes include texturing of the test surface (e.g., the friction of human hair is very anisotropic depending on the sliding direction) or the first stroke can develop a preferred surface orientation that resists sliding in the opposite direction. The investigator should therefore take special care to position the slider initially at the same position along the sliding path to begin each reciprocating test.

Chart recorders were once the first choice in recording frictional behavior, but availability of computer data acquisition systems make possible extremely detailed recording of friction data and the manipulation of data to obtain extensive statistical information about friction experiments. The author uses both methods. Chart recorders, even without memory modules, are handy for monitoring overall trends in friction behavior and are simple to set up. Chart outputs can be scanned or imported to a computer if needed. However, the mechanical inertia of the pen or stylus holder limits the responsiveness of these devices to record rapid transients or stick-slip behavior, and many a test record has been interrupted or lost by finicky pens or stuck paper. Current data loggers embody built-in storage media or disk drives but some of these are not very user-friendly and require hours of study and a dictionary of data acquisition terminology to set up for the first time. Data logger displays can employ a chart recorder mode so that real-time friction and other sensor outputs (e.g., temperature and contact electrical resistance) can be monitored in real time.

Any time that analog data is "chopped up" into data segments, filtered, and converted into digital data, there is a potential to lose information; however, the author has found that a data rate of about 100–200 Hz is adequate for most friction experiments. Less than this may lose information, and more than this will unnecessarily take up file space. Obviously, special cases, like high-speed sliding and high-frequency stick-slip studies, require a higher data rate. Usually a few scoping tests will reveal the appropriate data rate.

A comparison of chart recorder friction traces with digital data recordings for the same test was discussed by Blau and DeVore.[112] In Figure 3.31, chart recorder records for an alumina ball sliding on an aluminum disk at 0.98 N normal force

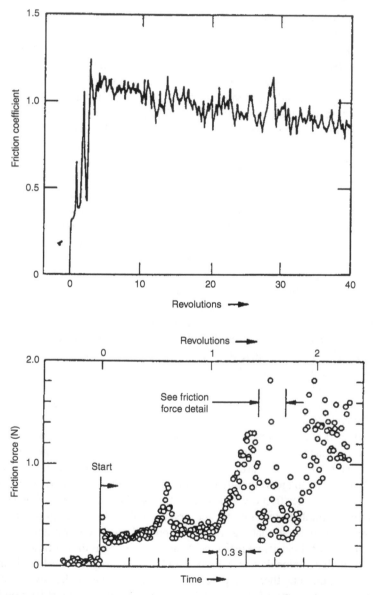

FIGURE 3.31 Comparison of chart recorder (analog) record with digital records for the same sliding experiment (alumina on aluminum, pin-on-disk, 0.98 N, 0.1 m/s, in air).

and 0.1 m/s sliding speed are compared with computer data acquisition system plots for the same test. The chart recorder shows the general trends of the data without much detail; the 100 Hz data show some features but lose information at the point when the friction force is varying too quickly to capture. At 1000 Hz, the data appear connected, suggesting no significant loss of detail.

The sampling rate also affects the distribution of friction values. Using data from the same experiment already discussed, one can plot histograms of the occurrence of friction coefficient values during the transient region of running-in at two different sampling rates, 100 counts/s and 1000 counts/s. Results are shown in Figure 3.32. Not

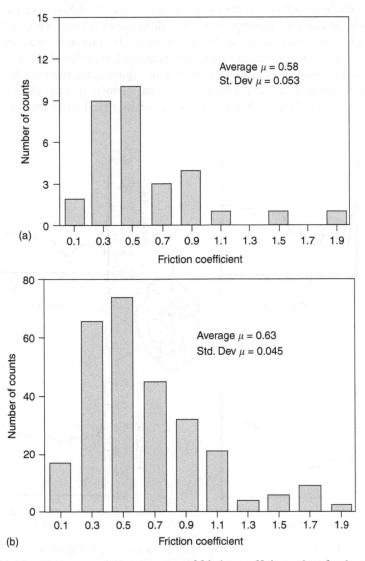

FIGURE 3.32 Histograms of the occurrence of friction coefficient values for the same test but with two data sampling rates: (a) 100 counts/s, (b) 1000 counts/s.

only is the distribution different for the two data samples, but also the averages and standard deviations are also different. Rabinowicz et al.[113] have discussed the statistical nature of friction, a subject relevant to design of tribometers. These authors concluded that at light loads, statistical variations in friction are to be expected; however, for well-lubricated surfaces, fluctuations tend to decrease as surface roughness decreases.

As discussed in Chapter 7 on running-in notes, surfaces may break-in differently at different locations. Therefore, the rate of friction data sampling can influence the picture one obtains of the physical process. An experiment was performed using a 52100 steel ring rotating against a fixed, annealed oxygen-free, high-conductivity (OFHC) copper block under a load of 4.9 N and with a speed of 20 cm/s in Ar gas at room temperature. The data-recording rate was synchronized at four times the rotation rate of the ring so that a friction value was obtained at every 90° on the ring circumference. The frictional histories of those four locations could be separately plotted, as shown in Figure 3.33, and each location on the ring circumference had a different frictional history. It is therefore advisable to sample friction frequently to ensure that the localized variations in behavior

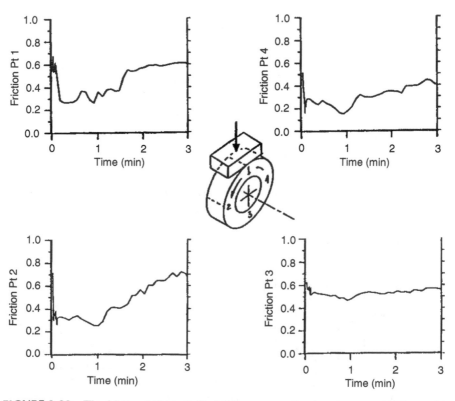

FIGURE 3.33 The frictional "histories" of different sampling locations on a block-on-disk test obtained at four locations during each rotation of the ring specimen.

are not lost and that the general behavior of the tribosystem is not misinterpreted by biased data sampling procedures.

The fact that friction can vary not only as a function of time but also from one place to another on the same surface at similar sliding times suggests that friction models that attempt to predict a single value of friction, and not a range or time-dependent distribution of friction values, may be overly simplistic. As Rabinowicz and others have discovered, the better lubricated the surface, the less likely it will be that large variations in friction will be observed, but in poorly lubricated or unlubricated tribosystems, the spatial and temporal behavior of friction cannot be ignored in fundamental studies of friction processes because it reflects the complexities of asperity interactions, especially when lubricants are absent.

The use of deadweight or pneumatic (hydraulic) loading systems for applying the normal force to the frictional contact is a consideration in designing tribometers and in interpreting results of experiments. In constant-load friction experiments, many investigators assume that the friction coefficient can be obtained directly from the friction force by dividing it with the value of the applied load. Deadweight-loaded systems may not provide a constant normal force, particularly if the sliding speed is increased to the point where contact vibrations occur (see Chapter 7 for additional details). Therefore, the *applied* load and the instantaneous *actual* load may not be the same. Mechanical instabilities can be compensated for by using hydraulic or pneumatic loading systems, but even these can exhibit inertial or response-time effects. The best way to ensure that the friction coefficient is correctly computed, especially in cases of high sliding speed and contact vibrations, is to monitor both the friction force and the normal force.

Other types of sensor outputs, such as contact temperature (by embedded thermocouples or other means), vibrations (by accelerometers mounted in various orientations on the tribometer), and contact resistance, have been employed to gain additional insights into frictional processes. For example, contact temperature changes may reflect the degree of running-in of the surfaces,[114] and vibration levels may indicate the degree of scuffing or interfacial transfer in the interface.

3.6   DESIGNING FRICTION EXPERIMENTS

The design of friction experiments follows the same rationale as would any systematic experimental design. Czichos[115] has provided a detailed discussion of designing friction and wear experiments in particular. Some of the factors involved in friction experiment design are summarized in Table 3.16.

One additional consideration in selecting experimental methods for friction is whether comparison of the current results with past results is of value. Then duplicating the previous experimental conditions becomes important. The following rules of thumb apply:

- The less effective the lubrication, the more replicate tests will be needed to characterize frictional behavior satisfactorily.

TABLE 3.16
Some Important Factors Involved in Designing Friction Experiments

Factor	Considerations
Purpose for friction testing	Simulation; basic research; preliminary screening of material combinations
Type of motion	Can affect the value of friction measured (e.g., unidirectional vs. reciprocating)
Macrocontact geometry	Conformal; nonconformal; affects the regime of lubrication, thermal conditions, and the management of debris
Load, speed, duration of testing	Running-in phenomena; transitional processes; simulation of service history; change of dominant frictional processes
Specimen preparation and cleaning	Simulation of an application; how long during the experiment does the initial surface condition persist?; ambient films and contaminants
State of lubrication	Flow of lubricants; how lubricants are supplied or replenished; thermal effects on lubrication
Number of tests per condition	Unlubricated tests often behave more variably than well-lubricated tests requiring more replicates; confidence increases with the number of tests; quality control requirements
Accuracy required	Affects choice of sensor system; method of data recording; how data are treated (statistics)

- Increasing or decreasing normal load periodically, as in "step" tests, will not necessarily produce the same frictional behavior at each load that would have occurred had the tests been run at constant load.
- Different frictional results may be obtained if starting at a high load and stepping it down versus starting at a low load and stepping it up, due to the effects of prior surface conditioning on friction.
- The same considerations as for load may be applied to changes in velocity.

Standards have been developed by the ASTM subcommittee G-2.50 on friction to help in reporting friction data.[116] Other ASTM standards have specific ways in which friction data are reported. In any case, including a description of the experimental test configuration, materials, surface preparation, applied conditions, and state of lubrication in detail is advisable so that the interpretation and usefulness of friction test results can be determined. At minimum, the average (nominal) static or kinetic friction coefficient is reported. Further information includes the starting friction coefficient (not necessarily the same as the static friction coefficient), maximum or minimum friction value during running-in, the time to reach steady state, the variation of friction force at steady state, and the presence and time over which friction transitions may occur.

Eiss[117] discussed the statistical reporting of friction and wear data, stating that it is the responsibility of the experimenter to measure and report the variability of data so that others can properly interpret it. He discusses the proper use of graphical representations and measures of variability, especially stressing the importance of reporting the number of data points used in determining the average value. The standard deviation s is the most usual measure of variability. The variance is the square of the standard deviation. Accordingly, the standard deviation is the square root of the average sum of the squares of the deviations of each value (x_i) from the mean value (x_{ave}). For N values

$$s = \sqrt{\sum_{i=1}^{n} \frac{(x_i - x_{ave})^2}{N}} \tag{3.19}$$

For a small number of observations (<14), N is replaced with $N - 1$. With a small number of data, a bias develops, causing s to deviate from the true standard deviation of the sample. Using $N - 1$ helps to correct this bias.

The confidence interval is a specified percentage, typically 90, 95, or 99%, which represents the range of values centered on the mean of the sample. A 90% confidence limit means that out of 100 points, 90 of them will be within the confidence limit that contains the true mean of the data. For sample sizes under 120, the t distribution is preferred. Given that the number of degrees of freedom is one less than the sample size, one can calculate the confidence interval from

$$I = x_{ave} \pm t\left(\frac{s}{\sqrt{N}}\right) \tag{3.20}$$

where t is a factor obtained from a table of t-distribution values listed by number of degrees of freedom. If, for example, the mean value of the friction coefficient for $N = 9$ was 0.60 and the standard deviation was 0.03, then the 95% confidence interval is

$$I = 0.60 \pm 2.262\left(\frac{0.03}{\sqrt{9}}\right) = 0.60 \pm 0.023 \tag{3.21}$$

where the value 2.262 was obtained from a table of 95% confidence limits with a degree of freedom of 8. Then, with 95% confidence, the true mean of the sample is between 0.577 and 0.623.

In designing friction tests, it is extremely difficult to vary just one thing and hold everything else constant. For example, let us assume sliding velocity is a variable. As sliding velocity is increased, the frictional temperature rise increases and tribochemical reactions may be promoted. Thus, changing the velocity also

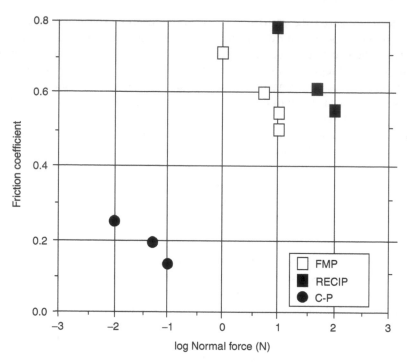

FIGURE 3.34 Illustration of the fact that the friction testing method can have a pronounced influence on the friction coefficient obtained for the same sliding materials (440C steel on IC-50).

changes temperature, and perhaps surface film formation kinetics. As another example, assume normal force is intended to be a variable. At low normal forces the contact pressure may not change surface roughness very much, but as load is increased, the surface may begin to wear more, thereby introducing the variable of surface roughness into the experiment. It is important to recognize that when plotting either friction force or friction coefficients versus a single test variable, the concomitant changes in other variables may be concealed within those results.

In summary, friction measurement requires careful consideration of the reasons why friction data are needed and how they might be used. The next chapter discusses the fundamentals of friction processes, especially with respect to the issue of contact size relative to the dominant frictional processes. It will support the contention that to simulate certain devices friction test methods should be selected with an understanding of the relative scale of operative phenomena. Since friction coefficients are characteristics of the materials in the context of the tribosystem, the reader should not be surprised to see a difference in the friction coefficients for the same material pair measured in different tribometers. The plot in Figure 3.34 is for reciprocating, sphere-on-flat sliding of stainless steel 440C on polished Ni_3Al alloy at room temperature.[118] Intelligent procedural design and complete documentation of the methodology are essential if tribometry is to provide meaningful friction data.

APPENDIX

TABLE A1
ISO Friction-Related Standards

Designation	Title	Comment
ISO 12156-1:2006	Diesel fuel—Assessment of lubricity using the high-frequency reciprocating rig (HFRR)—Part 1: Test method	Uses a specific commercial high-frequency reciprocating (HFRR) machine to assess the lubricity of diesel fuels and additives
ISO 12957-2:2005	Geosynthetics—Determination of friction characteristics—Part 2: Inclined-plane test	Frictional characteristics of geotextiles and geotextile-related products in contact with soils using an inclined-plane apparatus
ISO 12957-1:2005	Geosynthetics—Determination of friction characteristics—Part 1: Direct shear test	Friction characteristics of geotextiles and geotextile-related products in contact with a standard sand using a direct shear apparatus
ISO 14512:1999	Passenger cars—Straight-ahead braking on surfaces with split coefficient of friction—Open-loop test procedure	Test method for determining vehicle reactions during a straight-line braking maneuver on a surface having a split coefficient of friction (e.g., half wet, half dry road)
ISO 15113:2005	Rubber—Determination of frictional properties	A method for measuring the coefficient of friction of rubber against standard comparators, itself, or any other specified surface
ISO 15359:1999	Paper and board—Determination of the static and kinetic coefficients of friction—Horizontal plane method	Uses a horizontal plane to determine static coefficient of friction, and the static and kinetic coefficients of friction of paperboard after a specified amount of wear
ISO 16234:2006	Heavy commercial vehicles and buses—Straight-ahead braking on surfaces with split coefficient of friction—Open-loop test method	An open-loop test method for determining vehicle reactions during a straight-line braking maneuver on a surface having a split coefficient of friction
ISO 18904:2000	Imaging materials—Processed films—Method for determining lubrication	Method of determining the presence of a lubricant on photographic film after processing, applicable to both the emulsion and support sides of the film

(*Continued*)

TABLE A1 (Continued)

Designation	Title	Comment
ISO 20808:2004	Fine ceramics (advanced ceramics, advanced technical ceramics)—Determination of friction and wear characteristics of monolithic ceramics by ball-on-disk method	Ball (pin)-on-disk configuration under dry sliding conditions
ISO 21182:2005	Light conveyor belts—Determination of the coefficient of friction	Test method for determining the dynamic and static coefficients of friction for light conveyor belts
ISO 22653:2003	Footwear—Test methods for lining and insocks—Static friction	Two methods for assessing the frictional properties of lining and insocks
ISO 6601:2002	Plastics—Friction and wear by sliding—Identification of test parameter	Identifies parameters and test methods associated with friction and wear of plastics
ISO 7148-2:1999	Plain bearings—Testing of the tribological behavior of bearing materials—Part 2: Testing of polymer-based bearing materials	Tribotests of polymer-based plain bearing materials under specified load, sliding velocity, and temperature, with and without lubrication
ISO 7176-13:1989	Wheelchairs—Part 13: Determination of coefficient of friction of test surface	Sliding a rubber-faced block with definite speed over a rough test surface
ISO 8295:1995	Plastics—Film and sheeting—Determination of coefficients of friction	Method for determining the coefficients of starting and sliding friction of plastics film and sheeting for quality control
ISO 8349:2002	Road vehicles—Measurement of road surface friction	Friction force values for tires on paved surfaces. Procedures for determining peak- and slide-braking coefficients on actual test surfaces
ISO 5904:1981	Gymnastic equipment—Landing mats and surfaces for floor exercises—Determination of resistance to slipping	Procedure involves pulling, by means of a rope, a friction piece over the horizontal surface of the mat in the specified directions

TABLE A2
ASTM Standards Related to Friction Measurement

ASTM Designation	Materials	Type of test
C 808	Carbon and graphite against other materials	General, comprehensive test method with no specific apparatus described
D 1894	Plastic films and sheeting	Sled-type test
D 2047	Flooring materials against heels and soles	Articulated strut test (James machine)
D 2394	Wood and wood finished	Sled-type test
D 2714 (G 77)	Lubricated steel-on-steel	Calibration of the ring-on-block testing machine
D 3028	Plastic solids and sheeting	Pendulum device
D 3108	Yarns on various materials	Compound pulley-type device with a pendulum
D 3247	Corrugated and solid cardboard	Sled-type device
D 3248	Corrugated and solid cardboard	Inclined-plane device
D 3334	Woven fabrics	Inclined-plane device
D 3412	Yarn-on-yarn	Yarn wound around a cylinder
D 3702	Self-lubricating materials	Thrust-washer geometry
D 4103	Vinyl and wood tiles	General discussion, no specific type of machine
D 4917	Uncoated, self-mated printing and writing paper	Horizontal sled test for static and kinetic friction
D 4918	Uncoated, self-mated printing and writing paper	Inclined-plane test for static friction
D 5183	Lubricants	Upper ball spinning on a stationary triangular nest of three lower balls (four-ball test)
D 5707	Greases	Linearly reciprocating test also intended to measure antiwear characteristics
D 6205	Shoe leather on flooring	Calibration for static friction measurements using the James machine
D 6425	Extreme-pressure lubricant additives	Linearly reciprocating test also intended to measure antiwear characteristics
E 274	Tires on wet pavement	Locked wheel dragged over wet pavement at constant load and speed
E 303	Tires on road surfaces	"British Pendulum Tester" uses a swinging mass to skid on road surfaces
E 510	Rubber-on-pavement	Rotating axle containing two rollers
E 670	Tires on pavement	Trailer device containing full-sized tires
E 707	Rubber-on-pavement	Skid resistance test using a pendulum
E 1911	Rubber on wet pavement	Horizontal spinning disk with three spring-loaded sliders

(Continued)

TABLE A2 (Continued)

ASTM Designation	Materials	Type of test
F 489	Walking surfaces and heels and soles	"James machine" for flooring materials; uses an articulated strut mechanism
F 609	Walking surfaces and heels and soles	Horizontal pull, spring scale "slip-meter" uses a fixed load and constant speed
F 695	Walking surfaces and heels and soles	Articulated strut device for static slip resistance
F 732	Medical and surgical materials	Reciprocating pin-on-flat
F 1677	Walking surfaces and heels and soles	Portable version of the articulated strut device for wet surfaces
F 1679	Walking surfaces and heels and soles	Uses a variable incidence tribometer (VIT) powered by a CO_2 cartridge that enables measurements on inclines
G 99	Various materials	Pin-on-disk test
G 115	Various materials	General guide to measuring and reporting friction
G 133	Various materials	Reciprocating ball-on-flat wear test with optional friction measurement
G 164	Residual lubrication on flexible webs	Inclined-plane test
G 181	Engine materials	Reciprocating piston ring segment on cylinder liner material with lubrication
G 182	Rolling bearings	Inclined-plane test for breakaway friction

Note: Normally ASTM standards are designated with a letter, a number, and the year of the most recent modification or approval. The year has been omitted in the table because these are subject to frequent change. Readers should consult the ASTM Web site www.astm.org.

REFERENCES

1. R. P. Feynman, R. B. Leighton, and M. Sands (1963). *The Feynman Lectures in Physics*, Addison-Wesley, New York, pp. 12–15.
2. DIN 50 322 (1986). *Wear; Classification of Categories in Wear Testing*, Deutsches Institut für Normung, Berlin, Germany.
3. S. Hogmark and S. Jacobson (1992). Hints and guidelines for tribotesting and evaluation, *Lubr. Eng.*, 48(5), pp. 401–409.
4. ASLE (1976). *Friction and Wear Devices*, STLE, Park Ridge, IL.
5. *Glossary of Terms and Definitions in the Field of Friction, Wear and Lubrication (Tribology)* (1969). Research Group on Wear of Engineering Materials, Organization for Economic Cooperation and Development (OECD), Paris.
6. R. Voitik (1993). The tribological aspect number, in *Wear Test Selection for Design and Application*, eds. A. W. Ruff and R. G. Bayer, Spec. Tech. Pub. 1199, ASTM, Philadelphia, PA.
7. F. T. Barwell (1959). Friction and its measurement, *Metallurgical Rev.*, 4(14), pp. 141–177.
8. B. Bhushan, ed. (2001). *Fundamentals of Tribology: Bridging the Gap Between the Macro- and Micro/Nanoscales*, Kluwer, Dordrecht.
9. P. J. Blau (1991). Scale effects in steady state friction, *Tribol. Trans.*, 34, pp. 335–342.

10. J. F. Archard (1957). Elastic deformation and the laws of friction, *Proc. Royal Soc., Part A*, 243, pp. 190–205.
11. E. Rabinowicz (1992). Friction coefficients of noble metals over a range of loads, *Wear*, 159, pp. 89–94.
12. C. J. Studman and J. E. Field (1973). A simple method for measuring friction for spheres and cylinders, *Wear*, 24, pp. 243–246.
13. K. G. Budinski (1992). Laboratory testing methods for solid friction, in *ASM Handbook, Volume 18: Friction, Lubrication, and Wear Technology*, 10th ed., ASM International, Materials Park, OH, pp. 45–58.
14. L. J. Brendall, L. B. Johnson, Jr., and D. Kuhlmann-Wilsdorf (1987). Teaming measurements of the coefficient of friction and of contact resistance as a tool for the investigation of sliding interfaces, *Proceedings of International Conference on Wear of Materials*, ASME, New York, pp. 861–868.
15. H. Czichos, S. Becker, and J. Lexow (1987). Multilaboratory testing: Results from the Versailles Advanced Materials and Standards Program on wear test methods, *Wear*, 114, pp. 109–130.
16. H. Czichos, S. Becker, and J. Lexow (1989). International multilaboratory sliding wear tests with ceramics and steel, *Wear*, 135, pp. 171–191.
17. B. Stupp and J. Wright (1963). Dry films lubricant at elevated temperatures, *Lubr. Eng.*, 19(11), p. 463.
18. M. P. Wolverton, K. Talley, and J. F. Theberge (1992). Friction and wear of thermoplastic composites, in *ASM Handbook, Volume 18: Friction, Lubrication, and Wear Technology*, 10th ed., ASM International, Materials Park, OH, pp. 820–822.
19. Lewis Research, Lewes, Delaware, Model LRI-1a, is a computer-controlled thrust washer testing machine designed to conduct ASTM D3702.
20. F. P. Bowden and L. Leben (1939). The nature of sliding and the analysis of friction, *Proc. Royal Soc., Part A*, 169, pp. 371–391.
21. ASTM F732: Standard test method for wear testing of polymeric materials used in total joint prostheses, *Annual Book of Standards*, Vol. 13.01, ASTM International, West Conshohocken, PA.
22. L. K. Ives and P. J. Boyer (1979). Pin-and-V-block and ring-and-block bench wear tests for engine oil evaluation, in *Proceedings of Joint Conference on Measurements and Standards for Recycled Oil Systems and Durability*, National Bureau of Standards Special Pub. 584, Gaithersburg, Maryland.
23. P. J. Blau (1992). Rolling contact wear, in *ASM Handbook, Volume 18: Friction, Lubrication, and Wear Technology*, 10th ed., ASM International, Materials Park, OH, pp. 257–262.
24. J. P. O'Donaghue and A. Cameron (1966). Friction and temperature in rolling sliding contacts, *ASLE Trans.*, 9, pp. 186–194.
25. L. D. Wedeven (1997). *Method and Apparatus for Comprehensive Tribological Evaluation of Materials*, US Patent 5,679,883.
26. B. J. Briscoe, A. Winkler, and M. J. Adams (1992). *A Statistical Analysis of the Frictional Forces Generated Between Monofilaments During Intermittent Sliding*, Imperial College, London.
27. C. Latorre and B. Bhushan (2005). Nanotribological effects of hair care products and environment on human hair using atomic force microscopy, *J. Vacuum Sci. Technol.*, 23(4), pp. 1034–1045.
28. P. K. Gupta (1991). Simple method for measuring the friction coefficient of thin fibers, *J. Am. Ceram. Soc.*, 74(7), pp. 1692–1694.
29. ASTM G 143-03 (2004). Standard test method for measurement of web/roller friction characteristics, *Annual Book of Standards*, Vol. 03.02, ASTM International, West Conshohocken, PA.

30. K. G. Budinski (2001). Friction of plastic webs, *Tribol. Int.*, 34(9), pp. 625–633.
31. R. Hegel (1993). Hygroscopic effects on magnetic tape friction, *Tribol. Trans.*, 36(1), pp. 67–72.
32. G. J. Wenzloff, T. A. Hylton, and D. K. Matlock (1992). Technical note: A new test procedure for the bending under tension friction test, *J. Mater. Eng. Perform.*, 1(5), pp. 609–614.
33. T. H. C. Childs and D. Cowburn (1984). Contact observations on and friction of rubber drive belting, *Wear*, 100, pp. 59–76.
34. M. K. Ramasubramanian and S. D. Jackson (2003). A friction sensor for real-time measurement of friction coefficient on moving flexible surfaces, *Sensors*, 2003. *Proc. IEEE*, 1(22–24), pp. 152–157.
35. P. J. Blau, ed. (1993). *A Compilation of International Standards for Friction and Wear Testing of Materials*, VAMAS, Report No. 14, available through NIST, Gaithersburg, MD, 49 pp.
36. G. M. McClelland, C. M. Mate, R. Erlandsson, and S. Chiang (1988). Direct observation of friction at the atomic scale, in *Adhesion in Solids*, eds. D. M. Mattox, J. E. E. Baglin, R. J. Gottschall, and C. D. Batich, Materials Research Society, Pittsburgh, PA, pp. 81–86.
37. M. Salmeron (1993). Use of the atomic force microscope to study mechanical properties of lubricant layers, *Mater. Res. Soc. Bull.*, 18(5), pp. 20–25.
38. P. J. Blau (1992). Appendix: Static and kinetic friction coefficients for selected materials, in *ASM Handbook, Volume 18: Friction, Lubrication, and Wear Technology*, 10th ed., ASM International, Materials Park, OH, pp. 70–75.
39. R. Overney and E. Meyer (1993). Tribological investigations using friction force microscopy, *Mater. Res. Soc. Bull.*, 18(5), pp. 26–34.
40. E. Meyer, R. Overney, R. Luthi, D. Brodbeck, L. Howald, J. Frommer, H.-J. Guntherodt, O. Walter, M. Fujihara, H. Takano, and Y. Gotoh (1992). Friction force microscopy of mixed Langmuir-Blodgett films, *Thin Solid Films*, 220, pp. 132–137.
41. F. P. Bowden and T. P. Hughes (1939). *Proc. Royal Soc., Part A*, 172, p. 263.
42. J. Cumings and A. Zettl (2000). Low-friction nanoscale linear bearing realized from multiwall carbon nanotubes, *Science*, 289, pp. 602–604.
43. P. J. Blau (1990). Friction microprobe studies of composite surfaces, in *Tribology of Composite Materials*, eds. P. K. Rohatgi, C. S. Yust, and P. J. Blau, ASM International, Materials Park, OH, pp. 59–68.
44. P. J. Blau and C. E. Haberlin (1992). An investigation of the microfrictional behavior of C_{60} particle layers on aluminum, *Thin Solid Films*, 219, pp. 129–134.
45. P. J. Blau (1993). Friction microprobe investigation of particle layers affects on sliding friction, *Wear*, 162–164A, pp. 102–109.
46. V. V. Konchits (1981). Influence of electric current on frictional interaction of metals, *Treni i Iznos*, 2(1), pp. 170–176.
47. T. P. Weihs and W. D. Nix (1988). Direct measurements of the frictional resistance to sliding of a fiber in a brittle matrix, *Acta Metallurg.*, 22, pp. 271–275.
48. L. N. McCartney (1989). New theoretical model of stress transfer between fiber and matrix in a uniaxially-fiber-reinforced composite, *Proc. Royal Soc. London, Part A*, 425, pp. 215–244.
49. C.-H. Hsueh (1990). Interfacial debonding and fiber pull-out stresses of fiber-reinforced composites, *Mater. Sci. Eng.*, A123, pp. 1–11.
50. J. Hutchinson and J. T. Jensen (1990). Models of fiber debonding and pullout in brittle composites with friction, *Mech. Mater.*, 9(2), pp. 139–163.
51. B. J. Briscoe and T. A. Stolarski (1980). The influence of linear and rotating motions on the friction of polymers, *ASME* preprint 80-C2/Lub-44, p. 5.

52. B. J. Briscoe and T. A. Stolarski (1982). The effect of complex motion in the pin-on-disc machine on the friction and wear mechanism of organic polymers, in *Other Tribological Problems*, Vol. IV, eds. M. Hebda, C. Kajdas, and G. M. Hamilton, Elsevier, Amsterdam, pp. 80–99.
53. M. Turell, A. Wang, and A. Bellare (2003). Quantification of the effect of cross-path motion on the wear of ultra-high molecular weight polyethylene, *Wear*, 255, pp. 1034–1039.
54. N. Fujisawa, N. L. James, R. N. Tarrant, D. R. McKensie, J. C. Woodard, and M. V. Swain (2003). A novel pin-on-apparatus, *Wear*, 254, pp. 111–119.
55. V. Kragelskii, M. N. Dobychin, and V. S. Kombalov (1982). *Friction and Wear Calculation Methods*, Pergamon Press, New York, pp. 212–215.
56. Z. Rymuza (1989). *Tribology of Miniature Systems*, Elsevier, Amsterdam.
57. G. Barker (1946). The comprehensive laboratory testing of instrument lubricants, *ASTM Bull.*, March, p. 25.
58. P. Lacey, S. Gunsel, and R. A. Ward (2001). A new apparatus to evaluate friction modified additives for engine oils, *Lubr. Eng.*, 57(11), pp. 13–22.
59. D. J. Hargreaves and J. C. H. Tang (2006). Assessing friction characteristics of liquid lubricants, *Tribotest*, 12, pp. 309–319.
60. S. E. Hartfield-Wunch, S. C. Tung, and C. J. Rivard (1993). *Development of a Bench Wear Test for the Evaluation of Engine Cylinder Components and the Correlation with Engine Test Results*, SAE Technical paper # 932693, Society of Automotive Engineers, Warrendale, PA.
61. P. J. Blau (2001). *A Review of Sub-Scale Test Methods to Evaluate the Friction and Wear of Ring and Liner Materials for Spark- and Compression Ignition Engines*, Oak Ridge National Laboratory, tech. memorandum , ORNL/TM-2001/184, 19 pp.
62. R. A. Mutti and M. Priest (2005). Experimental evaluation of Piston-assembly friction under motored and fired conditions in a gasoline engine, *J. Tribol.*, 127, pp. 826–836.
63. M. Noorman, D. N. Assanis, D. J. Patterson, S. C. Tung, and S. I. Tseregounis (2000). *Overview of Techniques for Measuring Friction Using Bench Tests and Fired Engines*, Society of Automotive Engineers, SAE reprint no. 2000-1-1780, Warrendale, PA, 11 pp.
64. P. A. Gaydos and K. F. Dufrane (1992). *Studies of Dynamic Contact of Ceramics and Alloys for Advanced Heat Engines*, Final Report, Oak Ridge National Laboratory report ORNL/Sub/84-00216/1, 77 pp.
65. M. Cater, N. W. Bolander, and F. Sadeghi (2006). *A Novel Suspended Liner Test Apparatus for Friction and Side Force Measurement with Corresponding Modeling*, Society of Automotive Engineers (SAE) International, Paper number 2006-32-0041/20066541, 13 pp.
66. F. G. Rounds (1981). *Soots from Used Diesel-Engine Oils: Their Effects on Wear as Measured in 4-ball Wear Test*, SAE Paper 810499, Society of Automotive Engineers, Warrendale, PA, 16 pp.
67. F. Rounds (2006). Changes in friction and wear performance caused by interactions among lubricant additives, *Lubr. Sci.*, 1(4), pp. 333–363.
68. M. G. S. Naylor (1992). *Development of Wear-Resistant Coatings for Diesel Engine Components*, Oak Ridge National Laboratory subcontract final report, ORNL/Sub-87-SA581/1, 195 pp.
69. J. J. Truhan, J. Qu, and P. J. Blau (2005). A rig test to measure friction and wear of heavy duty diesel engine piston rings and cylinder liners using realistic lubricants, *Tribol. Int.*, 38, pp. 211–218.
70. J. J. Truhan, J. Qu, and P. J. Blau (2005). The effect of lubricating oil condition on the friction and wear of piston ring and cylinder liner materials in a reciprocating bench test, *Wear*, 259, pp. 1048–1055.

71. ASTM G 181-04 (2004). Standard test method for conducting friction tests of piston ring and cylinder liner materials under lubricated conditions, *Annual Book of Standards*, Vol. 03.02, ASTM International, West Conshohocken, PA.
72. T.-L. Ho and M. B. Peterson (1976). Wear formulation for aircraft brake materials sliding against steel, *Wear*, 43, pp. 199–210.
73. C. Cueva, A. Sinatora, W. L. Guesser, and T. P. Tschiptschin (2003). Wear resistance of cast irons used in brake disc rotors, *Wear*, 255, pp. 1256–1260.
74. P. J. Blau, R. L. Martin, M. H. Weintraub, H. Jang, and W. Donlon (1996). *Frictional Behavior of Automotive Brake Materials Under Wet and Dry Conditions*, ORNL Technical Report, ORNL/M-5824, p. 41.
75. M. Mosleh, P. J. Blau, and D. Dumitrescu (2004). Characteristics and morphology of wear particles from laboratory testing of disk brake materials, *Wear*, 256, pp. 1128–1134.
76. P. J. Blau (2004). *Research on Non-Traditional Materials for Friction Surfaces in Heavy Vehicle Disc Brakes*, Oak Ridge National Lab., Tech. Memorandum, ORNL/TM-2004/265, p. 32.
77. P. J. Blau and H. M. Meyer III (2003). Characteristics of wear particles produced during friction tests of conventional and non-conventional disc brake materials, *Wear*, 255, pp. 1261–1269.
78. P. J. Blau, J. Qu, B. C. Jolly, W. H. Peter, and C. A. Blue (2007). Tribological investigation of titanium-based materials for brakes, *Wear*, 263(7–12), pp. 1202–1211.
79. P. J. Blau and J. C. McLaughlin (2003). Effects of water films and sliding speed on the frictional behavior of truck disc brake materials, *Tribol. Int.*, 36(10), pp. 709–715.
80. L. Strawhorn (2000). *The Mystery of Aftermarket Brake Lining Selection*, Presented. at the SAE North American Truck and Bus Meeting, December 5.
81. Federal Motor Carrier Safety Standard (FMVSS) No. 121, Air Brake Systems, 49 CFR (10-1-99 Edition), National Highway Traffic Safety Administration, pp. 368–386.
82. *Aftermarket Brake Lining Classification*, Technology and Maintenance Council, Recommended Practice RP 628A, revised 3/1997.
83. P. G. Sanders, T. M. Dalka and R. H. Basch (2001). A reduced-scale brake dynamometer for friction characterization, *Tribol. Int.*, 34(9), pp. 609–615.
84. P. J. Blau (2001). *Compositions, Functions, and Testing of Friction Brake Materials and Their Additives*, Oak Ridge National Laboratory, Technical Memorandum, ORNL/TM-2001/64, p. 38.
85. A. E. Anderson, S. Gratch, and H. P. Hayes (1967). *A New Laboratory Friction and Wear Test for the Characterization of Brake Linings*, Society of Automotive Engineers, Warrendale, PA, Paper 670079.
86. G. Capps, O. Franzese, W. Knee, P. Blau, J. Massimini, J. Petrolino, and D. Rice (2006). *Heavy Single-Unit Truck Original Equipment and Aftermarket Brake Performance Characterization in Field, Test-Track and Laboratory Environments*, Final Report to the National Transportation Research Center, Inc. (NTRCI), by the Oak Ridge National Laboratory (ORNL), Oak Ridge, Tennessee.
87. C. E. Agudelo and E. Perro (2005). *Technical Overview of Brake Performance Testing for Original Equipment and Aftermarket Industries in the US and European Markets*, Link Technical Report FEV2005-01, p. 27.
88. K. Hartman (1988). *Development, Evaluation, and Application of Performance-Based Brake Testing Technologies*, Office of Motor Carriers Tech Brief, U.S. Federal Highway Administration Publication FHWA-MCRT-98-001, Washington, DC, p. 4.
89. U.S. Department of Energy and U.S. Environmental Protection Agency website, http://www.fueleconomy.gov/feg/atv.shtml, accessed April 2007.
90. Internet, http://www.wvdot.com/3_roadways/rp/facts/Chapter%202/Road%20Surface%20Type.pdf, accessed April 2007.

91. R. R. Hegmon (1993). A close look at road surfaces, *Public Roads* On-line, 57(1), pub. by the U.S. Federal Highway Administration, Summer 1993 edition.

92. R. Y. Liang (2003). *Blending Proportions of High Skid and Low Skid Aggregate*, Ohio Dept. of Transportation Report FHWA/OH-2003/014, p. 131.

93. S. Müller, M. Uchanski, and K. Hedrick (2003). Estimation of the maximum tire-road friction coefficient, *J. Dyn. Syst. Measure. Contr*, 125(4), pp. 607–617.

94. A. Norheim, N. K. Sinha, and T. J. Yager (2001). Effects of the structure and properties of ice and snow on the friction of aircraft tyres on movement area surfaces, *Tribol. Int.*, 34(9), pp. 617–623.

95. M. Marpet (2002). Issues in the development of modern walkway-safety tribometry standards: Required friction, contextualization of test results, and non-proprietary standards, in *Metrology of Pedestrian Locomotion and Slip Resistance, ASTM STP 1424*, eds. M. I. Marpet, and M. A. Sapienza, ASTM International, West Conshohocken, PA.

96. J. V. Durá, E. Alcántara, T. Zamora, E. Balaguer, and D. Rosa (2005). Identification of floor friction safety level for public buildings considering mobility disabled people needs, *Safety Sci.*, 43(7), pp. 407–423.

97. S. Kim, T. Lockhart, and H. Y. Yoon (2005). Relationship between age-related gait adaptations and required coefficient of friction, *Safety Sci.*, 43(7), pp. 425–436.

98. W. R. Chang, K. W. Li, Y. H. Huang, A. Filiaggi, and T. K. Courtney (2004). Assessing floor slipperiness in fast-food restaurants in Taiwan using objective and subjective measures, *Appl. Ergon.*, 35(4), pp. 401–408.

99. M. I. Marpet and R. J. Brungraber (1992). *Walkway Friction: Experiment and Analysis*, Presented at the National Educators Workshop, November 11–13, Oak Ridge, TN.

100. ASTM F 1694-04 (2004). Standard guide for composing walkway surface evaluation and incident report forms for slips, stumbles, trips, and falls, *Annual Book of Standards*, Vol. 15.07, ASTM International, West Conshohocken, PA.

101. H. Nagata (2002). An analysis of sliding properties on roofing surfaces, in *Metrology of Pedestrian Locomotion and Slip Resistance*, eds. M. Marpet and M. A. Sapienza, ASTM STP 1424, ASTM International, West Conshohocken, PA, pp. 1–9.

102. J. A. Schey (1983). *Tribology in Metalworking*, ASM International, Materials Park, OH, p. 736.

103. J. F. Lin, A. Y. Lee, and K. Y. Lee (1992). Friction evaluation in deep drawing using an instrumented blankholder, *Tribol. Trans.*, 35(4), pp. 635–642.

104. L. R. Sanchez (1999). Characterization of a measurement system for reproducible friction testing on sheet metal under plane strain, *Tribol. Int.*, 32(10), pp. 575–586.

105. G. Totten, ed. (2006). *Handbook of Lubrication and Tribology, Volume 1: Application and Maintenance, Second Edition*, CRC Press (Taylor and Francis Group LLC, Boca Raton, Florida)/STLE, p. 1068.

106. T. E. Tullis (1996). Rock friction and its implications for earthquake prediction examined via models of Parkfield earthquakes, *Proc. Natl. Acad. Sci.*, 93, pp. 3803–3810.

107. L. Knopoff, J. A. Landoni, and M. S. Abinante (1992). Dynamical model of an earthquake fault with localization, *Phys. Rev. A (Atomic, Molecular, and Optical Physics)*, 46(12), pp. 7445–7449.

108. C. H. Scholz (1990). *The Mechanics of Earthquakes and Faulting, Chapter 2, Rock Friction*, Cambridge University Press, Cambridge, pp. 53–54.

109. D. E. S. Middleton and W. Copeland (1996). The friction of currency bills, *Wear*, 193, pp. 126–131.

110. *Coefficient of Static Friction by Horizontal Plane*, TAPPI Standard T549-PM-90, Norcross, GA, 1990.

111. A. Dellah, P. M. Wild, T. N. Moore, M. Shalaby, and J. Jeswiet (2002). An embedded friction sensor based on a strain-gauged diaphragm, *J. Manu. Sci. Eng.*, 124(3), pp. 523–527.

112. P. J. Blau and C. E. DeVore (1989). Interpretations of the sliding friction break-in curves of alumina-aluminum couples, *Wear*, 129, pp. 81–92.

113. E. Rabinowicz, B. G. Rightmire, C. E. Tedholm, and R. E. Williams (1955). The statistical nature of friction, *Trans. ASME*, October, pp. 981–984.

114. M. J. Neale (1973). *Tribology Handbook*, Wiley, New York, Section B 30 "Running-in Procedures."

115. H. Czichos (1992). Design of friction and wear experiments, in *ASM Handbook, Volume 18: Friction, Lubrication, and Wear Technology*, 10th ed., ASM International, Materials Park, OH, pp. 480–487.

116. ASTM G-115-93 (1993). *Standard Guide for Measuring and Reporting Friction Data*, ASTM, Philadelphia, PA.

117. N. S. Eiss (1992). Back to basics—statistical reporting of data, *Lubrication Eng.*, 48(8), pp. 651–654.

118. P. J. Blau (1992). Scale effects in sliding friction—An experimental study, in *Fundamentals of Friction*, eds. I. L. Singer and H. M. Pollock, Kluwer, Dordrecht, pp. 523–534.

4 Fundamentals of Sliding Friction

The "father of friction research," Leonardo da Vinci, was probably the first who consciously investigated *pairs* of materials. From that time on, the unfortunate misunderstanding about a friction "coefficient" as a materials property has grown, and it is still deeply rooted in the mind of engineers. ... [However] the friction coefficient is just a convenience, describing a friction *system* and *not* a materials property.

G. Salomon
Mechanisms of Solid Friction, p. 3[1]

Solid friction is a difficult subject. ... A knowledge of the laws of friction requires extensive experimental research.

D. Mendeleev
Molecular Physics of Boundary Friction, p. ix[2]

Chapter 2 illustrated traditional introductory approaches to friction, and Chapter 3 described a variety of methods that were developed to measure friction coefficients. This chapter begins with a hypothetical experiment to illustrate increasing levels of complexity when one attempts to understand the fundamental basis for friction. It lays the groundwork for understanding the role of scale effects and why it is important to define the tribosystem before selecting a friction model. Section 4.2 discusses static friction and stick-slip and also the circumstances under which a relationship between adhesion and friction is established. Section 4.3 describes the basic concepts of sliding friction, and Section 4.4 addresses frictional heating.

Friction modeling involves a systematic process, which is very much like any process used to model a physical phenomenon. Eight stages of model development can be identified as follows:

1. *Observation.* Observe and describe the phenomenon (e.g., it is difficult to slide solid objects over one another; objects slide more easily when their surfaces are covered with grease).
2. *Definition.* Give the phenomenon a name and determine how general it is. (For example, resistance to sliding occurs in many places. Call it "friction," from the Latin *"fricare,"* or adopt the prefix "tribo-" from Greek.)
3. *Metrology.* Quantify the phenomenon so that its behavior can be measured (e.g., define friction force, normal force, and friction coefficient).
4. *Conceptualization.* Conceptualize the physical conditions that give rise to the phenomenon.

5. *Select variables.* Define and select quantities to be used in the model (e.g., define friction force, shear strength, hardness, angle of repose, asperity shape, layer thickness).
6. *Hypothesize a set of rules.* Establish relationships between those selected quantities (e.g., friction force is equal to shear stress times area of contact).
7. *Validate.* Experimentally test the validity and generality of the relationships.
8. *Refine or discard the model.* If the experiments and other evidence support the model, reveal it to the scientific or engineering community. Otherwise, re-evaluate the assumptions, hypotheses, or the choice of experiments used in validation. Refine the approach or discard the model.

One might ask, if the scientific method is so well established, why then are there so many different models for friction? The answer lies in steps 4 through 6. Contributions to friction vary from one tribosystem to another, and models should be tailored to fit each case. W. B. Hardy, who studied the physics of friction at Cambridge from about 1919 to 1933, said: "Friction phenomena are equally interesting for the physicist and the engineer: their investigation belongs to a most difficult field of boundary problems of physics."

When friction modelers use simplifying assumptions they risk neglecting influential variables. Some theorists introduce quantities with elegant-sounding names but which cannot be measured directly. These "adjustable models" confirm the notion that at least some of the basic physics of friction still remain elusive. While simplified models can adequately describe specific cases, their general applicability is limited. Therefore, the following three important elements are needed to develop accurate and predictive models for static or kinetic friction:

1. *Process identification*—knowledge of the dominant interfacial processes of friction and their relative stability over time
2. *Understanding of scale*—knowledge of the size scale at which these processes operate
3. *Definition of functional relationships*—identification of the rules that translate the external stimulus to the response of the tribosystem, consistent with interfacial processes and scale of interaction

At each level of interaction, different physical processes come into play, and a hierarchy of effects emerges (see Figure 4.1).[3] If the lubrication regime is adequate to promote full separation of the surfaces (zone I), then the frictional interaction can be effectively modeled using lubrication theories like those discussed in Chapter 6. If the conditions of lubrication are insufficient to avoid solid contact (zone II), then the characteristics of the solid bodies, surface structure and third bodies play a role. Should there be no effective lubrication in the system and if the shear forces generated by sliding cannot be fully accommodated by shearing the materials immediately adjacent to the interface, then these forces are transmitted to the fixtures holding the solid bodies together, and accounting must be made for the characteristics of the surrounding structure (e.g., its stiffness, natural frequency). The latter effects are depicted as zone III.

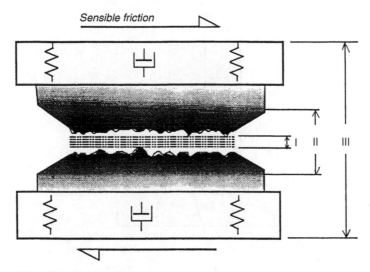

Sensible friction

Hierarchy of interaction levels

I-Interfacial media

    Lubrication theories
    Tribochemical films
    Triboparticulates

II-Bounding solids

    Deformation and fracture
    Fatigue
    Material/environment interaction
    White-layers and transfer

III-Machine and its environs

    Dynamics of the containment

FIGURE 4.1   Three-level hierarchy embodied in a frictional interface and its surroundings.

A hypothetical case. A simple illustration can be used to show how what might first appear to be a trivial problem can evolve to challenge the frontiers of understanding. As more and more attention is paid to the details of a friction problem, one's appreciation for its underlying complexity and for the difficulty of predictive modeling grows.

A solid block of "tribium" with a square face of apparent area A slides horizontally along a solid surface of another material that we will call "frictionite" (Figure 4.2). The magnitude of the normal force holding the tribium against the frictionite is N. The block slides with a constant velocity of v and experiences a friction force F. The investigator want to determine the fundamental causes for the observed behavior of F.

Under the first set of N and v, we might observe that the surface of the frictionite is not visibly altered by the passage of the block. Therefore, we could assume that

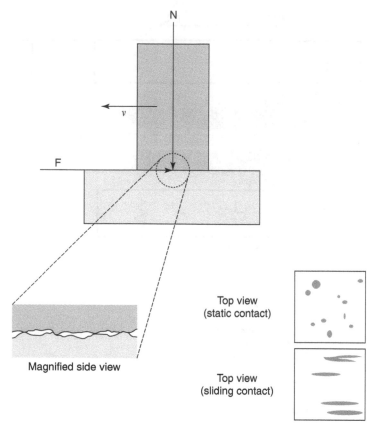

FIGURE 4.2 Hypothetical tribosystem consisting of a block of tribium sliding on frictionite.

the magnitude of **F** is due to phenomena that are too small to be observed with the naked eye. Using a surface roughness instrument, we measure the surface profiles of the tribium block and the frictionite transversely to the direction of motion and realize that they are not perfectly flat. Early investigators have developed models for friction based primarily on the roughness of the two contacting surfaces. Since the topographic features of the surfaces are unlikely to be identical, only a unique distribution of high spots will be in contact at a given instant. Consequently, the apparent area A of the face of the block, defined by its length and width, will in general be greater than the true area of contact, defined by the summation of the areas of all the contact spots. Abbott and Firestone[4] said it this way, over half a century ago: "When two newly machined surfaces are placed together, they touch only on the peaks of the highest irregularities, and the actual contact area is very small."

What, then, is the actual (true) area of contact (A^*)? To the layman, the difficulty of measuring A^* may not appear very challenging, but this central issue remains one of contention in tribology. There have been countless experimental attempts to determine A^*, and every one of them has physical limitations or makes debatable assumptions. If the surfaces behave plastically, like most common materials do above their yield strength, then we might argue that A^* is simply the ratio of **N** to the indentation hardness number H, which is by definition the load per unit area under static equilibrium. But of the several scales of hardness that use the units of load per unit area, which value or values should be selected? Different schemes of indenting the surface and measuring the results are used in hardness scales. Indenters differ in size and shape, and engineering handbooks frequently contain "conversion tables" for translating the numerical values between different hardness scales (such as Brinell, Vickers, and Knoop hardness) even though the units of measure are the same (pressure or stress). So for a given material, the numerical value of indentation hardness depends on how it is measured and calculated. To further complicate matters, the hardness of a metal that has been severely deformed (say, by sliding under a heavy load) may be two or three times higher than it was before sliding began. The Meyer hardness number (MHN), a scale rarely used outside materials research, reflects the ability of metals and alloys to work-harden. It is defined as the ratio of the applied load (**P**) to the projected area of the permanent indentation having diameter d,

$$\text{MHN} = \frac{\textbf{P}}{A_{\text{P}}} = \frac{4\textbf{P}}{\pi d^2} \tag{4.1}$$

MHN is affected by the diameter of the sphere used to indent the surface,[5] and thus the Meyer law is written as

$$\text{MHN} = k \left(\frac{d}{D} \right)^m \tag{4.2}$$

where D is the diameter of the spherical indenter, m is a constant related to the work-hardening ability of the material being tested, and k is a constant determined from the plot of $\log(d/D)$ against $\log$ MHN.

Models deriving from such constructs usually contain no explicit consideration of time-dependent changes in material properties, yet it is well known in the field of mechanical metallurgy that the properties of many metals and alloys exhibit strain-rate sensitivity, which changes their mechanical properties depending on how fast they are deformed. Metals, alloys, and polymers can all exhibit strain-rate effects. Furthermore, the yield stresses in very thin surface layers, such as near-surface, highly deformed shear zones in metals, or even very thin films of liquid lubricants (see Chapter 6), may be much different from those of a bulk material of the same composition. Strain-rate effects and differences in surface versus bulk properties are not considered in the majority of published macroscale friction models.

It seems reasonable that when one surface slides over another, the asperities on one or both surfaces might change shape. Early asperity models for friction made no accounting for this because it was too complicated to handle analytically. Furthermore, even though we might observe no change in the surfaces with the naked eye, is it not possible that change occurs on a microscale, and that the surface roughness we measure after the experiment differs from the initial surface roughness? *Running-in* and *wearing-in* are terms often used to describe the change in surface roughness during the initial stages of sliding. Running-in can occur on a microscale. The previous quote from Abbott and Firestone[4] continues as follows: "If the surfaces

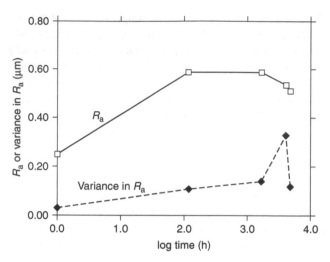

FIGURE 4.3 Changes in the surfaces roughness of a marine engine cylinder liner during normal operation.

are run-in under load, or otherwise fitted, the projecting irregularities are gradually removed, and the actual area of contact is increased. At first the wear is quite rapid, but it decreases as the contact area increases."

The concept of an *equilibrium surface roughness* was introduced by some schools of investigators to account for the nature of the frictional steady-state surface conditions,[6] but even after running-in, surfaces may continue to change their roughnesses, and one notable paper on the wear of marine engines[7] experimentally refutes the concept of an equilibrium surface roughness (see Figure 4.3).

We now begin to realize that each sliding contact has a "life" of its own. The instantaneous points of contact on imperfectly flat surfaces dance about wildly, changing shapes and sizes as time goes on. Evidence to support this notion was first obtained by investigators who observed hot spots generated during frictional heating.[8,9] More recent work by Wang and her colleagues[10] at Northwestern University has elegantly demonstrated the power of computers to portray changes in microcontacts and local pressures during sliding.

If adhesive bonds form at the instantaneous points of contact, might not the friction force be the sum of the forces needed to break those bonds to maintain sliding? Some friction modelers have taken that approach while others have refuted adhesion-based models, arguing that most real free surfaces are not perfectly clean and that mechanical interlocking of asperities can overwhelm adhesive contributions. Surfaces contain contaminants, adsorbed hydrocarbons, adsorbed moisture, and, in the case of nonnoble metals, oxides and tarnishes. Natural contaminants may be invisible to the naked eye, yet they may have lubricating properties. Consequently, frictional resistance may, at least in part, derive from the mechanical and chemical properties of surface films. As a result, we ask: How stable and mechanically important are those films? Can stable films form during sliding or will they quickly wear away?

Returning to our hypothetical example, if we slide the tribium block faster and faster, frictional heating will begin to change the mechanical and chemical properties

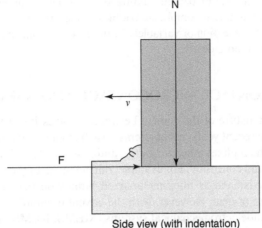

Side view (with indentation)

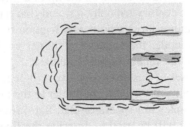

Top view (material is displaced)

FIGURE 4.4   Increases in tangential force due to digging-in of the block's leading edge.

of the surfaces. But if there is a viscous, continuous fluid film on the surface, and if the geometry of the contact has the right characteristics, the fluid might be channeled into the sliding interface in such a way that a hydrodynamic pressure develops, and this pressure can counteract the normal force, thus separating the surfaces and lowering the friction (refer to the explanation of hydrodynamic lubrication in Chapter 6).

If we further increase the load, at some point the front edge of the hard tribium block will begin to dig into the surface of the frictionite, analogous to the machining of a soft surface by a harder tool. In that case, the forces required to create cutting chips or plow through the surface will raise the total friction force. The geometry of the leading edge of the block will affect its tendency to cut or plow through the surface (Figure 4.4). In orthogonal metal-cutting literature, the angle of tilt from the normal to the surface is called the rake angle. The rake angle is taken to be negative when the top is tilted toward the sliding direction and positive in the opposite case. Mulhearn and Samuels[11] have shown that the critical rake angle for the transition between cutting and plowing varies between different metals (e.g., −5° for Al, −45° for Cu, and −35° for Pb), and it also varies with the type of heat treatment given to certain steels (0 to −80°). The mechanics of surface deformation and wear can therefore affect friction, arguing for the embedment of wear models into friction models if the definition of the given tribosystem requires it.

The hypothetical case of tribium sliding on frictionite was used to reveal the many interacting factors that complicate friction modeling and how physical observations influence the selection of variables. In the next section, scale effects on friction are discussed in more detail.

4.1 MACROCONTACT, MICROCONTACT, AND NANOCONTACT

Characterizing the nature of the contact between surfaces involves both geometry and feature size. In recent years, tribosystems have become miniaturized to the point where the use of the prefixes macro-, micro-, and nano- in describing them can generate ambiguities and confusion. For example, during much of the twentieth century, asperities were envisioned as micrometer-sized bumps on the surface of a larger body, say a bearing or gear. However, with the advent of atomic force microscopes and micro- and nanoelectromechanical systems (MEMS, NEMS) an entire contacting body can be a micrometer or less in dimension, smaller than one asperity in the historical sense. Individual atoms on the tip of a tiny probe have become the asperities of twenty-first century tribophysics, and that has created a schism in tribology modeling community: those who model friction on the molecular scale and those who continue to model the behavior of larger engineering components in which microcontacts are the smallest level of interest.

For the purposes of this discussion, we will begin with the more traditional connotations. The term *macrocontact* will refer to the overall shapes of the bodies making contact. Examples are sphere-on-flat, flat-on-flat, or cylinder-on-cylinder. The term *microcontact* will be used to describe the fine-scale interaction of asperities. Examples include using standard roughness parameters to describe surfaces, exploring the effects of the orientation (lay) of surface grooves on friction, assessing the development of tiny pits or flakes on worn surfaces, studying the nature of asperity shapes, and so forth. In recent years, considerable effort has been devoted to imaging and modeling contacting surfaces at the atomic level. We use the term *nanocontact* to describe phenomena at this scale; however, it should be recognized that the features associated with micro- and nanocontacts are not necessarily factors of a thousand, different in size. For example, Figure 4.5 shows the size scales involved in various tribological phenomena.

Macrocontacts and the stresses developed in them have been extensively modeled, particularly in regard to the design of bearings and gears. It is commonplace in such studies to assume that the properties of the materials in contact are homogeneous and isotropic, and for many practical purposes, those assumptions are justified. Macrocontact conditions are useful in friction models when there are thick lubricating films relative to the surface roughness and the effects of surface heterogeneities are of little importance.

Macrocontact approaches to a great many problems (e.g., in bearing and gear analysis) owe much to the ground-breaking work of Heinrich Hertz in the 1880s. Hertz's equations permit one to calculate the maximum compressive stresses and contact dimensions for bodies in elastic contact (i.e., no permanent plastic deformation). The parameters required to calculate these quantities and the algebraic equations used for simple geometries are given in Tables 4.1a and 4.1b.[12] Remember that

Phenomena	Applications
Molecular scale (10^{-10} to 10^{-8}m)	
Bonding/adhesion Order-disorder of interfacial layers Shear-thinning Surface diffusion Adsorption Asperity tip penetration Langmuir-Blodgett films Microchemistry Energy dissipative processes	Electrical contact friction Lubricant–surface interactions and catalysis Boundary lubricant development Fiber–matrix interface control in composites High–vacuum bearing applications Micromachining processes Lubrication of micromachines Coatings for magnetic storage media Ion implantation process development
Microstructural scale (10^{-7} to 10^{-5}m)	
Dislocation structures/twinning Highly deformed layer formation Micro-asperity interactions Metallurgical aspects of friction Grain deformation and fracture Transfer layers Wear debris particle effects Microabrasion	Carburizing, nitriding, etc. Solid lubricant application and development Surface roughness in bearings, gears, etc. Inclusion control for surface durability Surface patterning for friction control Various automotive applications Electroplating and thin thermal spray coats Flow of particles in material processing
Macroscale (10^{-4} to 10^{-2}m)	
Bulk mechanical properties effects Macrocontact geometry effects Surface texture (geometric) Macro-scale stick-slip Friction–vibration interactions	Friction textiles/fabrics/paper products Rail-wheel friction (railroads) Friction of shoes on flooring Thermally-sprayed coatings Mining and agricultural processes Forging, rolling, extrusion, etc. Weld overlays Flow in powders, sand, and fine gravel
Geologic scale (10^{-1} to $>10^{6}$ m)	
Earthquakes (grand scale stick-slip) Shear behavior of soils Shear behavior of rock masses	Design of building foundations Handling of ores and gravel Earthquake prediction Glacial flows Ice-breakers bow plate friction Plate tectonics

FIGURE 4.5   Typical sizes of phenomena affecting tribological response.

the Hertz's equations apply to the static or quasi-static, elastic case. In the case of sliding, plastic deformation, and the contact of rough surfaces, both the distribution of stresses and the contact geometry will be changed and more sophisticated analyses are required. "Hertzian contact" equations have been successfully used for engineering design and in friction and wear models in which the individual asperities are modeled as simple, elastic geometric contacts.

One of the most popular surface geometry models was developed by Greenwood and Williamson,[13] who modeled contacts based on a distribution of asperity heights, similar to the tail end of a Gaussian distribution. From those assumptions, contact

TABLE 4.1a

Equations for Calculating Elastic (Hertzian) Contact Stress (Definitions of Symbols)

Symbol	Definition
P	Normal force
p	Normal force per unit contact length
$E_{1,2}$	Modulus of elasticity for bodies 1 and 2, respectively
$v_{1,2}$	Poisson's ratios for bodies 1 and 2, respectively
D	Diameter of the curved body, if only one is curved
$D_{1,2}$	Diameters of bodies 1 and 2, where $D_1 > D_2$ by convention
S_c	Maximum compressive stress
a	Radius of the elastic contact
b	Width of a contact (for cylinders)
E^*	Composite modulus of bodies 1 and 2
A, B	Functions of the diameters of bodies 1 and 2

TABLE 4.1b

Equations for Calculating Elastic (Hertzian) Contact Stress and Contact Dimensions

$$E^* = \frac{1-v_1^2}{E_1} + \frac{1-v_2^2}{E_2}$$

$$A = \frac{D_1 + D_2}{D_1 D_2} \qquad B = \frac{D_1 - D_2}{D_1 D_2}$$

Geometry	Contact Dimension	Contact Stress
Sphere-on-flat	$a = 0.721(PDE^*)^{1/3}$	$S_c = 0.918(P/D^2E^{*2})^{1/3}$
Cylinder-on-flat	$b = 1.6(pDE^*)^{1/2}$	$S_c = 0.798(P/DE^*)^{1/2}$
Cylinder-on-cylinder (axes parallel)	$b = 1.6(pE^*/A)^{1/2}$	$S_c = 0.798(PA/E^*)^{1/2}$
Sphere in a spherical socket	$a = 0.721(PE^*/B)^{1/3}$	$S_c = 0.918[P(B/E^*)^2]^{1/3}$
Cylinder in a circular groove	$b = 1.6(pE^*/B)^{1/2}$	$S_c = 0.798(pB/E^*)^{1/2}$

between a rough surface and a smooth, rigid plane could be determined by three parameters: the asperity radius (R), the standard deviation of asperity heights (σ^*), and the number of asperities per unit area. To predict the extent of plastic deformation of asperities, a plasticity index (ψ), also a function of the hardness (H), elastic modulus (E), and Poisson's ratio (v), was introduced.

$$\psi = \left(\frac{E'}{H}\right)\left(\frac{\sigma^*}{R}\right)^{1/2} \tag{4.3}$$

where $E' = E/(1 - v^2)$. This basic formulation was subsequently refined by various investigators such as Whitehouse and Archard[14] to incorporate other forms of

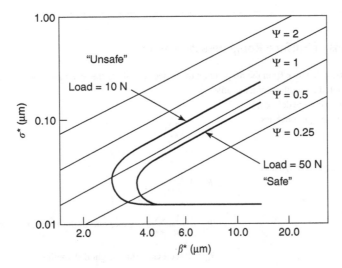

FIGURE 4.6   Scuffing diagram proposed by Hirst and Hollander in 1974.

height distributions and the incorporation of a distribution of asperity radii, represented by the so-called *correlation distance* β^*, produced higher contact pressures and increased plastic flow.[15] Thus,

$$\psi = \left(\frac{E'}{H}\right)\left(\frac{\sigma^*}{\beta}\right)^{1/2} \tag{4.4}$$

Hirst and Hollander[16] used the plasticity index to develop diagrams to predict the onset of scuffing (see Figure 4.6). Twenty years later, Park and Ludema[17] argued that the plasticity index was inadequate to explain scuffing behavior when liquid lubricants were present, as in many engineering situations.

Later, other roughness parameters, such as the average or root mean square (rms) slope of asperities, were incorporated into wear models to better account for such drawbacks.[18] Real engineering surfaces, particularly worn surfaces, are observed to be much more complex than simple arrangements of nesting spheres or spheres resting on flat planes. In fact, Greenwood acknowledged some of the problems associated with such simplifying assumptions about surface roughness.[19]

Those in the field of surface roughness measurement refer to the attributes of surface microgeometry as *texture*. National and international standards have been developed to define measures of surface roughness.[20] A comprehensive review of surface texture measurement methods was prepared by Song and Vorburger of the U.S. National Institute of Standards and Technology.[21] Some of the most commonly used roughness parameters are listed in Table 4.2. Of these, the arithmetic roughness, the rms roughness, and, to a lesser extent, the peak-to-valley roughness are the most widely used, but 10-point height is also gaining popularity for describing bearing surfaces. Skewness is useful for determining the lubricant retention qualities of surfaces, since it reflects the presence of crevices.

TABLE 4.2

Definitions of Surface Roughness Parameters

Let y_i = vertical distance from the ith point on the surface profile to the mean line and N = number of points measured along the surface profile.

Then, the following are defined:

Arithmetic average roughness

$$R_a = \frac{1}{N}\sum_{i=1}^{N}|y_i|$$

Root mean square roughness

$$R_a = \left[\frac{1}{N}\sum_{i=1}^{N}| y_i^2 |\right]^{1/2}$$

Skewness

$$R_{sk} = \frac{1}{NR_q^3}\sum_{i=1}^{N}y_i^3, \text{ a measure of the symmetry of the profile}$$

R_{sk} = 0 for a Gaussian height distribution

Kurtosis

$$R_{kurtosis} = \frac{1}{NR_q^4}\sum_{i=1}^{N}y_i^4, \text{ a measure of the sharpness of the profile}$$

$R_{kurtosis}$ = 3.0 for a Gaussian height distribution
$R_{kurtosis}$ < 3.0 for a broad distribution of heights
$R_{kurtosis}$ > 3.0 for a sharply peaked distribution

Surface metrologists have struggled with the development and selection of microgeometric parameters to make them as descriptive as possible of the actual contact. But one parameter alone cannot precisely model the microgeometry of surfaces, even relatively idealized ones. Figure 4.7 shows that it is possible to have the same average roughness (or rms roughness) for two quite differently appearing surfaces.

Even relatively small amounts of wear can change the roughness of surfaces on the microscale and obviously disrupt the nanoscale structure as well. Figure 4.8, which shows the wear surface of a Cu–Zn alloy subjected to unlubricated sliding in a pin-on-disk apparatus, dramatically highlights the shortcomings of models that use simple geometric shapes or statistically derived shape parameters in their contact stress calculations.

Geometry-based models for friction usually assume that the normal load remains constant. This assumption may be unjustified when sliding speeds are relatively high or when there are significant levels of vibration. Also, as sliding speed increases, frictional heating increases and localized thermal expansion at frictionally heated asperities can cause contact patches to form. The growth and subsequent wear of such patches has been called thermoelastic instability (TEI).[22] But TEI is only one way to stimulate vibrations and localized stress variations in sliding contacts. Another is the eccentricity of rotating shafts, excessive run-out, misalignment, and the transmission of vibrations to the contact from neighboring machinery.

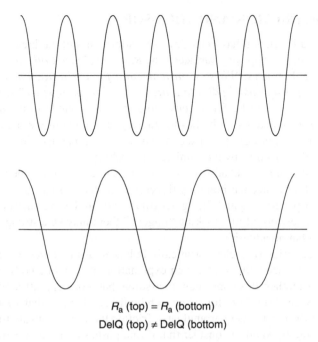

R_a (top) = R_a (bottom)

DelQ (top) ≠ DelQ (bottom)

FIGURE 4.7 Surfaces with different peak spacings and steepness can produce the same average roughness value.

FIGURE 4.8 Features of a Cu–10 wt% Zn alloy surface subjected to unidirectional sliding against a 52100 steel sphere in Ar gas. Magnification 190X.

4.2   STATIC FRICTION AND STICK-SLIP

Historically, static friction has been linked to adhesion and the breaking of bonds between atoms on the opposing surfaces. The *Random House Collegiate Dictionary* (1973) defines adhesion as: *"Physics, the molecular force of attraction in the area of contact between two unlike bodies that acts to hold them together."* If all possible causes for friction are to be fairly considered, it is reasonable to inquire whether there are other means to cause bodies to stay together without the requirement for molecular bonding. For example, a keeper bar holds to a magnet without adhesively bonding to it. Velcro works by mechanical interlocking.

Surfaces may *adhere*, but adherence is not the same as adhesion because there is no requirement for molecular bonding. If a certain material is cast between two surfaces and, after penetrating and filling irregular voids in the two surfaces, solidifies so as to form a network of interlocking "fingers," there may be a strong mechanical joint produced but no adhesion.

Invoking adhesion (i.e., electrostatically balanced attraction/chemical bonding) in friction theory meets the need for an explanation of how one body can transfer shear forces to another. It is convenient to assume that molecular attraction is strong enough to allow the transfer of force between bodies, and in fact that assumption has led to many of the most widely promulgated friction theories. From another perspective, is it not equally valid to consider that if one pushes two rough bodies together so that their asperities penetrate and then attempts to move those bodies tangentially, that the atoms may approach each other closely enough to repel strongly, thus causing a "back force" against the bulk materials and away from the interface? (See the schematic illustration in Figure 4.9.) The repulsive force parallel to the sliding direction

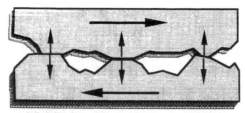

Adhesive forces impede relative movement

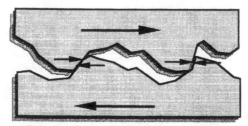

Repulsive forces impede relative movement

FIGURE 4.9   Two apparently contradictory arguments for the mechanism of static friction. At the top, the adhesive forces hold surfaces together at the tips of the asperities and resist relative (shearing) motion. At the bottom, horizontal components of repulsive forces at the sides of interpenetrating asperities resist relative motion.

must be overcome to move the bodies tangentially, whether accommodation occurs by asperities climbing over one another or by deforming one another. In the latter conception, it is repulsive forces and not adhesive bonding that produce sliding resistance.

Solid adhesion experiments have been conducted for many years. Ferrante et al.[23] have provided a comprehensive review of the subject. In addition, an excellent discussion of adhesion and its relationship to friction can be found in the book by Buckley.[24]

Atomic force microscopes came into prominence during the closing decades of the twentieth century. They permit investigators to study and map adhesion and lateral forces between surfaces on the molecular scale. Operators of such instruments are familiar with the tendency of the fine probe tip to jump to the surface as the approach distance becomes very small. The force required to shift the two bodies tangentially must overcome whichever bonds are holding the surfaces together. In the case of dissimilar metals with a strong bonding preference, the shear strength of the interfacial bonds can exceed the shear strength of the weaker of the two metals, and the static friction force ($\mathbf{F}_S$) will depend on the shear strength of the weaker material (τ_m) and the area of contact (A). In terms of the static friction coefficient μ_s,

$$\mathbf{F}_S = \mu_s \mathbf{P}^* = \tau_m A \qquad (4.5)$$

or

$$\mu_s = \left(\frac{\tau_m A}{\mathbf{P}^*} \right) A \qquad (4.6)$$

where $\mathbf{P}^*$ is the normal force that is the sum of the applied load and the adhesive force normal to the interface. Under specially controlled conditions, such as friction experiments with clean surfaces in a high vacuum, the static friction coefficients can be much greater than 1.0, and the experiment becomes more of a measurement of the shear strength of the solid materials than of interfacial friction.

Historically, scientific understanding and approaches to modeling friction have been strongly influenced by our concepts of solid surfaces and by the instruments available to study them. Atomic force microscopes and scanning tunneling microscopes have permitted views of surface atoms with amazing resolution and detail. Among the first to study nanocontact frictional phenomena were McClellan et al.,[25,26] who used the instrument diagrammed in Figure 4.10. A tungsten wire with a very fine tip was brought down to the surface of a highly oriented, cleaved basal plane of pyrolytic graphite as the specimen was oscillated at about 10 Hz using a piezoelectric driver. The cantilevered wire was calibrated so that its spring constant was known (2500 N/m) and the normal force could be determined by measuring the deflection of the tip with a reflected laser beam. As normal force was decreased, the contributions of individual atoms to the tangential force became increasingly apparent (see Figure 4.11). At the same time, it appeared that the motion of the tip became less uniform, exhibiting what some might call *nanoscale stick-slip*.

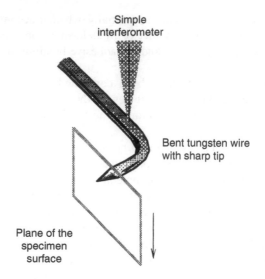

FIGURE 4.10 Principle of the atomic force microscope used by McClelland et al. (Adapted from McClelland, G.M., Mate, C.M., Erlandsson, R., and Chiang, S., *Adhesion in Solids*, Materials Research Society, Pittsburgh, PA, 1988.)

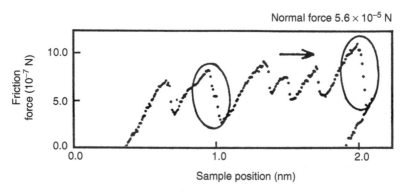

FIGURE 4.11 Atomic-scale periodicity of the friction force. (Adapted from McClelland, G.M., Mate, C.M., Erlandsson, R., and Chiang, S., *Adhesion in Solids*, Materials Research Society, Pittsburgh, PA, 1988.)

Thompson and Robbins[27] discussed the origins of nanocontact stick-slip when analyzing the molecular behavior in thin fluid films trapped between flat surfaces of face-centered cubic solids. At that scale, stick-slip was believed to arise from the periodic phase transitions between ordered static and disordered kinetic states. Immediately adjacent to the surface of the solid, the fluid assumed a regular, crystalline structure, but this was disrupted during slip events. Experimental plots of friction force per unit area versus time exhibited extremely uniform, "classical" stick-slip appearance. Once slip occurs, all the kinetic energy must be reconverted into potential energy. In subsequent papers[28,29] the authors used this argument to calculate the critical velocity v_c below which the stick-slip occurs:

$$v_c = c \left(\frac{\sigma \mathbf{F}_S}{M} \right)^{1/2} \tag{4.7}$$

where σ is the lattice constant of the wall, $\mathbf{F}_S$ the static friction force, M the mass of the moving wall, and c a constant.

Friction might simply be defined as *the resistance to relative motion between two contacting bodies parallel to a surface that separates them.* Motion at the atomic scale is invariably unsteady. One struggles to define the physical boundaries of the "surface that separates them." In nanocontact, accounting for the tangential components of thermal vibrations of the atoms thus affects our ability to clearly define "relative motion" between surfaces. Furthermore, under some conditions it may be possible to translate the surface laterally while the adhesive force between the probe tip and the opposite surface exceeds a small externally applied tensile force. When the sense of the applied normal force is reversed, should the friction coefficient of necessity take on a negative value? Progress in instrumentation and experimental sophistication forces us to consider whether the term "friction" has a traditional meaning at the atomic scale. Perhaps a new term, such as "frictomic forces," should be introduced to avoid such concerns.

A parallel development, molecular dynamics (MD) modeling on high-speed computers, has further broadened the possibilities of modeling friction on the finest scale.[30] Landman et al.[31] reviewed progress in this field, drawing together developments in experimental techniques as well as computer simulations. By conducting MD simulations of an Ni tip and a flat Au surface, Landman et al. illustrated how the tip can attract atoms from the surface simply by close approach and without actual indentation. A connective neck or "bridge" of surface atoms was observed to form as the indenter was withdrawn. The neck can exert a force to counteract the withdrawal force on the tip, and the MD simulations indicate the tendency to transfer material between opposing asperities under pristine surface conditions. Landman et al. subsequently conducted numerous other MD simulations, including complete indentation and indentation in the presence of organic species between the indenter and the substrate. Belak and Stowers[32], using a material volume containing 43,440 atoms in 160 layers, simulated many of the deformational features associated with metals, such as edge dislocations, plastic zones, and point defect generation. Calculated shear stresses for a triangular indenter passing along the surface exhibited erratic behavior, not unlike that observed during metallic sliding under clean conditions (Figure 4.12). Pollock and Singer[33] have compiled a fascinating series of papers on atomic-scale approaches to friction. Readers interested in detailed discussions of the relationship between adhesion and friction may want to consult the annotated references in Table 4.3.

While MD simulations and atomic-scale experiments continue to provide fascinating insights into frictional behavior, albeit under somewhat idealized conditions, most engineering tribosystems are, in comparison, quite "dirty." Not only

TABLE 4.3
Discussions of the Relationships between Adhesion and Friction

M. E. Sikorski (1964). The adhesion of metals and factors that influence it, in *Mechanisms of Solid Friction*, eds. P. J. Bryant, L. Lavik, and G. Salomon, Elsevier, Amsterdam, pp. 144–162.

Describes macroscopic "twist-compression" experiments (two methods) to compare the measured coefficients of adhesion to measured friction coefficients for various metals. Concludes that both mechanical and physicochemical factors affect the correlation. Correlations of adhesion with other properties, such as melting point and Vickers hardness are presented.

E. Rabinowicz (1965). *Friction and Wear of Materials*, Wiley, New York, NY, pp. 62–67.

Describes and critiques the adhesion-based theories for friction. The critique addresses five issues: (a) inconsistencies between adhesion theories and the effects of surface roughness on friction, (b) whether the conditions in sliding are sufficient to weld surfaces together, (c) lack of experimental evidence for strong adhesive forces between surfaces in ambient environments, (d) consideration of the behavior of brittle nonmetals, and (e) the fact that experimental friction coefficients are two or three times greater than the ratio of the shear strength of the weaker material to the contact pressure.

K. L. Johnson, K. Kendall, and A. D. Roberts (1971). Surface energy and the contact of elastic solids, *Proc. R. Soc. Lond.*, A324, pp. 301–313.

This classic paper is probably the most-cited paper concerning the effects of adhesion on contact between elastic solids. First describing what is now referred to as the JKR model after the last names of the authors, it lays the conceptual basis for the analysis of numerous studies in nanotribology in which surface forces play a role in dissipating energy associated with both static and moving contacts, with and without liquids present in confined interfaces.

J. J. Bikerman (1976). Adhesion in friction, *Wear*, 39, pp. 1–13.

The author disputes the importance of adhesion in friction, addressing four hypotheses regarding such relationships. He argues that the adhesive force between solids under normal conditions is so small as to play a negligible role in friction, especially in the presence of air. He also refutes the adhesive junction formation concepts often used to explain frictional behavior. Bikerman concludes: "As long as the term friction is employed in its common meaning adhesion does not cause friction and has no influence on frictional phenomena."

D. Tabor (1985). Future directions in adhesion and friction—status of understanding, in *Tribology in the '80s*, Noyes, Park Ridge, NJ, pp. 116–137.

This materials-oriented review covers metals, ceramics, elastomers, and polymers. Distinctions are made between the thermodynamic concept of adhesion and adhesive strength, as measured by pull-off testing. The roles of interfacial bonding, surface energies, and energy dissipation are discussed, emphasizing differences between various classes of materials.

F. P. Bowden and D. Tabor (1986). *The Friction and Lubrication of Solids*, Clarendon Press, Oxford, pp. 299–314, 358–360.

Discusses experiments on the role of lubricant films on adhesion between both metals and nonmetals. Describes several studies attempting to link adhesion with friction. In particular authors discuss steel sliding on indium. Before gross macroslip occurs, the friction coefficient μ and the adhesion coefficient v are well represented over a range of experiments by $0.3v^2 - \mu^2 = 0.3$, or alternately $\mu = 0.3(v^2 - 1)^{1/2}$.

Q. Quo, J. D. J. Ross, and H. M. Pollock (1988). What part do adhesion and deformation play in fine-scale static and sliding contact? in *New Materials Approaches to Tribology: Theory and Applications*, eds. L. E. Pope, L. L. Fehrenbacher, and W.O. Winer, Vol. 140, Materials Research Society, Pittsburgh, PA, pp. 57–66.

Describes single-asperity experiments, junction growth, surface forces contributions; "maps" based on energy-balance, contact mechanics, and allows prediction of conditions for adhesion-induced plastic deformation.

TABLE 4.3 (Continued)

J. N. Israelachvili (1992). Adhesion, friction, and lubrication of molecularly smooth surfaces, in *Fundamentals of Friction: Macroscopic and Microscopic Processes*, eds. H. M. Pollock and I. L. Singer, Kluwer, Dordrecht, The Netherlands, pp. 351–381.

   Provides an extensive review of energy dissipation mechanisms in molecular interfaces. Includes reversible and irreversible adhesion, contact theories, adhesive hysteresis, molecular relaxations, capillarity, surface forces, boundary friction, stick-slip, and very thin films.

K. L. Johnson (1998). Mechanics of adhesion, *Tribol. Int.*, 31(8), pp. 413–418.

   Twenty-seven years after the introduction of the JKR model, one of its creators reviews the mechanics of adhesion between spherical surfaces, especially with regard to magnetic storage devices. It offers a conceptual map with nondimensional axes: the ratio of adhesive force to applied load and the elastic deformation to the range of surface forces.

FIGURE 4.12 Molecular dynamics simulation of a wedge cutting through Cu and generating an edge dislocation that extends diagonally into the bulk material. (Courtesy of J. Belak, Lawrence Livermore Laboratory.)

are surfaces not atomically flat, but the materials are not homogeneous, and surface films and contaminant particles of many kinds (orders of magnitude larger than the atomic scale) may influence interfacial behavior.

   Static friction coefficients measured experimentally under ambient or contaminated conditions probably will not equate to values obtained in controlled research environments. In a series of carefully conducted experiments on the role of adsorbed oxygen and chlorine on the shear strength of metallic junctions, Wheeler[34] showed

TABLE 4.4
Static Friction Coefficients for Clean Metals in Helium Gas at Two Temperatures

Material Combination	Static Friction Coefficient	
	300 K	80 K
Fe (99.9%) on Fe (99.99%)	1.09	1.04
Al (99%) on Al (99%)	1.62	1.60
Cu (99.95%) on Cu (99.95%)	1.76	1.70
Ni (99.95%) on Ni (99.95%)	2.11	2.00
Au (99.98%) on Au (99.98%)	1.88	1.77
Ni (99.95%) on Cu (99.95%)	2.34	2.35
Cu (99.95%) on Ni (99.95%)	0.85	0.85
Au (99.98%) on Al (99%)	1.42	1.50
Fe (99.9%) on Cu (99.95%)	1.99	2.03

Source: Adapted from Arkarov, A.M. and Kharitonova, L.D. in *Tribology Handbook*, Mir, Moscow, 1986, 84.

clearly how μ_s can be reduced in the presence of adsorbed gases. Figure 4.13 (Ref. 34, p. 14) summarizes his results. However, static friction coefficients for pure, well-cleaned metal surfaces in the presence of nonreactive gases such as He can be relatively high, as the data in Table 4.4 indicate. It is interesting to note that the friction of copper on nickel and the friction of nickel on copper are not the same. This is not an error but rather a demonstration of the fact that reversing the materials of the sliding specimen and the counterface surface can affect the measured friction, confirming the assertion that friction is a property of the tribosystem and not only of the materials in contact. A special "cryotribometer" was used to obtain the data in Table 4.4.[35]

The length of time that two solids are in contact can also affect the relative role that adhesion plays in establishing the value of the static friction coefficient. Two distinct possibilities can occur: (a) if the contact becomes contaminated with a lower shear-strength species, the friction will decline; and (b) if the contact is clean and a more tenacious interfacial bond develops, the static friction will tend to increase. Akhmatov[36] demonstrated over half a century ago, using freshly cleaved rock salt, that the formation of surface films over time lowers the static friction (see Figure 4.14).

The opposite effect has been demonstrated for metals[37] (see Figure 4.15). A first approximation of rising static friction behavior is given by

$$\mu_{s(t)} = \mu_{s(t=\infty)} - [\,\mu_{s(t=\infty)} - \mu_{s(t=0)}]e^{-ut} \tag{4.8}$$

where $\mu_{s(t)}$ is the current value of the static friction coefficient at time t, $\mu_{s(t=\infty)}$ the limiting value of the static friction coefficient at long times, $\mu_{s(t=0)}$ the initial static

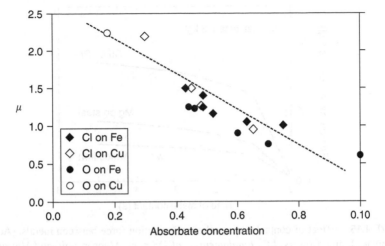

FIGURE 4.13 Relationship of the static friction coefficient of Fe and Cu sliders to the fractional coverage of the surface by O and Cl. (Adapted from Wheeler, D.R., *The Effect of Adsorbed Chlorine and Oxygen on the Shear Strength of Iron and Copper Junctions*, NASA TN D-7894, 1975.)

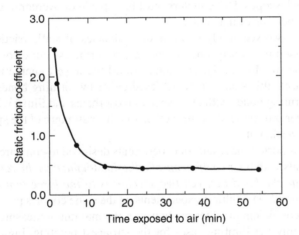

FIGURE 4.14 Effect of time of exposure to air on the static friction coefficient of cleaved salt surfaces. (Adapted from Akhmatov, A.S., *Trudy Stankina*, 1939; as cited by Kragelski, I.V. in *Friction and Wear*, Butterworths, London, 1965.)

friction coefficient, and u a rate constant. In contrast to exponential dependence on time, Buckley showed, using data for tests of single-crystal Au touching Cu–5% Al alloy, that junction growth (contact creep) can cause the adhesive force to increase linearly with time.

When two bodies are pressed together, the atoms on their surfaces will begin to interact. The extent of interaction will depend on the contact pressure, the temperature, and the degree of chemical reactivity that the species have for each other, so it should be no surprise that static friction can change with the duration of contact.

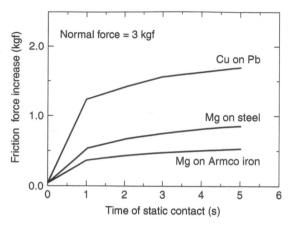

FIGURE 4.15 Effect of contact time on the static friction force between metals. (Adapted from Belak, J. and Stowers, J.F., *Fundamentals of Friction: Macroscopic and Microscopic Processes*, Kluwer, Dordrecht, The Netherlands, 1992.)

This phenomenon is of great importance to the intermittent operation of remotely controlled mechanisms, such as the antennas and other moving parts in earth-orbiting satellites and telescopes. Drive motors must be capable of overcoming starting friction after long periods of inactivity.

Despite the two seemingly opposite dependencies of static friction on time of contact, discussed previously, observations are nevertheless quite consistent from a thermodynamic standpoint. Systems tend toward the lowest energetic state. In the case of interfaces, this state can be achieved either by forming bonds between the solids or by forming bonds with other species (adsorbates and films) within the interface. The former process tends to strengthen the shear strength of the system, and the latter tends to weaken it.

Sikorski[38] reported the results of experiments designed to compare friction coefficients of metals with their *coefficients of adhesion* (*defined as the ratio of the force needed to break the bond between two specimens to the force that initially compressed them together*). Multiple experiments on the same couples provided different values for the coefficient of adhesion; however, the maximum measured coefficients of adhesion imply the limiting cases for the strongest bonding. Figure 4.16 shows Sikorski's data relating the coefficient of adhesion and friction coefficients (in laboratory air environments) for eight cubic metals. Under those conditions, the friction coefficients seem remarkably high; however, upon further analysis, one must conclude that the investigator was not measuring resistance to sliding in an interface but rather the shear strength of the bulk material, which was lower than the strength of the interfacial bonds.

Rabinowicz[39] conducted a series of simple, tilting-plane tests with milligram- to kilogram-sized specimens of a variety of metals. Results demonstrated a trend for the static friction coefficient to increase as slider weight (normal force) decreased (see Figure 4.17). For metal couples such as Au/Au, Au/Pd, Au/Rh, Ag/Ag, and Ag/Au, as the normal force increased over about six orders of magnitude (1 mg to 1 kg), the static friction coefficients tended to decrease by nearly one order of magnitude.

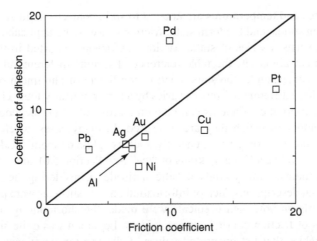

FIGURE 4.16 Relationship between the coefficient of adhesion and the coefficient of friction, based on the experiments of Sikorski. (Adapted from Sikorski, M.E., *Mechanisms of Solid Friction*, Elsevier, Amsterdam, 1964.)

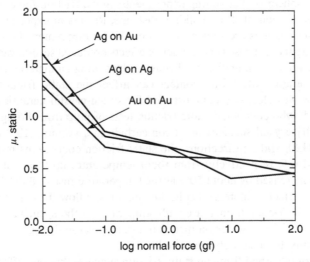

FIGURE 4.17 Increase in static friction of noble metals with decreasing normal force. (Plotted from the data of Rabinowicz, E., *Wear*, 159, 89, 1992.)

Therefore, in designing or analyzing machines in which there are very lightly loaded contacts, static friction coefficient values obtained by more macroscopic experiments should not be used as a basis for material selection.

Under low contact pressures, surface chemistry effects can play a relatively large role in governing static friction behavior. However, under more severe contact conditions, such as extreme pressures and high temperatures, other factors, more directly related to bulk properties of the solids, dominate static friction behavior. When very high pressures and temperatures are applied to solid contacts, solid-state welds can form between solids, and the term static friction ceases to be applicable.

Table 4.5 lists a series of static friction coefficients reported in the literature. Note that in certain cases, the table references list quite different values for these coefficients. For example, the static friction coefficient of aluminum on aluminum reported in three different references varies by a factor of nearly four. This reinforces the idea that friction coefficients can take on a range of values depending strongly upon the manner in which they are measured. In many cases, a reference might describe the materials only very generally, such as "glass on metal." Likewise, there are many kinds of "steel," many kinds of "ice" (see Section 5.6), and ceramists will confirm that there are many kinds of "silicon nitride." Considering the prevalence of vague material descriptions, lack of information on how surfaces were prepared, and missing details on how measurements were made, the shortcomings of handbook compilations of friction coefficients are evident. Feynman was quite right when he said: "The tables that list purported values of the friction coefficient of 'steel on steel' or 'copper on copper' and the like, are all false, because they ignore the factors that really determine friction."[40]

The effect of surface roughening by wear on reducing static friction has been studied by Wiid and Beezhold.[41] Friction and contact resistance were measured in a series of hemisphere-on-flat sliding experiments. Static friction force was measured in increments of about 0.05 mm sliding distance. Results of experiments with several metal combinations are shown in Figure 4.18. Sliding removed initial films and altered the roughness. These two interactive effects produced several variations of the static friction versus sliding distance behavior, depending on material combination.

The temperature of sliding contact can affect the static friction coefficient. This behavior was elegantly demonstrated for single-crystal ceramics by Miyoshi and Buckley,[42] who conducted static friction tests of pure iron sliding on carefully cleaned {0001} crystal surfaces of silicon carbide in a vacuum (10^{-8} Pa). For both ⟨1010⟩ and ⟨1120⟩ sliding directions, the static friction coefficients remained about level (0.4 and 0.5, respectively) from room temperature up to about 400°C; then each of them increased by about 50% as the temperature rose to 800°C. The authors attributed this effect to increased adhesion and plastic flow. Just as some situations may cause static friction to increase with temperature, others, such as the enhanced formation of lubricious oxides on metals in air, may cause the opposite trend. Each tribosystem must be separately evaluated.

The role of adsorbed films on static friction suggests that one effective strategy for alleviating or reducing static friction is to introduce a lubricant or other surface treatment to impede the formation of adhesive bonds between mating surfaces. Contamination of surfaces from exposure to the ambient environment performs essentially the same function but is usually less reproducible. Campbell[43] demonstrated how the treatment of metallic surfaces by oxidation or tarnishing solutions can reduce the static friction coefficient. Oxide films were produced by heating metals in air. Sulfide films were produced by immersing the metals in sodium sulfide solution. Except for the film on steel, which was colorless, film thicknesses were estimated to be 100–200 nm, judging from interference colors. Results of 10 tests, using a

TABLE 4.5
Static Friction Coefficients for Metals and Nonmetals (Dry or Unlubricated Conditions)

Material Combination			Table Reference
Fixed Specimen	Moving Specimen	μ_s	Number[a]
Metals and alloys on various materials			
Aluminum	Aluminum	1.05	1
		1.9	2
		0.57	3
	Ice (dry, at 0°C)	0.35	2
	Ice (wet, at 0°C)	0.4	2
	Steel, mild	0.61	1
	Titanium	0.54	3
Al, 6061-T6	Al, 6061-T6	0.42	4
	Copper	0.28	4
	Steel, 1032	0.35	4
	Ti-6Al-4V	0.34	4
Brass	Steel, mild	0.53	1
Cast iron	Cast iron	1.10	1
Cadmium	Cadmium	0.79	3
	Iron	0.52	3
Chromium	Cobalt	0.41	3
	Chromium	0.46	3
Cobalt	Cobalt	0.56	3
	Chromium	0.41	3
Copper	Cast iron	1.05	1
	Chromium	0.46	3
	Cobalt	0.44	3
	Copper	1.6	2
		0.55	3
	Glass	0.68	1
	Iron	0.50	3
	Nickel	0.49	3
	Zinc	0.56	3
Gold	Gold	2.8	2
		0.49	3
	Silver	0.53	3
Iron	Cobalt	0.41	3
	Chromium	0.48	3
	Iron	0.51	3
		1.2	2
Iron	Tungsten	0.47	3
	Zinc	0.55	3
Indium	Indium	1.46	3

(*Continued*)

TABLE 4.5 (Continued)

Material Combination			Table Reference
Fixed Specimen	Moving Specimen	μ_s	Number[a]
Lead	Cobalt	0.55	3
	Iron	0.54	3
	Lead	0.90	3
	Silver	0.73	3
Magnesium	Magnesium	0.60	1
		0.69	3
Molybdenum	Iron	0.69	3
	Molybdenum	0.8	2
Nickel	Chromium	0.59	3
	Nickel	0.50	3
		1.10	1
		3.0	2
Niobium	Niobium	0.46	3
Platinum	Platinum	3.0	2
		0.55	3
Silver	Copper	0.48	3
	Gold	0.53	3
	Iron	0.49	3
	Silver	1.5	2
		0.5	2
Steel	Cast iron	0.4	2
Steel, hardened	Steel, hardened	0.78	1
	Babbitt	0.42, 0.70	1
	Graphite	0.21	1
Steel, mild	Steel, mild	0.74	1
	Lead	0.95	1
Steel, 1032	Aluminum	0.47	4
	Copper	0.32	4
	Steel, 1032	0.31	4
	Ti-6Al-4V	0.36	4
Steel, stain. 304	Copper	0.33	4
Tin	Iron	0.55	3
	Tin	0.74	3
Titanium	Aluminum	0.54	3
	Titanium	0.55	3
Tungsten	Copper	0.41	3
	Iron	0.47	3
	Tungsten	0.51	3
Zinc	Cast iron	0.85	1
	Copper	0.56	3
	Iron	0.55	3
	Zinc	0.75	3
Zirconium	Zirconium	0.63	3

TABLE 4.5   (Continued)

Material Combination			Table Reference
Fixed Specimen	Moving Specimen	μ_s	Number[a]
Glass and ceramics on various materials			
Glass	Aluminum, 6061-T6	0.17	4
	Glass	0.94	1
		0.9–1.0	2
	Metal (clean)	0.5–0.7	2
	PTFE	0.10	4
	Nickel	0.78	1
Sapphire	Sapphire	0.2	2
Tungsten carbide	Copper	0.35	1
	Graphite (outgassed)	0.62	2
	Graphite (clean)	0.15	2
	Steel	0.5	1
	Tungsten carbide	0.2	1
Polymers on various materials[b]			
Acetal	Acetal	0.06	5
Nylon 6/6	Nylon 6/6	0.06	5
	Polycarbonate	0.25	5
PMMA	PMMA	0.80	5
PTFE	Ice (dry)	0.02	2
	Ice (wet)	0.05	2
	Nickel	0.15	4
	PTFE	0.04	1
	Steel	0.04	1
	Steel, 1032	0.27	4
Rubber	Metal	1–4	2
Miscellaneous material combinations			
Brick	Wood	0.6	6
Cotton	Cotton	0.3	6
Diamond	Diamond	0.1	6
Graphite	Graphite	0.10	2
Ice	Ice: at 0°C	0.05–0.15	2
	Ice: at −71°C	0.5	2
Leather	Metal	0.6	2
Mica	Mica: fresh-cleaved	1.00	2
	Mica: contaminated	0.2–0.4	2
Oak	Oak	0.54–0.62	1
Paper, copier	Paper, copier	0.28	4
Silk	Silk	0.3	2

(*Continued*)

TABLE 4.5 (Continued)

Material Combination			Table Reference
Fixed Specimen	Moving Specimen	μ_s	Number[a]
Ski wax	Ice (dry, at 0°C)	0.04	2
	Ice (wet, at 0°C)	0.1	2
Wood	Wood (clean)	0.25–0.5	2

[a] Table references: (1) From the compilation in Table 2.1, B. Bhushan and B. K. Gupta (1991). *Handbook of Tribology*, McGraw Hill, New York, New York. (2) (1967). R. C. Weast ed. *Handbook of Chemistry and Physics*, 48th Ed., CRC Press. (3) E. Rabinowicz (1971). *ASLE Trans.*, 14, p. 198; plate sliding on inclined plate at 50% relative humidity. (4) *Friction Data Guide* (1988). General Magnaplate Corp., Ventura, California 93003, TMI Model 98-5 Slip and Friction Tester, 200 g load, ground specimens, 54% relative humidity, average of five tests. (5) Lubricomp® internally reinforced thermoplastics and fluoropolymer composites, *ICI Advanced Materials Bulletin, Exton, PA*, 254–688; thrust washer apparatus, 0.28 MPa, 0.25 m/s, after one revolution running-in. (6) F. P. Bowden and D. Tabor (1986). *The Friction and Lubrication of Solids*, Clarendon Press, Oxford, Appendix IV; method unspecified.

[b] PTFE, polytetrafluoroethylene (Teflon); PMMA, polymethylmethacrylate.

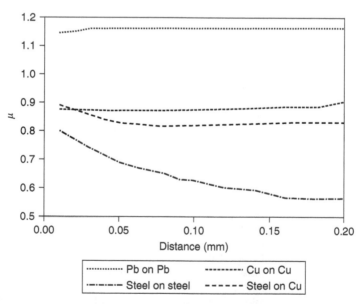

FIGURE 4.18 Incremental-distance sliding tests exhibited different effects on the static friction coefficient for different material combinations. (Adapted from Arkarov, A.M. and Kharitonova, L.D., *Tribology Handbook*, Mir, Moscow, 1986.)

three-ball-on-flat plate apparatus, were averaged to obtain static friction coefficients. In addition to producing oxides and sulfides, Campbell also tested oxide and sulfide films with "Acto" oil. The results of this investigation are shown in Table 4.6. For copper, the static friction coefficient ($\mu_s = 1.2$, with no film) decreased markedly

TABLE 4.6

Reduction of Static Friction by Surface Films

Material Combination	μ_s, No Film	μ_s, Oxide Film	μ_s, Sulfide Film
Copper on copper	1.21	0.76	0.66
Steel on steel	0.78	0.27	0.39
Steel on steel	078	0.19[a]	0.16[a]

[a] With both film and oil.

Source: Adapted from Campbell, W.E. in *Proceedings of M.I.T. Conference on Friction and Surface Finish*, MIT Press, Cambridge, MA, 1940, 197.

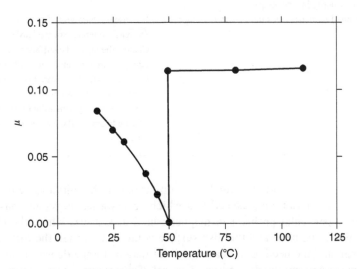

FIGURE 4.19 Hardy[44] observed a discontinuous change in μ_s with temperature in studies of palmitic acid-covered quartz surfaces.

when the sulfide film thickness was increased from 0 to about 300 nm, after which the static friction coefficient remained about constant at 0.75.

The extent to which a solid lubricant can reduce static friction may be dependent on temperature, as illustrated by Figure 4.19, adapted from Hardy's studies on the static friction of palmitic acid films on quartz.[44] Between 20 and 50°C the static friction coefficient decreases until melting occurs, at which time the lubricant looses its effectiveness.

Table 4.7 summarizes possible causes for static friction behavior and corresponding strategies to help reduce static friction. In some cases, changing the chemical species in the interface may help, but in others, mechanical techniques, such as physically vibrating the contact, may work better.

Stiction is phenomenologically similar to static friction. The phenomenon is sometimes associated with fine particles becoming jammed between contacting surfaces so as to prevent or impede motion during the start-up of sliding. It is not quite

TABLE 4.7
Possible Causes and Strategies for Controlling Static Friction

Causes	Strategies
Chemical (adhesion):	
Formation of bonds between surfaces	Select less chemically compatible materials
	Introduce a lubricating species
	Apply a parting compound to reduce adhesion
	Increase surface roughness
	Vibrate the contact continuously
	Change the temperature of the system
Mechanical (asperity interlocking):	
Penetration of hard asperities into a softer counterface	Increase the hardness of the softer material
	Decrease roughness of the harder material
	Change the lay of the surface roughness
	Apply a hard coating to one or both surfaces
	Decrease contact pressure to decrease the amount of penetration
	Apply a thick layer of solid lubricant or grease to spread the load and reduce shear strength

the same as classical static friction but is similar in that the initial force required to initiate relative motion may exceed that which occurs after movement has started. Stiction is a particular problem in magnetic hard disk drives where the clearance between the flying head and the disk surface is far smaller than the thickness of a human hair, and the need to increase the information storage density is driving the gap to smaller and smaller dimensions. Control of paniculate contamination involves the use of clean-room assembly areas for components. Other strategies, such as controlling surface roughness by special etching or burnishing treatments and controlling the orientation of fine finishing marks (lay) with respect to the sliding direction, are also used to alleviate stiction problems. Some computer hard disks that spin at over 3600 rpm are designed for 10,000–50,000 or more start–stop cycles over their service lifetimes. Surfaces that are too flat can "wring" together when the disk stops, and controlled surface relief is used to prevent this. Various lubricating films and coatings, some only 2–3 nm thick, and special curing treatments for those films have been successful in reducing stiction in magnetic disk applications.[45]

Stiction has also been associated with the formation of corrosion deposits between cast iron brake drums and linings when a vehicle has been parked in moist environments for an extended period of time.[46] This phenomenon results in high initial torque to break the rust bond and, due to the damage done to the surface finish, uneven braking response thereafter.

Stick-slip, a term coined by Bowden and Leben in 1938–1939,[47] is a commonplace phenomenon exemplified by the crying of chalk on blackboards, the vibration and squealing of brakes as an aircraft comes to a halt, and the chatter of windshield wipers on partly wet window glass. Stick-slip can be a simple annoyance like the squeaking of a door hinge, or it can be the source of pleasure, as in the bowing of stringed instruments.[48] Stick-slip has important technological implications. For example, Hammerschlag[49] described the requirement to eliminate stick-slip from certain mechanical parts of a high-resolution solar telescope: preloaded screws that fasten the support structure, spherical roller bearings in the right ascension and declination axes, and involute spur gears in the drive train.

Stick-slip has been referred to as a "relaxation–oscillation phenomenon," and consequently, some degree of elasticity is needed in the sliding contact for stick-slip to occur. As Holm[50] pointed out:

> A condition for regular periodicity of stick-slip is that the process is the same in every period. Plastic flow can obliterate this condition, and make the phenomenon very irregular. Another change occurs when a certain critical speed is attained. Then the motion proceeds relatively smoothly.

Like frictional phenomena in general, stick-slip occurs over a wide range of times and scales of interaction. Frequencies of oscillation can range from within the audible range (say, ~2–5 kHz, inferred from the spacing of tiny ridges on surfaces, as shown in Figure 4.20) to earthquakes[51] with less than one slip in every 200 years $(6.2 - 10^{-10}$ Hz). Table 4.8 lists famous geological "slips" and their dire consequences, recognizing that a major earthquake may release the energy equivalent to over 180 million metric tons of trinitrotoluene (TNT).[52] Stick-slip phenomena on many scales may exhibit similar types of relaxation–oscillation characteristics; however, the fundamental causes for each case of stick-slip behavior cannot be generalized.

Israelachvilli[53] considered stick-slip on a molecular level, as measured with surface forces apparatus. He considers the order–disorder transformations described by Thompson and Robbins[28,29] in the context of computer simulations. Most classical treatments of stick-slip, however, take a mechanics approach, considering that the

---

TABLE 4.8
Some Major Earthquakes and Their Consequences

Epicentral Region	Magnitude	Year	Estimated Loss of Human Life
Central China	—	1556	830,000
Hokaido, Japan	—	1730	130,000
Calcutta, India	—	1737	300,000
San Francisco, California	8.3	1906	503
Gansu, China	8.6	1920	200,000
Nan-Shan, China	8.3	1927	200,000
Quetta, India (now Pakistan)	7.5	1935	60,000
Chimbote, Peru	7.7	1970	66,794
Tangshan, China	8.2	1976	240,000
Southcentral Mexico	—	1985	7,000

Source: Adapted from *World Book Encyclopedia*, World Book, Chicago, 1988, 33.

FIGURE 4.20 Fine-scale "chatter marks" on the surface of the worn Cu alloy surface shown in Figure 4.8 (scanning electron micrograph). Marks are spaced about 7.0 μm apart.

behavior in unlubricated solid sliding is caused by making and breaking adhesive bonds.

In stick-slip, friction force versus time behavior is portrayed as a sawtooth curve (Figure 4.21, top). Most investigators agree that for a system to exhibit stick-slip behavior, a characteristic of decreasing friction coefficient with velocity is needed. Moore[54] has described the characteristic decrease in friction coefficient (μ) with sliding velocity (v) by approximating the slightly nonlinear behavior with a linear expression of the form $\mu = \mu_o - q_B v$, as defined in Figure 4.22. During deceleration, the friction force rises until relative motion stops. Then, when the static friction force F_S is exceeded, stored potential energy in the system (analogous to a spring) is released and a rapid acceleration occurs, resulting in a lower sliding faction coefficient. As the rapidly slipping surface catches up to the opposite surface and slows down to match the externally imposed sliding velocity, the friction rises again, and the process repeats. In the simple elastic case, the friction force rises linearly with time.

As Kragelski[37] observed, the value of the static friction force F_S can be a function of time in certain material systems. Therefore, it should not be surprising to see other types of stick-slip friction curves in which the friction force assumes other than a linear rise before the slip (see Figure 4.21, center panel).

As Holm suggested, stick-slip curves need not be perfect sawteeth or exponential rises to the static value but can be quite irregular. The irregularities in behavior enter

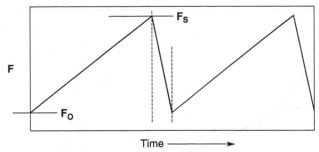

Stick segment: linear elastic, no relative displacement
Slip segment: complete recovery with $v \gg v_{ave}$

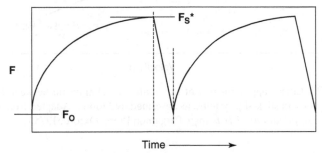

Stick segment: μ_S time dependent, possible creep
Slip segment: complete recovery with $v \gg v_{ave}$

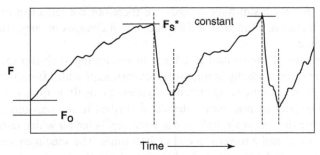

Stick segment: Interrupted with some relative displacement
Slip segment: incomplete recovery caused by interruptions

FIGURE 4.21 Various manifestations of stick-slip behavior: classical sawtooth behavior of the tangential force (top), and fine-scale sticks and slips superimposed or a lower frequency process (bottom).

when, for example, (a) heterogeneities in the contact surface produce momentary contact yielding before the complete rise to F_S or (b) the slip portion of the process is disrupted by premature engagement of the rapidly slipping surface with asperities on the opposing surface (see Figure 4.21, bottom panel).

When considering the time dependence of tangential force during stick-slip, it should be recognized that the instantaneous values of the force cannot in general be used to determine the friction coefficient as a function of time. During a linear rise,

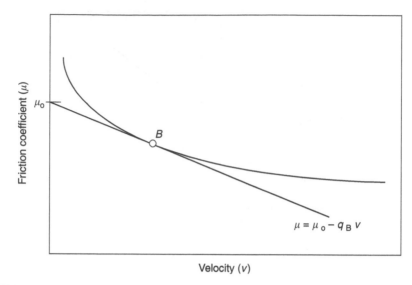

FIGURE 4.22   Linear approximation of the nonlinear relationship between friction coefficient and velocity in stick-slip systems, as described by Moore. (Adapted from Moore, D.F., *Principles and Applications of Tribology*, Pergamon Press, Oxford, 1975.)

for example, the tangential force does not fit the definition of the friction force until the maximum value just before slip (corresponding to μ_s). During slip there may be temporary loss of surface contact, and one should not use values of **F** during the slip to determine μ_k. Stick-slip force records are therefore not reflective of changes in μ (static or kinetic) as a function of time but only of changes in tangential force as a function of time.

The elasticity of the contact is critical to permitting stick-slip to occur. Blau and Yust[55] observed stick-slip in microscale experiments with 1.0-mm-diameter silicon nitride spheres sliding on sputtered molybdenum disulfide films (~1.5 μm thick) on silicon. The friction microprobe device, described in the previous chapter, was used to measure the fine-scale, ball-on-flat frictional behavior with a normal force of 98 mN (10 g load) and a traverse speed of 10.0 μm/s. The specimen stage was suspended on elastic webs to provide high sensitivity to the tangential force, but this increased elasticity promoted stick-slip. Figure 4.23 shows this behavior. Examination of the sliding surface with interference contrast microscopy indicated that the behavior was not caused by film failure, as might have been expected, but rather by the periodic sinking of the slider sphere into the soft film's surface followed by its climbing out of the impression and settling into the subsequent one. Other material-specific causes for stick-slip have been proposed, such as the viscoelastic behavior of organic polymers described by Briscoe.[56]

Certain types of sliding friction variations can look like stick-slip, but they do not conform to the standard definition. For example, a localized irregularity such as a small patch of transferred material on the surface of a rotating journal can cause periodic friction rises and falls, yet there is no significant change in the relative sliding velocity with time, and the maximum value of the friction trace may not be equal

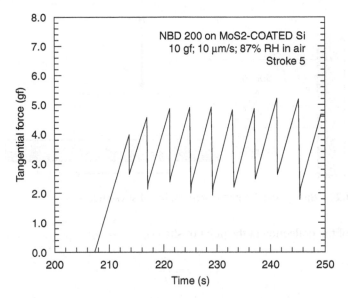

FIGURE 4.23 Appearance of stick-slip behavior recorded during a microfriction experiment of a 1.0 mm diameter silicon nitride ball sliding at 10 μm/s over an MoS$_2$ sputter-coated surface of Si. The spacing of events corresponded to the spacing of shallow impressions in the film, suggesting a sink-in/pull-out type of behavior. The film remained intact.

to the static friction for that material couple. Acoustic or vibrational frequency spectra of mechanical sliding components can be used to discover certain characteristic frequencies; however, the fact that characteristic frequencies are identified does not guarantee the occurrence of stick-slip in the traditional sense.

Stick-slip behavior can be modeled in several ways. Generally, the system is represented schematically as a spring-loaded contact, sometimes including a dashpot element to account for viscoelastic response (Figure 4.24). A more complete discussion of the effects of time-dependent material properties on stick-slip behavior of metals can be found in the paper by Kosterin and Kragelski.[57] Bowden and Tabor's analysis[58] considers a free surface of inertial mass m being driven with a uniform speed v in the positive x direction against an elastic constant k. Then the instantaneous resisting force **F** over distance x equals $-kx$. With no damping of the resultant oscillation,

$$ma = -kx \qquad (4.9)$$

where acceleration $a = (d^2x/dt^2)$. The frequency n of simple harmonic motion is given by

$$n = \left(\frac{1}{2}\pi\right)\left(\frac{k}{m}\right)^{1/2} \qquad (4.10)$$

Under the influence of a load **P** (mass **W** acting downward with the help of gravity **g**), the static friction force **F$_S$** can be represented as

$$\mathbf{F}_S = \mu_s \mathbf{P} \qquad (4.11)$$

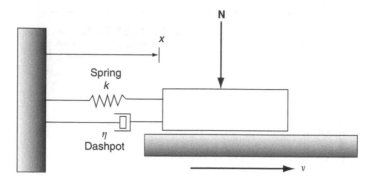

FIGURE 4.24   Spring and dashpot representation of stick-slip.

In terms of the deflection at the point of slip (x),

$$x = \frac{\mathbf{F_S}}{k} \qquad (4.12)$$

If the kinetic friction coefficient μ is assumed to be constant during slip, a departure from Moore's treatment of friction coefficient as a linear function of velocity (see Figure 4.22), then

$$ma - \mu\mathbf{P} = -kx \qquad (4.13)$$

Letting time $= 0$ at the point of slip (where $x = \mathbf{F_S}/k$), and the forward velocity $v \ll$ the velocity of slip, then,

$$x = \frac{\mathbf{P}}{k}[(\mu_s - \mu)\cos \omega t + \mu] \qquad (4.14)$$

in which $\omega = (k/m)^{1/2}$. In this case, the magnitude of slip δ is

$$\delta = \frac{\mathbf{P}(2\mu_s - 2\mu)}{k} \qquad (4.15)$$

From this equation, the larger the μ relative to μ_s, the less the effects of stick-slip, and when they are equal, the sliding becomes completely steady. Note that this first-order treatment neglects damping and other possible influences on behavior.

Kudinov and Tolstoy[59] derived a critical velocity above which stick-slip could be suppressed. This critical velocity v_c was directly proportional to the difference in the static and kinetic friction coefficients $\Delta\mu$ and inversely proportional to the square root of the product of the relative dissipation of energy during oscillation ($\psi = 4\pi\tau$), the stiffness of the system k, and the slider mass m. (Note: These authors' choice of the symbol ψ should not be confused with the plasticity index.) Thus,

$$v_c = \frac{\Delta\mu N}{\sqrt{\psi km}} \qquad (4.16)$$

where N is a factor of safety. The authors report several characteristic values of $\Delta\mu$ for slideways on machine tools: cast iron on cast iron, 0.08; steel on cast iron, 0.05; bronze on cast iron, 0.02; and Teflon™ on cast iron, 0.04.

System resonance within limited stick-slip oscillation ranges was discussed by Bartenev and Lavrentev,[60] who cited experiments in which an oscillating normal load was applied to a system in which stick-slip was occurring. The minimum in stick-slip amplitude and friction force occurred over a range of about 1.5–2.5 kHz, the approximate value predicted by $(1/2\pi)(k/m)^{-1/2}$. Using an oscillator and a loud-speaker to match the pitch, Blau and Whitenton[61] observed a frequency of about 2 kHz for unlubricated block-on-ring sliding of ferrous alloys.

Rabinowicz's article on stick and slip[62] is one of the most-cited reviews of the subject and serves as an excellent introduction to the subject. To treat stick-slip problems, Rabinowicz[63] suggested two possible strategies in his book, published nine years later:

1. Decrease the slip amplitude or slip velocity by increasing contact stiffness, increasing system damping, or increasing inertia (see Figure 4.24).
2. Lubricate or otherwise form a surface film to ensure a positive μ versus velocity relationship.

As Rabinowicz pointed out, the latter solution requires that effective lubrication be maintained, and stick-slip can return if the lubricant is depleted. The fact that stick-slip is associated with a significant difference between static and kinetic friction coefficients suggests that strategies that lower the former or raise the latter can be equally effective. Stick-slip is clearly one of the most fascinating and mechanistically diversified aspects of frictional behavior, and each problem requires careful analysis to determine the cause-and-effect relationships involved. Once each stick-slip problem is correctly defined in terms of its particular operative mechanisms, solution strategies become more obvious.

4.3 SLIDING FRICTION

Like static friction, sliding friction has played a role in the technological development of evolving societies—from the manual twirling of a stick on another stick to start a fire in an ancient cave to the design of high-speed bearings for the space shuttle's main turbopump. Hailing[64] estimated the changes in the drag force-to-weight ratio that occurred as technology has evolved over the past 10,000 years (Figure 4.25). Eons ago, single individuals or small groups of primitive beings would labor to drag heavy objects about on the ground. Later, even before the first wheels, water and animal fats were used as lubricants on rollers made from tree logs. Then spoked wheels and smooth, rigid wheels were invented and less and less tangential force was needed to initiate and sustain relative motion. With modern, well-lubricated bearings, friction coefficients are lower than 0.001. This section describes concepts and models for unlubricated sliding friction. The subject of lubricated friction is discussed in Chapter 6.

The diversity of the occurrences of kinetic friction makes the study of its fundamental nature a considerable challenge, and relatively subtle differences in the

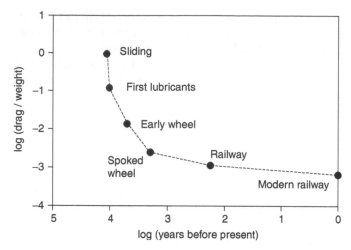

FIGURE 4.25 Ratio of drag force to weight for various tribosystems over several thousand years. (After Kosterin, J.I. and Kragelski, I.V., *Wear*, 5, 190, 1962.)

conditions of rubbing can alter behavior. In fact, it may be inaccurate to use the phrase "fundamental nature of friction," but rather one should say "the fundamental natures of friction" because the causes of observed frictional behavior vary from one tribosystem to another. At this point in the discussion, it will be helpful to differentiate between "friction processes" and "friction mechanisms":

Friction process—a set of interfacial mechanisms that, when acting together in an interface, produces a characteristic resistance to sliding. Examples of friction processes include wear debris layer formation, surface roughness changes due to fatigue and fracture, the formation of a friction polymer, and localized adhesive transfer of material from one surface to another. More than one process can operate simultaneously to produce variations in macroscopic (measured) friction force.

Friction mechanism—a physical phenomenon that can be modeled or qualitatively depicted on a first-principles level, and which acts in an additive or synergistic combination with other phenomena to result in the operation of a friction process. Examples of friction mechanism include interdiffusion of atomic species, nucleation of subsurface microfractures, and the generation and movement of crystallographic point defects (interstitial atoms), line defects (dislocations), or volume defects (void clusters).

Use of the phrases "fundamental mechanism(s) of friction" and "friction mechanism" is commonplace in the literature of tribology. However, in light of the foregoing distinction, these uses can be misleading, and use of the term "friction process(es)" is advocated.

Kinetic friction models, other than empirical ones, can be grouped into five categories:

1. Plowing and cutting-based models
2. Adhesion, junction growth, and shear models

3. Single- and multiple-layer shear models
4. Debris layer and transfer layer models
5. MD models

Each type of model was developed to explain frictional phenomena of some kind. Some of the models are based on observations that contact surfaces contain grooves that are suggestive of a dominant contribution from plowing. Others rely on observations that material from one surface transfers (sticks) to the other, suggesting adherence or adhesion. Certain models are built on the interpenetration of rigid asperities. Single-layer models rely on a view of the interface showing flat surfaces separated by a layer whose shear strength controls friction. Some models involve combinations, such as adhesion plus plowing. Recent friction models emerging from the physics community involve molecular-level phenomena.

As depicted in Figure 4.1, lubrication-oriented models and the debris-based models describe phenomena that take place in zone I, whereas most of the classical models for solid friction are based on zone II phenomena. Few models take into account the effects of both the interfacial properties and the surrounding mechanical system, represented in Figure 4.1 as zone III. Rather, models that address the mechanical properties of the surrounding system tend to ignore the properties and characteristics of the materials in the interface, preferring instead to depict the system as a series of springs, dashpots, and masses (i.e., a mysterious "black box" with inputs and responses). The remainder of this section describes models from the listed five categories.

4.3.1 MODELS FOR SLIDING FRICTION

This section summarizes the most common historical approaches that have been used in developing sliding friction models. Each of these falls into one or more of the five categories.

4.3.1.1 Plowing Models

Plowing models assume that the dominant contribution to friction is the energy required to displace material ahead of a rigid protuberance or protuberances moving along a surface. One of the simplest models for plowing is that of a rigid cone of slant angle θ plowing through a surface under a normal load $\mathbf{P}$ (see Figure 4.26a).[65] If we assign a groove width w (i.e., twice the radius r of the circular section of the penetrating cone at surface level), the triangular projected area A_p that is swept out as the cone moves along is

$$A_p = \frac{1}{2}w(r\tan\theta) = \frac{1}{2}(2r)(r\tan\theta) = r^2\tan\theta \tag{4.17a}$$

The friction force $\mathbf{F}_p$ for this plowing contribution to sliding is found by multiplying the swept-out area by the compressive strength p. Thus, $\mathbf{F}_p = (r^2\tan\theta)p$, and the friction coefficient, if this were the only contribution, is $\mu_p = \mathbf{F}_p/\mathbf{P}$. From the definition of the compressive strength p as force per unit area, we can write:

$$p = \frac{\mathbf{P}}{\pi r^2}$$

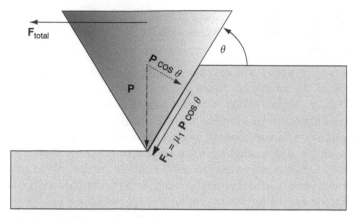

$$F_{total} = F_{displacement} + F_{interface}$$
$$\mu_{total} = \mu_{displacement} + \mu_i \cos^2 \theta$$

(a)

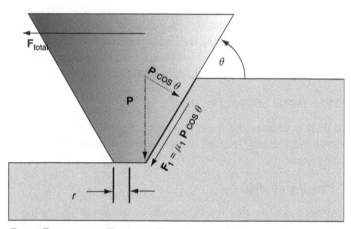

$$F_{total} = F_{displacement} + F_{interface} + F_{circle}$$
$$\mu_{total} = \mu_{displacement} + \mu_i \cos^2 \theta + \frac{\tau}{p}(\pi r^2)$$

(b)

FIGURE 4.26 Simple model for a cone cutting through a material: (a) sharp-pointed cone and (b) truncated cone.

and

$$\mu_p = \frac{F_p}{P} = \frac{(r^2 \tan \theta)\, p}{\pi r^2 p} = \frac{\tan \theta}{\pi} \tag{4.18}$$

Equation 4.18 can also be written in terms of the apex angle of the cone α ($=90 - \theta$)

$$\mu_p = \frac{2 \cot \alpha}{\pi} \tag{4.19}$$

Note that μ_p is for the plowing of a hard asperity and is not necessarily the same as the friction coefficient of the material sliding along the sides of the conical surface. This suggests that there is a difference between the magnitude of the macroscopic friction coefficient and that which occurs on the slopes of individual asperities. Surprisingly, some researchers who purportedly work on scale effects in friction do not appreciate this distinction.

4.3.1.2 Adhesion, Junction Growth, and Shear Models

Adhesion, junction growth, and shear (AJS) models' interpretations of friction are based on a scenario in which two rough surfaces are brought close together, causing the highest peaks (asperities) to touch. As the normal force increases, the contact area increases and the peaks are flattened. Asperity junctions grow until they are able to support the applied load. Adhesive bonds form at the contact points. When a tangential force is applied, these bonds must be broken, and overcoming the shear strength of the bonds results in the friction force. Early calculations comparing bond strengths to friction forces obtained in experiments raised questions as to the general validity of such models. Observations of material transfer and similar phenomena suggested that the adhesive bonds might be stronger than the softer of the two mated materials, and that the shear strength of the softer material, not the bond strength, should be used in friction models.

Traditional friction models, largely developed for metal-on-metal sliding, have added the force contribution due to the shear of junctions to the contribution from plowing, giving the extended expression:

$$\mu = \frac{\tau A_r}{P} + \frac{\tan \theta}{\pi} \qquad (4.20)$$

where A_r is the real area of contact and τ is the shear strength of the material being plowed. This expression has met with relatively widespread acceptance and has been used as the basis for other sliding friction models. But if the tip of the cone wears down, as depicted in Figure 4.26b, not two but three contributions to the plowing process can be identified: the force needed to displace material from in front of the cone, the friction force along the leading face of the cone (i.e., the component in the macroscopic sliding direction), and the friction associated with the shear of the interface along the worn frustum of the cone. It follows that friction on two scales is involved: the macroscopic friction force for the entire system and the friction forces associated with the flow of material along the face of the cone and across its frustum. This situation is analogous to the interpretation of orthogonal cutting of metals in which the friction force of the chip moving up along the rake face of the tool and friction along the wear land are not the same as the cutting force for the tool as a whole.[66] Considering the three contributions to the friction of a flat-tipped cone gives

$$\mu = \left(\frac{\tau}{P} \right)(\pi r^2) + \mu_i \cos^2 \theta + \frac{\tan \theta}{\pi} \qquad (4.21)$$

where r is now redefined as the radius of the top of the worn cone and μ_i is the friction coefficient of the cone against the material flowing across its face.

Equation 4.21 helps explain why the friction coefficients for ceramics and metals sliding on faceted diamond films are up to 10 or more times higher than the friction coefficients reported for smooth surfaces of the same materials sliding against smooth surfaces of diamond (i.e., $\mu \gg \mu_i$).[67] When the rake angle θ is small, $\cos^2 \theta$ is close to 1.0, and the second term is only slightly less than μ_i (0.02–0.12 typically). If one assumes that the friction coefficient for the material sliding across the frustum of the cone is the same as that for sliding along its face (μ_i), then Equation 4.21 can be rewritten as

$$\mu = 2\mu_i + \frac{\tan \theta}{\pi} \qquad (4.22)$$

thus implying that the friction coefficient for a rigid sliding cone is more than twice that for sliding a flat surface of the two same materials. It is interesting to note that Equation 4.22 does not account for the depth of penetration, a factor that seems critical for accounting for the energy required to plow through the surface (displace the volume of material ahead of the slider), and at $\theta = 90°$, which implies infinitely deep penetration of the cone, it would be impossible to move the slider at all! Evidently this approach has limits.

In consideration of the complexities of surface finish and all the many ways that metrologists have devised to measure it, it seems remarkable that Equations 4.16 and 4.17, which depend on a single quantity [(tan θ)/π], should be able to predict the friction coefficient with any degree of accuracy. For one thing, the model is based exclusively on a single conical asperity cutting through a surface. It does not account for multiple contacts or for the possible differences in contact angle for those contacts. The model is also based on a surface's relatively ductile response to a perfectly rigid asperity and can neither account for fracture during wear nor for the change in the groove geometry that one might expect to happen for multiple passes over the same surface.

In 1962, Mulhearn and Samuels[11] published a paper on the transition between abrasive asperities cutting through a surface and plowing through it. Their results suggest that there exists a critical rake angle for that type of transition. (Note: The rake angle is the angle between the normal to the surface and the leading face of the asperity, with negative values indicating a tilt toward the direction of travel.) If plowing occurs only up to the critical rake angle, then we may compute the maximum contribution to friction due to plowing from the data of Mulhearn and Samuels and Equation 4.19 (see Table 4.9). This approach suggests that the maximum contribution of plowing to the friction coefficient of aluminum or nickel is about 0.03 in contrast to copper, whose maximum plowing contribution is 0.32. Since the sliding friction coefficient for aluminum can be quite high (over 1.0 in some cases), the implication is that factors other than plowing, such as the shearing of strongly adhering junctions, would be by far the major contributor. Examination of unlubricated sliding wear surfaces of both Al and Cu often reveals a host of ductile-appearing features not in any way resembling cones, and despite the similar appearances in the microscope of worn Cu and Al, one finds from the first and last rows in Table 4.9 that the contribution of plowing to friction should be different by a factor of 10. Again, something is wrong with the simple cone model.

TABLE 4.9

Estimates of the Maximum Plowing Contribution to Friction

Metal	Critical Rake Angle[a] (Degrees)	μ_p
Aluminum	−5	0.03
Nickel	−5	0.03
Lead	−35	0.22
α-Brass	−35	0.22
Copper	−45	0.32

[a] For a cone, the absolute value of the critical rake angle is 90 minus angle θ (defined in Figure 4.26).

Source: Adapted from Mulhearn, T.O. and Samuels, L.E., *Wear*, 5, 478, 1962.

Hokkirigawa and Kato[68] carried the analysis of abrasive contributions to sliding friction even further using observations of single hemispherical sliding contacts (quenched steel, tip radius 26 or 62 μm) on brass, carbon steel, and stainless steel in a scanning electron microscope. They identified three modes: (a) plowing, (b) wedge formation (prow formation), and (c) cutting (chip formation). The tendency of the slider to produce the various modes was determined to be related to the degree of penetration, D_p. Here, $D_p = h/a$, where h is the groove depth and a is the radius of the sliding contact. Aided by the work of Challen and Oxley,[69] the sliding friction coefficient was modeled in three ways depending upon the regime of sliding. Three parameters were introduced, namely: (1) $f = p/\tau$; (2) $\theta = \sin^{-1}(a/R)$; and (3) β, the angle of the stress discontinuity (shear zone) from Challen and Oxley's analysis. In terms of these parameters, where p is the contact pressure, τ is the bulk shear stress of the flat specimen, and R is the slider tip radius, the friction coefficient was given as follows for each mode:

Cutting mode:

$$\mu = \tan\left(\theta - \frac{\pi}{4} + \frac{1}{2}\cos^{-1}f\right) \tag{4.23}$$

Wedge-forming mode:

$$\mu = \frac{\{1 - \sin 2\beta + (1 - f^2)^{1/2}\}\sin\theta + f\cos\theta}{\{1 - \sin 2\beta + (1 - f^2)^{1/2}\}\cos\theta + f\sin\theta} \tag{4.24}$$

Plowing mode:

$$\mu = \frac{A\sin\theta + \cos(\cos^{-1}f - \theta)}{A\cos\theta + \sin(\cos^{-1}f - \theta)} \tag{4.25}$$

where

$$A = 1 + \frac{\pi}{2} + \cos^{-1} f - 2\theta - \sin^{-1} \frac{\sin \theta}{(1-f)^{-1/2}}$$ (4.26)

For unlubricated conditions, the transitions between the various modes were experimentally determined using the scanning electron microscope. Table 4.10 summarizes these results.

Results of the work by Kokkirigawa and Kato showed that the analytical form of the frictional dependence on the sharpness of asperities cannot ignore the mode of surface deformation, but the crystal structure of the material being plowed can also affect tangential force. For example, Bowden and Brookes[70] conducted experiments using diamond cones with tip angles of 60°, 120°, and 170° sliding on a single-crystal (001) diamond surface in specific orientations relative to the crystal structure. As Table 4.11 shows, the sharper the tip, the more pronounced were the effects of crystallography on friction. Data like these support the argument given in Section 3.3

TABLE 4.10
Critical Degree of Penetration (D_p) for Unlubricated Friction Mode Transitions

	Value of D_p for the Transition	
Material	Plowing to Wedge Formation	Wedge Formation to Cutting
Brass	0.17 (tip radius 62 μm)	0.23 (tip radii 62, 27 μm)
Carbon steel	0.12 (tip radius 62 μm)	0.23 (tip radius 27 μm)
Stainless steel	0.13 (tip radii 62, 27 μm)	0.26 (tip radius 27 μm)

Source: Adapted from Hokkirigawa, K. and Kato, K., *Tribol. Int.*, 21, 51, 1988.

TABLE 4.11
Effects of Crystallographic Anisotropy of Kinetic Friction Coefficient (μ)

Sliding Direction	μ, 60° cone	μ, 120° cone	μ, 170° cone
[001]	0.17	0.11	0.07
[110]	0.06	0.06	0.055
[100]	0.15	0.11	0.07
[110]	0.05	0.05	0.05
[010]	0.17	0.11	0.07
Max μ – Min μ	0.12	0.06	0.02

Note: Data taken from Figure 2 of Bowden, F.P. and Brookes, C.A., *Proc. R. Soc. Lond.*, A295, 244, 1966.

regarding the need for caution when representing scratch test results as if they were friction coefficients.

In summary, the foregoing treatments of the plowing contribution to friction assumed that asperities were regular geometric shapes; however, rarely do such shapes appear on actual sliding surfaces (notable exceptions are the pyramidal surface features that can be produced on chemical vapor-deposited diamond thin films or ion milled on silicon surfaces). The asperities on sliding surfaces tend to be irregularly shaped, as viewed in optical or scanning electron microscopes. Experimental data that further illustrate the effects of contact area, surface roughness, and load on friction are given in Chapter 7.

4.3.1.3 Plowing with Debris Generation

Even when the predominant contribution to friction is initially from cutting and plowing of hard asperities through the surface, the gradual generation of wear debris that submerges the asperities can reduce the severity of plowing. Consider, for example, the friction coefficients obtained by stroke-by-stroke sliding of bearing steel, an aluminum alloy, and polymethylmethacrylate sliding against silicon carbide abrasive papers.[71] Table 4.12 shows that starting with multiple hard "asperities" of the same geometric characteristics produced different initial and steady-state friction coefficients for the three slider materials. Wear debris accumulation in the contact region affected the frictional behavior. In the case of abrasive papers and grinding wheels, this is called *loading*. Avoidance of loading is important in maintaining the efficiency of a grinding process, and a great deal of effort has been focused on dressing grinding wheels to improve their material removal efficiency. An increase in the tangential grinding force or an increase in the power drawn by the grinding spindle suggests that loading or abrasive grit wear is occurring.

As wear progresses, the wear debris accumulates between the asperities and alters the effectiveness of the cutting and plowing action by covering the active points. If the cone model is to be useful at all for other than pristine surfaces, the effective value of θ must be given as a function of time or number of sliding passes. Not only is the wear rate affected, but the presence of debris affects the interfacial shear strength, as is explained in another section in this chapter in regard to third-body particle effects on friction. The notion that wear debris can accumulate and raise friction has led investigators to try patterning surfaces to create pockets where

TABLE 4.12
Effects of Material Type on Friction During Abrasive Sliding

Slider Material	24 μm Grit Size		16 μm Grit Size	
	Starting μ	Ending μ	Starting μ	Ending μ
AISI 52100 steel	0.47	0.35	0.45	0.29
2014-T4 aluminum	0.69	0.56	0.64	0.62
PMMA	0.73	0.64	0.72	0.60

Note: Normal force, 2.49 N; sliding speed, 5 mm/s; multiple strokes, 20 mm long.

debris can be collected (e.g., see Ref. 72). The orientation and depths of the ridges and grooves in a surface affect the effectiveness of the debris-trapping mechanism.

4.3.1.4 Plowing with Adhesion

Traditional models for sliding friction have historically been developed with metallic materials in mind. Classically, the friction force $\mathbf{F}$ is said to be an additive contribution of adhesive ($\mathbf{S}$) and plowing forces ($\mathbf{F}_{pl}$):[73]

$$\mathbf{F} = \mathbf{S} + \mathbf{F}_{pl} \qquad (4.27)$$

The adhesive force derives from the shear strength of adhesive metallic junctions that are created when surfaces touch one another under a normal force. Thus, by dividing by the normal force we find that $\mu = \mu_{adhesion} + \mu_{plowing}$. If the shear strength of the junction is τ and the contact area is A, then

$$\mathbf{S} = \tau A \qquad (4.28)$$

The plowing force $\mathbf{F}_{pl}$ is given by

$$\mathbf{F}_{pl} = pA' \qquad (4.29)$$

where p is the mean pressure required to displace the metal from the sliding path and A' is the cross section of the track presented to the moving hard asperities. While helpful in understanding the results of experiments in the sliding friction of metals, this approach involves several applicability-limiting assumptions, for example, that adhesion between the surfaces results in bonds that are continually forming and breaking, that the protuberances of the harder of the two contacting surfaces remain perfectly rigid as they plow through the softer counterface, and perhaps most limiting of all, that the friction coefficient for a tribosystem is determined only from the shear strength of the materials.

As pointed out in a review by Briscoe and Tolarski,[74] polymers, particularly rubber and other elastomers, represent a special case of applying the two-term model for friction. Instead of including a plowing term, a frictional hysteresis term is introduced. Thus,

$$\mu = \mu_{adhesion} + \mu_{hysteresis} \qquad (4.30)$$

where $\mu_{hysteresis}$ is calculated from the asperity density factor γ, the average wavelength of the surface topography λ, the energy dissipated per asperity contact E_d, and the contact pressure p as follows:

$$\mu_{hysteresis} = \left(\frac{4\gamma}{\lambda^3}\right)\left(\frac{E_d}{p}\right) \qquad (4.31)$$

The unique characteristics of polymer friction are discussed further in Chapter 5.

4.3.1.5 Single-Layer Shear Models

Single-layer shear (SLS) models for friction depict an interface as a layer whose shear strength determines the friction force and, hence, the friction coefficient. The layer

TABLE 4.13

Measured Values for the Shear Stress Dependence on Pressure

Material	τ_o (kgf/mm^2)	α
Aluminum	3.00	0.043
Beryllium	0.45	0.250
Chromium	5.00	0.240
Copper	1.00	0.110
Lead	0.90	0.014
Platinum	9.50	0.100
Silver	6.50	0.090
Tin	1.25	0.012
Vanadium	1.80	0.250
Zinc	8.00	0.020

Source: Adapted from Kragelskii, I.V., Dobychin, M.N., and Kombalov, V.S. in *Friction and Wear Calculation Methods*, Pergamon Press, Oxford, 1982, 178–180.

can be a separate film, such as a solid lubricant, or simply the near-surface zone of the softer material that is shearing during friction. The friction force **F** is the product of the contact area A and the shear strength of the layer:

$$\mathbf{F} = \tau A \qquad (4.32)$$

The concept that the friction force is linearly related to the shear strength of the interfacial material has a number of useful implications, especially as regards the role of thin lubricating layers, including oxides and tarnish films. It is known from the work of Bridgman[75] on the effects of pressure on mechanical properties that τ is affected by contact pressure, p:

$$\tau = \tau_o + \alpha p \qquad (4.33)$$

Table 4.13 lists several values for the shear stress and the constant α.[76] Further data for pressure effects on the shear strength of solid lubricating layers are given in Chapter 6.

There is an interesting similarity of the form of Equation 4.33 with what is known as *Byerlee's law* for the friction of rock at high pressures. Byerlee[77] conducted extensive studies of the frictional shear stress of a wide range of types of rock as a function of the applied normal stress, σ_n. The data could be fitted with two straight lines. Below a normal pressure of 200 MPa, $\tau = 0.85\sigma_n$. Above that pressure, $\tau = 50 + 0.6\sigma_n$. The Byerlee's law is observed to hold over a large range of hardness and ductility and is sometimes used to estimate the strength of natural faults.

4.3.1.6   Multiple-Layer Shear Models

SLS models presume that the sliding friction coefficient is determined by the shear strength of a single weak layer interposed between solid surfaces. Evidence revealed by the examination of frictional surfaces suggests that shear can occur at various positions in and adjacent to the interface, for example, at the upper interface between the solid and the debris layer, within the entrapped debris or transfer layer itself, at the lower interface, or below the original surfaces where extended delaminations may occur. One might envision sliding friction that involves several layers acting in parallel. Intuitively one would expect the predominant frictional contribution to come from the lowest shear strength in the "stack." Yet the shear forces transmitted across that weakest interface may still be sufficient to induce nonnegligible displacement in other layers above and below it, particularly if the difference in shear strengths of the various layers is small. Consider, for example, a deck of cards that can be sheared between two surfaces, leaving a parallelogram-shaped stack. Inclined-plane static experiments by the author on fresh playing cards (using the apparatus shown in Figure 3.6) revealed that slip occurs first on the weakest interface(s) near the top of the deck where the normal force is lowest. The average μ_s was 0.40 for a total of 90 experiments using different numbers of cards from 1 to 53 interfaces. However, when the clamping force was increased and the tangential velocity was higher, multiple planes slipped within the deck. Therefore, modelers who assume that slip occurs only on one interface should be cognizant of the "playing card phenomenon."

Interestingly, a discussion of deformation on multiple planes in or adjacent to a frictional interface can be found in the wear research publications of Godet's group at INSA in France under the topical headings: "third-body effects" and "velocity accommodation" (e.g., Refs 78 and 79).

4.3.1.7   Molecular Dynamics Models

Spurred on by advances in high-speed computing, there have been relatively recent developments in MD simulations of frictional interactions. When coupled with information from nanoprobe instruments, such as the atomic force microscope, the scanning tunneling microscope, the surface forces apparatus, and the lateral force microscope, such studies have made possible intriguing insights into the behavior of pristine surfaces on the atomic scale. Reviews of this subject are not presented here. Instead, the reader is referred to the special nanotribology issue of the *Materials Research Society Bulletin*[80] and the book *Fundamentals of Friction: Macroscopic and Microscopic Processes.*[81]

MD models of friction for assemblages of even a few hundred atoms tend to require millions upon millions of individual, iterative computations to predict frictional interactions taking place over only a fraction of a second in real time. As computers improve these models will undoubtedly improve as well. Because they begin with very specific arrangements of atoms, usually in single-crystal form with a specific sliding orientation, results are often periodic with sliding distance. Some of the calculations' results are remarkably similar to certain types of behavior observed in real materials, simulating such phenomena as dislocations (localized slip on preferred planes) and the adhesive transfer of material to the opposing counterface. For example, Rigney's group at The Ohio State University has developed models that predict how surfaces roughen and mechanically mix during sliding.[82] Figure 4.27 is a single frame from a sliding wear model depicting two clean, solid surfaces. Arrows at the top and bottom show the directions of shear. Periodic roughness

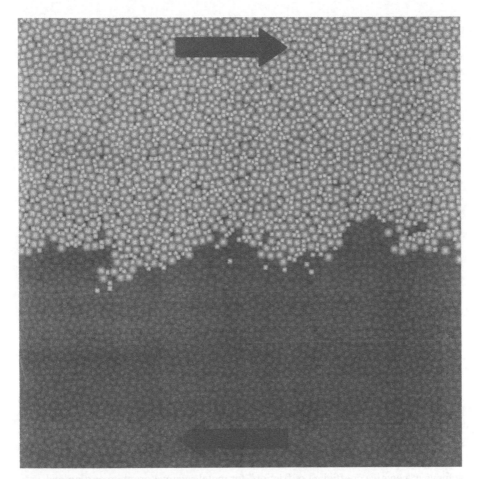

FIGURE 4.27 Model for interfacial mixing during sliding friction. (Courtesy of D.A. Rigney and X.Y. Fu, The Ohio State University.)

features develop, followed by the intermixing of atomic species. From visualizations like these, one might expect frictional behavior to change with time, and studies of running-in phenomena by a variety of investigators support that conclusion.

MD models are challenged by phenomena such as fracture, surface fatigue induced delaminations, wear debris particles agglomerating and deforming in the interface, high-strain-rate phenomena, work-hardening of near-surface layers, deformation twinning, pressure-induced phase transformations, recrystallization, amorphization, and other microstructural responses of bearing materials. The convergence of metallurgy and materials science with MD modeling is yet to be achieved.

4.3.1.8 Stimulus–Response Dynamical Friction Models

There is a certain group of friction modelers who take the approach that friction can be represented by one or more analytical expressions that relate inputs, such as velocity and fluctuations in normal forces, to outputs, such as the time dependence of the friction force. The tribological interface is assigned certain properties such as stiffness, damping capacity, viscosity, and the like, and then the model is applied to predict the frictional response. The sophistication and rigor of such approaches is exemplified in the book by Persson.[83]

In stimulus–response dynamical models where the surface morphology is included, the configuration is usually idealized, and there is little or no accounting for the heterogeneities observed on actual bearings, gears, brakes, and seals. For example, these models rarely, if ever, account for the fact that the two mating surfaces may have different initial roughness, waviness, and lay. The models also tend to ignore wear or plastic deformation, focusing instead on elastic or viscoelastic conditions. The complicating effects of material transfer, third bodies, hard particle embedment, and running-in are also usually ignored.

Stimulus–response models are sometimes tested using carefully designed apparatus and clean, smooth, pure test materials. Typically, quite low loads and sliding velocities are applied (e.g., Ref. 84). Although such models may adequately portray the responses of materials under pristine conditions, their usefulness to designers in selecting surface finishes, coatings, lubricants, and material compositions for practical engineering applications is limited. One exception may be in the design and analysis of ultrahigh-precision bearings, micromechanical systems, or disc drives where the lubricant layers are extremely thin and the tribosystem characteristics more closely approach those portrayed in such models.

4.3.1.9   Ultralow Friction and "Superlubricity"

It is tantalizing to envision the limiting state of zero friction, and in recent years this concept has been explored both philosophically and experimentally. In fact, it is not impossible to have friction coefficients of less than zero, as shown by Skinner and Gane[85] in elegant adhesion experiments conducted during the early 1970s. The two investigators measured adhesive forces, opposite in sense to the normal force, to in effect achieve a "negative coefficient of friction" for metals against graphite or diamond.

More recently, there has been theoretical and experimental work to explore nearly frictionless conditions and "superlubricity." In 2000, Erdemir et al.[86] reported sliding friction coefficients for diamond films in the range of 0.001–0.005 against steel and ceramic surfaces in dry nitrogen. It was subsequently found, in pin-on-disk tests, that the surrounding environment was key to achieving such low values and that the tribochemistry (hydrogenation) of the surfaces was key in achieving such behavior.[87] Table 4.14 lists low friction coefficients reported by Andersson et al. for self-mated films of hydrogenated carbon (HC) under three environmental conditions. The baseline vacuum was 5×10^{-5} Pa and 1.3 kPa of either water, oxygen, or hydrogen gases were introduced. The challenge in such work is to find the means to achieve long-term, sustained low friction in practical applications.

The term "superlubricity" was introduced by Hirano and Shinjo[88,89] about a decade before nearly frictionless diamond films were first reported. Models used to explain superlubricity were based on the fit between crystal lattices on opposing, incommensurate [sic] surfaces.[90,91] In 2007, a compilation of work on models and experimental work on superlubricity was published,[92] and the phenomenon remains a subject of active interest, especially in the tribophysics community.

TABLE 4.14
Friction Coefficients for Self-Mated HC Films in Three Environments (Load = 1 N and Velocity 0.03–0.07 m/s)

Environment	μ (Initial)[a]	μ (After 80 Disk Revolutions)[a]
Water	0.10	0.045
Oxygen	0.21	0.030
Hydrogen	0.21	0.030
Vacuum	0.19	0.025

[a] Estimated from traces presented in Figure 4 of Andersson, J., Erck, R.A., and Erdemir, A., *Wear*, 254, 1070, 2003.

4.3.1.10 Selecting Friction Models

The friction models described in this chapter range from fundamental treatments using interatomic potentials with idealized surfaces to empirical models that are force-fit to the frictional behavior of specific tribosystems under specific operating conditions. One might conclude the following about such eclectic approaches:

1. No friction model is capable of accounting for all possible variables that can affect friction in diverse tribosystems.
2. Simple friction models may adequately describe behavior under limited ranges of conditions, but applicability is limited.
3. Predictive, tribosystem-level friction models that explicitly account for changing interface geometry, materials properties, lubrication aspects, thermal, chemical, and external mechanical system responses, all in a time-dependent context, do not now exist nor are they likely to exist in the foreseeable future.
4. Friction models should be selected based on the size scale of frictional processes that operate in the given tribosystem.
5. Historically, materials-based friction models were focused on producing a single value for the friction force or friction coefficient; yet in practical engineering systems—especially under dry or boundary-lubricated conditions—friction fluctuates.
6. Models that predict not a single value but rather the time dependence and range of frictional fluctuation over time will be of great value to hardware designers and those who estimate the energy utilization of machinery.
7. In certain cases, friction models will need to include submodels for wear, especially when modeling friction during break-in or end-of-life transitions.

An alternative to friction modeling is to portray frictional behavior using graphical approaches, like those described in the next section.

4.3.2 Phenomenological, Graphical, and Statistical Approaches

Phenomenological approaches, graphical representations, or statistical methods have been used in place of first-principles models to estimate friction levels.

Phenomenological methods use observations of contact surfaces and machine behavior to develop an understanding of the physical characteristics of materials, surfaces, or surface treatments that are known to influence friction. Some consultants rely more on their career experience of what has worked in the past than on analytical modeling. Expert systems in the future may combine mathematical and phenomenological approaches to friction modeling.

Graphical methods are based on trends in experimental data that cover the range of conditions within the tribosystem of interest. Results are usually plotted on two or three axes in which regions of safe or acceptable frictional conditions are identified. Typical plot axes are load or contact pressure and sliding speed, but other variables such as temperature or surface finish have also been used.

Suh[93] developed so-called "friction spaces" in which adhesion, wear particle penetration, and surface roughness are displayed. Graphical approaches can be useful for engineering work, but they are limited by the need to restrict the number of variables to a manageable number, and a measure of interpolation is required to make up for the fact that no group can practically conduct enough variable combinations to fill the volume of the plot. Other types of graphical methods, including the friction process diagrams (FPDs) developed by the author, are described in Chapter 8 on running-in and transitions. FPDs display relative contributions of frictional processes such as interfacial transfer, debris coverage, and metallic sliding rather than applied parameters such as load, speed. In essence, FPDs combine phenomenological and graphical approaches.

Statistical approaches involve the design of experiments to investigate systematically the interactions of certain key variables on frictional behavior. Careful statistical design of friction experiments can provide useful empirical models for the given friction system. One common example of such approaches is in the formulation of multicomponent friction materials for brake linings.[94] However, like graphical approaches, an errant factor that was not identified as a variable in the original test matrix could influence the results.

Statistical approaches to modeling friction are a substitute for a lack of understanding of exactly what is going on in complex tribosystems. The use of statistics is helpful in engineering analysis but falls short of providing a complete insight into the underlying physics.

4.3.3   FRICTION MODELS THAT INCLUDE WEAR

Friction in many practical tribosystems is accompanied by wear. Wear can change the surface roughness, the nominal pressure and area of contact in nonconformal contacts, and by transfer, mechanical mixing, and third-body formation, the composition and properties of the materials in the interface as well. Any friction model that does not include considerations of wear implicitly presumes that a steady state of some kind exists in the contact—whether it is a pristine, nonworn surface, or a sliding condition in which there is no significant change in the properties or interfacial conditions with time. From that point of view, any friction model that fails to include wear falls short of simulating a great many practical engineering systems—especially those that operate under boundary lubrication or with no added lubrication.

The inclusion of wear considerations into friction models is not new. The two-term expression that includes for adhesion plus plowing, developed by Bowden and Tabor,[58] in effect recognized that deformation and its consequences (e.g., wear) can contribute to friction. Suh picked up on this theme when conceptualizing friction spaces,[93] and Liu et al.[10] included plasticity arguments in their computer models. A future challenge will be to extend the incorporation of wear models to account for changing wear processes during the evolution of friction surfaces, from running-in, through steady state, and onward to ultimate wear-out.

4.4   FRICTIONAL HEATING

Heat generation and rising surface temperatures are intuitively associated with friction. In elementary school, children are taught to rub their hands together to illustrate how friction can produce heat. When a friction force $\mathbf{F}$ moves through a distance x, an amount of work $\mathbf{F}x$ is produced. The laws of thermodynamics require that the energy so produced be dissipated to the surroundings. At equilibrium, the energy into a system $\mathbf{U}_{in}$ equals the sum of the energy output to the surroundings $\mathbf{U}_{out}$ (dissipated externally) and the energy accumulated $\mathbf{U}_{accumulated}$ (consumed or stored internally):

$$\mathbf{U}_{in} = \mathbf{U}_{out} + \mathbf{U}_{accumulated} \tag{4.34}$$

The rate of energy input in friction is the product of $\mathbf{F}$ and the sliding velocity v whose units work out to energy per unit time (e.g., N m/s). This energy input rate at the frictional interface is balanced almost completely by heat conduction and mechanical energy transmission (phonons) moving away from the interface, either into the contacting solids or to the surroundings. Although the partition of energy during friction under various conditions is yet to be clearly determined, as speed increases, less and less frictional energy, perhaps as little as 5%, is consumed or stored in the material as microstructural defects such as dislocations, the energy to produce phase transformations, surface energy of new wear particles and propagating subsurface cracks, etc. The rest is dissipated as heat, and there may be sufficient heat to promote chemical reactions, alter the microstructure, or melt the sliding interface.

Energy that cannot readily be removed from the interface raises the temperature locally. Assuming that the proportionality of friction force $\mathbf{F}$ to normal force $\mathbf{P}$ (i.e., $\mathbf{F} = \mu\mathbf{P}$) holds over a range of normal forces, we would expect the temperature to rise linearly with $\mathbf{P}$ in a constant-velocity sliding system. This result is presented in Figure 4.28 wherein temperature rise data for 20 sliding experiments are presented.[95] The steady-state temperatures were measured using thermocouples embedded in flat blocks of various copper alloys sliding unlubricated against rotating AISI 52100 steel rings in Ar gas at 50 mm/s and under a range of normal forces. The least-squares fitted line does not go through zero because the thermocouple measured only the bulk temperature rise in the block specimen, and further, its tip was about 2 mm from the actual contact interface. Nevertheless, in the absence of more detailed information, one can estimate the relative temperature rises of different sliding materials on the same counterface by scaling them directly to the differences in the measured friction forces.

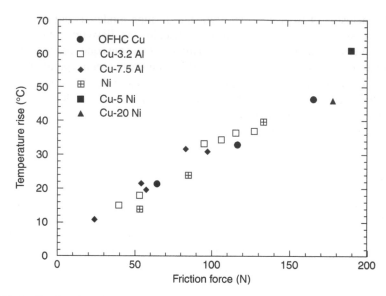

FIGURE 4.28 Steady-state bulk temperature rise was proportional to friction force in block-on-ring tests using a thermocouple embedded in the block specimen (tests in Ar gas at 5.0 cm/s).

It is convenient to distinguish between two temperatures, the *flash temperature* and the *mean surface temperature*. The former is localized, whereas the latter is averaged over the nominal contact zone.

Flash temperature—the maximum friction-induced temperature of the tips of interacting asperities

Mean surface temperature—the average temperature across a frictionally heated surface [Note: As described later, Ashby et al. used the term "bulk temperature" to represent the mean temperature at the surface. That usage should not be confused with the temperature of the bulk material well away from the zone of frictional heating.]

Since sliding surfaces touch at only a few locations at any instant, the energy is concentrated there and the heating is particularly intense—thus, the name flash temperature. The combined effect of many such "flashes" dissipating their energy in the interface under steady state is to heat a near-surface layer to an average temperature that is determined by the energy transport conditions embodied in Equation 4.35 given earlier. Blok[96] discussed the concept and calculation of flash temperature in a review article in the journal *Wear*.

The early work of Blok[97] and Jaeger[98] is still cited as a basis for more recent work, and it has been reviewed in a simplified form by Bowden and Tabor.[99] Basically, the temperature rise in the interface is given as a function of the total heat developed, Q:

$$Q = \frac{\mu \mathbf{W} \mathbf{g} v}{J} \qquad (4.35)$$

TABLE 4.15

Temperature Rise during Sliding—Jaeger Development

Conditions	Temperature Rise[a] $(T = T_o)$
Circular junction of radius a	$= \dfrac{Q}{4a(k_1 + k_2)}$
Square junction of side $= 2l$, at low speed	$= \dfrac{Q}{4.24l(k_1 + k_2)}$
Square junction of side $= 2l$, at high speed wherein the slider is being cooled by the incoming surface of the flat disk specimen	$= \dfrac{Qx^{1/2}}{3.776l[k_1(1/v)^{1/2} + 1.125x^{1/2}k_2]}$, where $x = (k_1/\rho_1 c_1)$, thermal diffusivity, for the disk material

[a] T, steady-state junction temperature; T_o, initial temperature; $k_{1,2}$, thermal conductivity of the slider and flat bodies; ρ, density; c, specific heat.

where μ is the sliding friction coefficient, **W** the load, **g** the acceleration due to gravity, v the sliding velocity, and J the mechanical equivalent of heat (4.186 J/cal). Expressions for various heat flow conditions are based on Equation 4.35. As Table 4.15 shows, the expressions become increasingly complicated when the cooling effects of the incoming, cooler surface are accounted for.

Rabinowicz[100] published a very rough rule of thumb for estimating the flash temperature rise in sliding:

$$\theta_m = \frac{v}{2} \ (\pm \ \text{a factor of } 2 - 3) \qquad (4.36)$$

where v is the sliding velocity (ft/min) and θ_m is the estimated surface flash temperature (°F). For curiosity, the author compared the results of using Equation 4.36 with those from more complicated models for frictional heating and found it gave surprisingly similar results. While such results may have been fortuitous, it highlights the uncertainties in using frictional heating models that contain many variables, the values of which must be estimated or empirically "adjusted" to give sensible results.

In general, nearly all expressions for flash temperature or mean temperature rise contain the friction force–velocity product. Sometimes the friction force is written as the product of the normal force and friction coefficients in such expressions, but it is present nevertheless.

A review of frictional heating calculations was prepared by Cowan and Winer[101] and included representative materials properties data. Their approach involves the use of two *heat partition coefficient* (γ_1 and γ_2) that describe the relative fractions of the total heat that go into each of the contacting bodies, such that $\gamma_1 + \gamma_2 = 1$. The *Fourier modulus F_o*, a dimensionless parameter, is introduced to establish whether or not steady-state conditions have been reached at each surface. For a contact radius a, an exposure time t, and a thermal diffusivity for body i of D_i

$$F_o = \frac{D_i t}{a^2} \tag{4.37}$$

The Fourier modulus is taken to be 100 for a surface at steady state. Another useful parameter grouping is the Peclet number P_e, defined in terms of the density of the solid ρ, the specific heat c_p, the sliding velocity v, the thermal conductivity k, and the characteristic length L_c:

$$P_e = \frac{\rho c_p v L_c}{k} \tag{4.38}$$

The characteristic length is the contact width for a line contact or the contact radius for a circular contact. The Peclet number relates the thermal energy removed by the surrounding medium to that conducted away from the region in which frictional energy is being dissipated. Since $D_i = (\rho c_p / k)$,

$$P_e = \frac{v L_c}{D_i} \tag{4.39}$$

The Peclet number is sometimes used as a criterion for determining when to apply various forms of frictional heating models. Cowan and Winer treat two cases in their review: (a) line contact with two bodies in motion and (b) circular contact with one body in motion. The reader is referred to that reference for their derivation.

It is instructive to compare several models for flash temperature. Certain symbols used by various authors have been changed for the sake of consistency. Further, it is important to note that the forms of these relationships will change for different sliding velocity regimes. The purpose of the comparison is to illustrate differences in the forms of frictional heating models, not to discuss their detailed derivation. For the latter, the reader should consult the original references. Four treatments for a pin moving along a stationary flat specimen are compared: Rabinowicz's derivation based on surface energy considerations, one of the cases from Cowan and Winer's review, Kuhlmann-Wilsdorf's model, and the model of Ashby et al.

Based on considerations of junctions of radius r and surface energy of the softer material Γ, Rabinowicz[102] arrived at the following expression:

$$T_f = \frac{3000 \pi \mu \Gamma v}{J(k_1 + k_2)} \tag{4.40}$$

where J is the mechanical equivalent of heat, v the sliding velocity, μ the friction coefficient, and k_1 and k_2 the thermal conductivities of the two bodies. The constant 3000 is obtained from the calculation of the effective contact radius r in terms of the surface energy Γ of the circular junctions and their hardness H (i.e., $r = 12,000\Gamma H$) and the load carried by each asperity ($\mathbf{P} = \pi r^2 H$). Thus, the numerator equates to $\mathbf{F}v$ expressed in terms of surface energy.

TABLE 4.16

Effects of Deformation Type and Peclet Number on Flash Temperature Calculation

Type of Deformation	Peclet Number (P_e)	Average Flash Temperature[a]
Plastic		
	<0.02	$T_f = \mu P^{0.5} v \dfrac{\sqrt{\pi p}}{8k}$
	>200	$T_f = 0.31\pi P^{0.25} v^{0.5} \left[\dfrac{(\pi p)^{0.75}}{(k\rho c)^{0.5}} \right]$
Elastic		
	<0.02	$T_f = 0.13 \mu P^{0.667} v \left(\dfrac{1}{k} \right) \left(\dfrac{E_v}{R} \right)^{0.333}$
	>200	$T_f = 0.36 \mu P^{0.5} v^{0.5} \left(\dfrac{1}{\sqrt{k\rho c}} \right) \left(\dfrac{E_v}{R} \right)^{0.5}$

[a] μ, friction coefficient; P, load; v, velocity; k, thermal conductivity; π, density; c, heat capacity; E_v, the reduced elastic modulus = $E/(1-v^2)$, where v is Poisson's ratio; p, flow pressure of the softer material; ρ, material density; and R, the equivalent radius of the sliding contact.

Source: Adapted from Cowan and Winer, Ref. 101.

The equation provided by Cowan and Winer, based on an earlier treatment by Holm, for the case of a circular contact with one body in motion is

$$T_f = \frac{\gamma_1 \mu P v}{\pi a k_1} \tag{4.41}$$

where γ_1 is the heat partition coefficient, described earlier, P the normal force, a the radius of contact, and k_1 is as defined earlier. The computation of γ_1 takes various forms depending on the specific case. The presence of elastic or plastic contact can also affect the form of the average flash temperature, as Table 4.16 shows. Here, the exponents of normal force and velocity are not necessarily 1.0.

Kuhlmann-Wilsdorf[103] considered an elliptical contact area as the planar moving heat source. The flash temperature is given in terms of the average temperature in the interface T_{ave}:

$$T_{ave} = \frac{\pi q r}{4 k_1} \tag{4.42}$$

where q is the rate of heat input per unit area (related to the product of friction force and velocity), r the contact spot radius, and k_1 the thermal conductivity, as defined earlier. Then

$$T_f = \frac{T_{ave}}{(1/ZS) + (k_1/S_o)} \tag{4.43}$$

where Z is a velocity function and S and S_o are contact area shape functions (both = 1.0 for circular contact). At low speeds, where the relative velocity of the surfaces $v_r < 2$ ($v_r = v/P_e$), Z can be approximated by $1/[1 + (v_r/3)]$. In this treatment, functions to account for the shape of the contact region and nonlinear dependence of sliding velocity on heat flow are introduced.

As a last example, we consider the form described by Ashby et al.[104] for flat-ended pin-on-disk geometry. Beginning with the expression for q, the heat generated per unit nominal contact area A_n per unit time,

$$q = \frac{\pi \mathbf{P} v}{A_n} \tag{4.44}$$

these authors developed the following expression for flash temperature:

$$T_f = T_b + \frac{\mu \mathbf{P} v}{A_n}\left[\frac{1}{(k_1/l_{1f}) + (k_2/l_{2f})}\right] \tag{4.45}$$

where T_b is the heat sink temperature of sliding solids and l_{1f} and l_{2f} are effective thermal diffusion distances. Here, an account is made for the distance that heat has to travel from the interface to reach the heat sink temperature of the two solids. The nominal contact area, rather than the true contact area of asperity junctions, is used. Ashby et al. computed sliding temperature maps for several metal combinations in which isotemperature contours, for both flash and mean temperatures (which they call "bulk" temperatures), are plotted on axes of normalized velocity and normalized pressure. These maps delineate regions in which thermal effects may seriously alter the wear behavior of the tribosystem.

The foregoing models for frictional heating assume that the heat source is located at the contact surface. Strictly speaking, a surface is a mathematical construct that has area but no thickness (no volume). From that point of view, it makes no physical sense that heat should be generated at a two-dimensional surface. Frictional heat is actually generated within a volume in which the conversion of frictional work to thermal energy occurs. Taking that more rigorous approach, Malkin and Marmur[105] calculated that the frictional temperature rises estimated using the assumption of planar heat sources could lead to significant overestimates.

In summary, models for frictional heating of unlubricated solid contacts differ with respect to the following attributes:

- They assume different shapes for the heat source on the surface.
- The assume different contact geometries (sphere-on-plane, flat-on-flat, cylinder-on-flat, etc.).
- They use different ways to partition the flow (dissipation) of heat between sliding bodies.
- They use different ways to incorporate the thermophysical properties of materials.
- Most assume heat is produced from a planar source, but some use a region of heat generation (volume).
- They use different velocity criteria for transitions in thermal regimes.

Software packages continue to be developed for use by component designers interested in frictional heat flows. However, these usually depend on handbook friction data or estimated boundary conditions. More rigorous models need to accommodate changes in friction with temperature, speed, or operating time.

It has been the experience of the author in comparing the temperature rises predicted by different models that for low sliding speeds the values predicted by simpler models fall very close to those predicted by the more sophisticated models, and the differences are small compared with the uncertainties in the values of the numeric variables that go into the calculations. At higher speeds, the predictions become more, and more unreliable since materials properties change as a function of temperature, and there is less likelihood that the interface will reach steady state. Furthermore, we recognize that as sliding speed increases in many tribosystems, the vibration and the variations in normal force also increase, negating many of the assumptions that go into the frictional heating calculations.

Experimental studies have provided very useful information in validating the forms of frictional heating models. Experimenters have often used embedded thermocouples in one or both members of the sliding contact to measure surface temperatures (as in the case of Figure 4.28), and others sometimes made thermocouples out of the contacts themselves;[106] however, techniques using infrared sensors have been used as well. Dow and Stockwell[107] used infrared detectors with a thin, transparent sapphire blade sliding on a 15 cm diameter ground cylindrical drum to study the movements and temperatures of hot spots. Later, Griffioen et al.[108] and Quinn and Winer[8] used an infrared technique with a sphere-on-transparent sapphire disk geometry. A similar arrangement was also developed and used by Furey and his students with copper,[109] iron,[110] and silver[111] spheres sliding on sapphire, and Enthoven et al.[112] used an infrared system with a ball-on-flat arrangement to study the relationship between scuffing and the critical temperature for its onset.

Piston ring and land temperatures in a firing engine have been measured by Chang et al.[113] using infrared methods. They designed a series of special, 5 mm diameter sapphire windows in the wall of a cylinder block. Infrared radiation was monitored using an infrared emission microscope through the windows, and data were recorded in a storage oscilloscope. Temperature transients ranging from about 1 to 10 ms could be observed.

Frictional heating is important because it changes the shear strengths of the materials in the sliding contact, promotes reactions of the sliding surfaces with chemical species in the environment, enhances diffusion of dissolved elements to the surface (surface segregation), and in the case of lubricated sliding can result in the breakdown or failure of the lubricant to perform its functions.[114] In fact, one of the functions of lubricants is to carry away frictional heat, but in some situations this function is not always achieved and frictional heat can still cause problems. Under extreme conditions, such as in rotating bands on artillery rounds[115] or kinetic energy weapons (rail guns)[116], frictional heating can result in molten layer formation, which can reduce friction and serve as a liquid lubricant.

As a final note, frictional heating is sometimes used as a condition monitor for the state of a machine. During run-in, the initial surface conditioning process, the frictional heating may be high while surface conformation (wear-in) occurs.

Monitoring temperature is one way of determining the degree to which components have run in successfully.[117] The subject of running-in is discussed in more detail in Chapter 8.

REFERENCES

1. G. Salomon (1964). *Mechanisms of Solid Friction*, eds. P. J. Bryant, L. Lavik, and G. Salomon, Elsevier, Amsterdam, p. 3.
2. D. Mendeleev (1966). Cited by A. S. Ahkmatov, *Molecular Physics of Boundary Friction*, Daniel Davy, New York, NY, p. ix.
3. P. J. Blau (1991). Scale effects in sliding friction, *Tribol. Trans.*, 34, pp. 335–342.
4. E. J. Abbott and F. A. Firestone (1933). Specifying surface quality, *Mech. Eng.*, 55, p. 569.
5. P. C. Jindal and J. Gurland (1973). An evaluation of the indentation hardness of spheroidized steels, in *The Science of Hardness Testing and Its Research Applications*, eds. J. H. Westbrook and H. Conrad, ASM International, Materials Park, OH, pp. 99–108.
6. V. S. Kombalov (1980). Status and prospects of the study of the influence of roughness on the friction characteristics of friction pairs, *Trente i Iznos*, 1(3), pp. 440–452.
7. V. A. Valentov (1983). Marine diesel cylinder-piston friction surface microgeometry change during operation, *Trente i Iznos*, 4(6), pp. 1104–1107.
8. T. F. J. Quinn and W. O. Winer (1987). An experimental study of the 'hot spots' occurring during the oxidational wear of tool steel on sapphire, *J. Tribol.*, 109(2), pp. 315–320.
9. T. A. Dow (1980). Thermoelastic effects in a thin sliding seal—A review, *Wear*, 59, pp. 31–52.
10. T. Liu, G. Liu, Q. Xie, and Q. Wang (2005). Two-dimensional adaptive-surface elasto-plastic asperity contact model, *J. Tribol.*, 128(4), pp. 898–903.
11. T. O. Mulhearn and L. E. Samuels (1962). The abrasion of metals: A model of the process, *Wear*, 5, p. 478.
12. W. C. Young (1989). *Roark's Formulas for Stress and Strain*, 6th ed., McGraw-Hill, New York, NY.
13. J. A. Greenwood and J. B. P. Williamson (1966). Contact of nominally flat surfaces, *Proc. R. Soc. Lond. A*, 295, pp. 300–319.
14. D. J. Whitehouse and J. F. Archard (1970). The properties of random surfaces of significance in their contact, *Proc. R. Soc. Lond. A*, 316, pp. 97–121.
15. R. A. Onions and J. F. Archard (1973). The contact of surfaces having a random structure, *J. Phys. D: Appl. Phys.*, 6, pp. 289–304.
16. W. Hirst and A. E. Hollander (1974). Surface finish and damage in sliding, *Proc. R. Soc. Lond. A*, 233, pp. 379–394.
17. K. B. Park and K. C. Ludema (1994). Evaluation of the plasticity index as a scuffing criterion, *Wear*, 175(1–2), pp. 123–131.
18. J. McCool (1986). Comparison of models for the contact of rough surfaces, *Wear*, 107, pp. 37–60.
19. J. A. Greenwood (1992). Problems with surface roughness, in *Fundamentals of Friction: Macroscopic and Microscopic Processes*, eds. I. L. Singer and H. M. Pollock, Kluwer, Dordrecht, The Netherlands, pp. 57–76.
20. ANSI Standard B46.1 (1985). *Surface Texture (Surface Roughness, Waviness, and Lay)*, American National Standards Institute.
21. J. F. Song and T. V. Vorburger (1992). Surface texture, in *ASM Handbook, Volume 18: Friction, Lubrication, and Wear Technology*, 10th ed., ASM International, Materials Park, OH, pp. 334–345.

22. R. A. Burton, ed. (1980). *Thermal Deformation in Frictionally-Heated Systems*, Elsevier, Lausanne, Switzerland, p. 290.
23. J. Ferrante, G. H. Bozzolo, C. W. Finley, and A. Banerjea (1988). Interfacial adhesion: Theory and experiment, in *Adhesion in Solids*, eds. D. M. Mattox, J. E. E. Baglin, R. J. Gottschall, and C. D. Batich, Materials Research Society, Pittsburgh, PA, pp. 3–16.
24. D. F. Buckley (1981). *Surface Effects in Adhesion, Friction, Wear, and Lubrication*, Elsevier, New York, NY, pp. 245–313 (Chapter 5).
25. G. M. McClelland, C. M. Mate, R. Erlandsson, and S. Chiang (1988). Direct observation of friction at the atomic scale, in *Adhesion in Solids*, eds. D. M. Mattox, J. E. E. Baglin, R. J. Gottschall, and C. D. Batich, Materials Research Society, Pittsburgh, PA, pp. 81–86.
26. G. M. McClelland, C. M. Mate, R. Erlandsson, and S. Chiang (1987). Atomic scale friction of a tungsten tip on a graphite surface, in *Phys. Rev. Lett.*, 59, pp. 1942–1945.
27. P. A. Thompson and M. O. Robbins (1990). Origin of stick-slip motion in boundary lubrication, *Science*, 250, pp. 792–794.
28. M. O. Robbins and P. A. Thompson (1991). The critical velocity of stick-slip motion, *Science*, 253, p. 916.
29. M. O. Robbins, P. A. Thompson, and G. S. Grest (1993). Simulations of nanometer-thick lubricating films, *Mater. Res. Soc. Bull.*, XVIII(5), pp. 45–49.
30. L. O'Conner (1992). Molecular simulations open the friction frontier, *Mech. Eng.*, September, pp. 60–61.
31. U. Landman, W. D. Luetke, N. A. Burnham, and R. J. Colton (1990). Atomistic mechanisms and dynamics of adhesion, nanoindentation, and fracture, *Science*, 248, pp. 454–461.
32. J. Belak and J. F. Stowers (1992). The indentation and scraping of a metal surface: A molecular dynamics study, in *Fundamentals of Friction: Macroscopic and Microscopic Processes*, eds. H. M. Pollock and I. L. Singer, Kluwer, Dordrecht, The Netherlands, pp. 511–520.
33. H. M. Pollock, and I. L. Singer, eds. (1992). *Fundamentals of Friction: Macroscopic and Microscopic Processes*, Kluwer, Dordrecht, The Netherlands, p. 621.
34. D. R. Wheeler (1975). *The Effect of Adsorbed Chlorine and Oxygen on the Shear Strength of Iron and Copper Junctions*, NASA TN D-7894.
35. A. M. Arkarov and L. D. Kharitonova (1986). Friction at low temperatures, in *Tribology Handbook*, ed. I. V. Kragelski, Mir, Moscow, p. 84.
36. A. S. Akhmatov (1939). Some items in the investigation of the external friction of solids, *Trudy Stankina*; as cited by I. V. Kragelski (1965) in *Friction and Wear*, Butterworths, London, p. 159.
37. I. V. Kragelski (1965). *Friction and Wear*, Butterworths, London, p. 200.
38. M. E. Sikorski (1964). The adhesion of metals and factors that influence it, in *Mechanisms of Solid Friction*, eds. P. J. Bryant, L. Lavik, and G. Salomon, Elsevier, Amsterdam, pp. 144–162.
39. E. Rabinowicz (1992). Friction coefficients of noble metals over a range of loads, *Wear*, 159, pp. 89–94.
40. R. P. Feynman, R. B. Leighton, and M. Sands (1963). *The Feynman Lectures on Physics*, Addison-Wesley, Reading, MA, p. 12.
41. D. H. Wiid and W. F. Beezhold (1963). The influence of wear on the coefficient of static friction in the case of hemispherical sliders, *Wear*, 6, pp. 383–390.
42. K. Miyoshi and D. H. Buckley (1981). Anisotropic tribological properties of silicon carbide, in *Proceedings of Wear of Materials*, ASME, New York, NY, pp. 502–509.
43. W. E. Campbell (1940). Remarks printed in *Proceedings of M.I.T. Conference on Friction and Surface Finish*, MIT Press, Cambridge, MA, p. 197.

44. W. B. Hardy (1936). *Collected Scientific Papers*, Cambridge University Press, Cambridge, UK.

45. H. J. Lee, R. Zubeck, D. Hollars, J. K. Lee, A. Chao, and M. Smallen (1993). Enhanced tribological performance of rigid disk by using chemically bonded lubricant, *J. Vac. Sci. Technol. A*, 11(3), pp. 711–714.

46. A. E. Anderson (1992). Friction and wear of automotive brakes, in *ASM Handbook, Volume 18: Friction, Lubrication, and Wear Technology*, ASM International, Materials Park, OH, pp. 569–577.

47. F. P. Bowden and L. Leben (1939). The nature of sliding and the analysis of friction, in *Proc. R. Soc. Lond. A*, 169, pp. 371–391.

48. M. E. McIntyre and J. Woodhouse (1986). Friction and the bowed string, *Wear*, 113, pp. 175–182.

49. R. H. Hammerschlag (1986). Friction and stick-slip in a telescope construction, *Wear*, 113, pp. 17–20.

50. R. Holm (1958). *Electrical Contacts Handbook*, Springer-Verlag, Berlin, pp. 234–235.

51. A. Ruina (1986). Unsteady motions between sliding surfaces, *Wear*, 113, pp. 83–86.

52. *World Book Encyclopedia* (1988), Vol. 6, World Book, Chicago, p. 33.

53. J. N. Israelachvilli (1992). Adhesion, friction, and lubrication of molecularly smooth surfaces, in *Fundamentals of Friction: Macroscopic and Microscopic Processes*, eds. H. M. Pollock and I. L. Singer, Kluwer, Dordrecht, The Netherlands, pp. 351–381.

54. D. F. Moore (1975). *Principles and Applications of Tribology*, Pergamon Press, Oxford, p. 152.

55. P. J. Blau and C. S. Yust (1993). Microfriction studies of model self-lubricating composite surfaces, Presented at International Conference on Metallurgical Coatings and Thin Films, San Diego, CA, April 19–23.

56. B. J. Briscoe (1992). Friction of organic polymers, in *Fundamentals of Friction: Macroscopic and Microscopic Processes*, eds. H. M. Pollock and I. L. Singer, Kluwer, Dordrecht, The Netherlands, pp. 167–181.

57. J. I. Kosterin and I. V. Kragelski (1962). Rheological phenomena in dry friction, *Wear*, 5, pp. 190–197.

58. F. P. Bowden and D. Tabor (1986). *The Friction and Lubrication of Solids*, Clarendon Press, Oxford, pp. 106–107.

59. V. A. Kudinov and D. M. Tolstoy (1986). Friction and oscillations, in *Tribology Handbook*, ed. I. V. Kragelski, Mir, Moscow, p. 122.

60. G. M. Bartenev and V. V. Lavrentev (1981). *Friction and Wear of Polymers*, Elsevier, New York, NY, pp. 53–61.

61. P. J. Blau and E. P. Whitenton (1980–1985). Unpublished studies at the U.S. National Bureau of Standards, Gaithersburg, MD.

62. E. Rabinowicz (1956). Stick and slip, *Sci. Am.*, 195(5), pp. 109–118.

63. E. Rabinowicz (1965). *Friction and Wear of Materials*, Wiley, New York, NY, pp. 99–102.

64. J. Hailing (1976). *Introduction to Tribology*, Springer-Verlag, New York, NY, p. 2.

65. E. Rabinowicz (1965). *Friction and Wear of Materials*, Wiley, New York, NY, p. 69.

66. P. H. Black (1961). *Theory of Metal Cutting*, McGraw-Hill, New York, NY, pp. 45–72 (Chapter 5).

67. P. J. Blau, C. S. Yust, R. E. Clausing, and L. Heatherly (1989). Morphological aspects of the friction of hot-filament-grown diamond thin films, in *Mechanics of Coatings*, eds. D. Dowson, C. M. Taylor, and M. Godet, Elsevier, Amsterdam, pp. 399–407.

68. K. Hokkirigawa and K. Kato (1988). An experimental and theoretical investigation of ploughing, cutting and wedge formation during abrasive wear, *Tribol. Int.*, 21(1), pp. 51–57.

69. J. M. Challen and P. L. B. Oxley (1979). An explanation of the different regimes of friction and wear using asperity deformation models, *Wear*, 53, pp. 229–243.
70. F. P. Bowden and C. A. Brookes (1966). Frictional anisotropy in nonmetallic crystals, *Proc. R. Soc. Lond.*, A295, pp. 244–258.
71. P. J. Blau, E. P. Whitenton, and A. Shapiro (1988). Initial frictional behavior during the wear of steel, aluminum, and poly(methylmethacrylate) on abrasive papers, *Wear*, 124, pp. 1–20.
72. N. P. Suh (1986). *Tribophysics*, Prentice-Hall, Englewood Cliffs, NJ, pp. 416–424.
73. F. P. Bowden and D. Tabor (1986). *The Friction and Lubrication of Solids*, Oxford Science Press, England, pp. 90–121 (Chapter V).
74. B. J. Briscoe and T. A. Tolarski (1993). Sliding friction, in *Characterization of Tribological Materials*, ed. W. A. Glaeser, Butterworth-Heinemann, Stoneham, MA, pp. 46–48.
75. P. W. Bridgman (1931). *The Physics of High Pressure*, Macmillan Press, New York, NY.
76. I. V. Kragelskii, M. N. Dobychin, and V. S. Kombalov (1982). *Friction and Wear Calculation Methods*, Pergamon Press, Oxford, pp. 178–180.
77. J. D. Byerlee (1978). Friction of rocks, *Pure Appl. Geophys.*, 116, pp. 615–626.
78. M. Godet (1984). The third-body approach: A mechanical view of wear, *Wear*, 100, pp. 437–452.
79. Y. Sun, Y. Berthier, B. Fantino and M. Godet (1993). A quantitative investigation of displacement accommodation in third-body contact, *Wear*, 165, pp. 123–131.
80. J. Belak, ed. (1993). Nanotribology (special issue), in *Mater. Res. Soc. Bull.*, 18(5), May issue.
81. I. L. Singer and H. M. Pollock (1992). *Fundamentals of Friction: Macroscopic and Microscopic Processes*, Kluwer, Boston, MA.
82. D. A. Rigney (2007). Flow and mechanical mixing in near surface layers, Plenary address at 15th International Conference on Wear of Materials, Montreal, Canada (also published in *Wear*).
83. B. N. J. Persson (1998). *Sliding Friction: Physical Principles and Applications*, Springer-Verlag, Berlin, p. 462.
84. F. Heslot, T. Baumberger, B. Perrin, B. Caroli, and C. Caroli (1994). Creep, stick-slip, and dry friction dynamics: Experiments and a heuristic model, *Phys. Rev. E*, 49(6), pp. 4973–4987.
85. J. Skinner and N. Gane (1972). Sliding friction under a negative load, *J. Phys. D: Appl. Phys.* 5, pp. 2087–2094.
86. A. Erdemir, O. L. Eryilmaz, I. B. Nylufer, and G. R. Fenske (2000). Synthesis of super-low friction carbon films from highly hydrogenated methane plasmas, *Surf. Coat. Technol.*, 133–134, pp. 448–454.
87. J. Andersson, R. A. Erck, and A. Erdemir (2003). Friction of diamond-like carbon films in different atmospheres, *Wear*, 254, pp. 1070–1075.
88. M. Hirano and K. Shinjo (1990). Atomistic locking and friction, *Phys. Rev. Lett.* B47, pp. 11837–11851.
89. K. Shinjo and M. Hirano (1993). Dynamics in friction: Superlubric state, *Surf. Sci.*, 283, pp. 473–478.
90. M. Hirano, H. Kuwano, and J. W. Weijtmans (1997). Superlubricity mechanism for microelectromechanical systems, *Proc. IEEE MEMS '97*, pp. 436–441.
91. M. Hirano (2003). Superlubricity: A state of vanishing friction, *Wear*, 254, pp. 932–940.
92. A. Erdemir and J.-M. Martin, ed. (2007). *Superlubricity*, Elsevier, Oxford, UK, p. 524.
93. N. P. Suh (1986). *Tribophysics*, Prentice-Hall, Englewood Cliffs, NJ, pp. 85–87.

94. R. C. D. Lavista (2000). *Development of a Friction Material Formulation Without Metals By Taguchi Design of Experiments*, Society of Automotive Engineers, Paper number 2000-01-2754.
95. P. J. Blau (1979). Interrelationships Among Wear, Friction, and Microstructure in the Unlubricated Sliding of Copper and Several Single-Phase Binary Copper Alloys, PhD dissertation, Ohio State University, Columbus, OH, p. 341.
96. H. Blok (1963). The flash temperature concept, *Wear*, 6, pp. 483–494.
97. H. Blok (1937). General discussion on lubrication, *Inst. Mech. Eng.*, 2, p. 222.
98. J. C. Jaeger (1942). Moving sources of heat and the temperature of sliding contacts, *J. Proc. R. Soc. N. South Wales*, 76, p. 203.
99. F. P. Bowden and D. Tabor (1986). *The Friction and Lubrication of Solids*, Clarendon Press, Oxford, pp. 52–57.
100. E. Rabinowicz (1965). *Friction and Wear of Materials*, Wiley, New York, NY, p. 89.
101. R. S. Cowan and W. O. Winer (1992). Frictional heating calculations, in *ASM Handbook Volume 18: Friction, Lubrication, and Wear Technology*, 10th ed., ASM International, Materials Park, OH, pp. 39–44.
102. E. Rabinowicz, op cit, Ref. 87.
103. D. Kuhlmann-Wilsdorf (1987). Demystifying flash temperatures I. Analytical expressions based on a simple model, *Mater. Sci. Eng.*, 93, pp. 107–117.
104. M. F. Ashby, J. Abulawi, and H. S. Kong (1991). Temperature maps for frictional heating in dry sliding, *Tribol. Trans.*, 34, pp. 577–587.
105. S. Malkin and A. Marmur (1977). Temperatures in sliding and machining processes with distributed heat sources in the subsurface, *Wear*, 42(2), pp. 333–340.
106. O. S. Dine, C. M. Ettles, S. J. Calabrese, and H. A. Scarton (1993). The measurement of surface temperature in dry or lubricated sliding, *J. Tribol.*, 115(1), pp. 78–83.
107. T. A. Dow and R. D. Stockwell (1977). Experimental verification of thermoelastic instabilities in sliding contact, *J. Lubr. Technol.*, 99(3), p. 359.
108. J. A. Griffioen, S. Bair, and W. O. Winer (1985). Infrared surface temperature in a sliding ceramic-ceramic contact, in *Mechanisms of Surface Distress*, eds. D. Dowson et al., Butterworths, London, pp. 238–245.
109. C. A. Rogers (1981). *An Experimental Investigation of the Effect of Subdivision of Contact Area on Surface Temperatures Generated by Friction*, M.S. thesis, Virginia Polytechnic Institute and State University, Blacksburg, VA.
110. S. C. Moyer (1982). *Infrared Radiometric Measurements of Surface Temperatures Generated by Friction of Sliding Iron-on-Sapphire*, M.S. thesis, Virginia Polytechnic Institute and State University, Blacksburg, VA.
111. E. L. Hollowell (1983). *Infrared Measurements of Steady-State Temperatures and Average surface Temperature Distribution for Silver Sliding on Sapphire*, M.S. thesis, Virginia Polytechnic Institute and State University, Blacksburg, VA.
112. J. C. Enthoven, P. M. Cann, and H. A. Spikes (1993). Temperature and scuffing, *Tribol. Trans.*, 36(2), pp. 258–266.
113. H. S. Chang, R. Wayte, and H. A. Spikes (1993). Measurement of piston ring and land temperatures in a firing engine using infrared, *Tribol. Trans.*, 36(1), pp. 104–112.
114. J. Fohl and H. Vetz (1976). Failure criteria in thin film lubrication—influence of temperature on seizing, wear, and reaction layer formation, *Wear*, 36, pp. 25–36.
115. R. S. Montgomery (1976). Surface melting of rotating bands, *Wear*, 38, pp. 235–243.
116. D. J. Laird and A. N. Palazotto (2003). Effect of temperature on the process of hypervelocity gouging, *AIAA J.*, 41(11), pp. 2251–2260.
117. M. J. Neale (1973). *Tribology Handbook*, Section B30 on Running-in Procedures, Halsted Press, New York, NY.

5 Solid Friction of Materials

Before the dawn of recorded history, human beings undoubtedly noticed that when they rubbed different kinds of materials together, they felt unequal degrees of sliding resistance. Later, engineers learned to use the unique sliding characteristics of certain material combinations to create wagon wheel brakes, pivot bearings, carriage axles, ship launching rails, and stone arches. More recently, A. J. Morin, who worked in Paris in the 1830s, experimented with various materials and compiled lists of friction coefficients. Later, when results from a growing number of investigations were compared, a problem emerged. Friction coefficients for ostensibly the same material pairs differed. Who was right and who was wrong?

The confusion over which friction data to use when designing machinery exists to this day. Since friction is a characteristic of both the materials and the tribosystem, the extent to which the properties of materials affect friction coefficients depends on a range of variables. One cannot state with certainty that the kinetic friction coefficient for Type 52100 bearing steel against 70–30 brass is always exactly 0.830. Therefore, although this chapter exemplifies the frictional characteristics of a number of different materials, it is not possible to span the entire range of situations in which such materials might come into contact. As a result, the graphs and tables of data presented here are intended for illustrative purposes, given the stated testing conditions.

5.1 FRICTION OF WOOD, LEATHER, AND STONE

Wood, leather, and stone were among the earliest materials used for bearing surfaces. Civilization's keen interest in issues such as the transportation of goods and people, the launching of ships, the grinding of grain, and the building of stone archways prompted studies into the behavior of frictional interactions between materials. Observations by early investigators laid the groundwork upon which more elegant theories of friction were built.

Table 5.1 provides data on the frictional behavior of wood, leather, and stone. In some of the references, the methods of friction measurement were not well-documented; however, a study of the magnitudes of the friction coefficients is nevertheless informative since, rightly or wrongly, designers have relied heavily on such data in the past.

As Table 5.1 shows, most of the reported friction coefficient values for wood, leather, and stone against various materials lie between about 0.2 and 0.6. It is not surprising to observe that the friction of wood on wood can be affected by the relationship of the sliding direction to the grain of the wood. Most of these data originated hundreds of years ago, yet it is rare to find recent work on the friction of these materials in the tribology literature.

TABLE 5.1
Friction Coefficients of Wood, Leather, and Stone

Material Pair	Conditions	μ_s	μ_k	Table Reference Number[a]
Wood				
Hardwood on hardwood	28–728 psi pressure	—	0.129	1
Oak on oak	Parallel to the grain	0.62	0.48	2
Oak on oak	Perpendicular to the grain	0.54	0.32	2
Cast iron on oak	—	—	0.49	2
Brick on wood	—	0.6	—	3
Clean wood on metals	Various	0.2–0.6	—	3
Leather				
Leather on iron	0.7–28.4 psi pressure	—	0.25	1
Leather on oak	Parallel to the grain	0.61	0.52	2
Leather on cast iron	—	—	0.56	2
Leather on metal	—	0.6	—	1
Stone				
Sandstone on sandstone	—	—	0.364	1
Granite on granite	—	—	0.303	1

[a] Table references: (1) G. Rennie's data, as cited by D. Dowson (1979). *History of Tribology,* Longman, London, p. 229. (2) E. R. Booser, ed. (1984). *Handbook of Lubrication*, Vol. II, CRC Press, Boca Raton, FL, compilation pp. 46–47. (3) Ref. 4, Appendix IV.

Most recent studies of rock friction have been done not in the context of construction materials, as was the early work, but rather from an interest in the movements and behavior of rock masses within the earth. A review of the latter subject can be found in the book by Scholz.[1] The composition and hardness of the rock seem to have relatively little effect on friction. Instead, the general trend is that with very smooth surfaces (say, <2 μm in surface roughness), the sliding friction coefficients of many kinds of rock are about equal to 0.15, and as surface roughness increases (to greater than about 4–5 μm roughness), the friction coefficient tends toward 0.4–0.6. These values are similar to those for structural ceramics, many of which have compositions similar to that of rocks and minerals.

5.2 FRICTION OF METALS AND ALLOYS

Many bearing components are made from metals and their alloys. Metal parts are relatively easy to fabricate, and they have the strength and stiffness to support heavy loads. Bearing alloys like high-chrome bearing steel (AISI 52100) do not usu-ally perform well without lubrication, whereas some alloys like gray cast iron with its graphite flakes have lubricative phases built into the microstructure. However, most metals do not have "built-in" lubricants and some form of added lubrication is required.

There are some situations where lubricants would contaminate chemically sensi-tive tribosystems. Examples include food, beverage, and pharmaceutical processing equipment where lubricants could contaminate the product. A common example of unlubricated sliding at light load is the stainless steel plates that form luggage deliv-ery carousels in airport terminals. They must slide back and forth as the carousel rotates around corners, producing a characteristic, triangular scuffing pattern. Any lubricants used here could transfer onto luggage items, adding yet another annoy-ance to the airport experience.

Since the friction of unlubricated metals is affected by a variety of tribosystem-related variables, it is impossible to generalize that the sliding friction coefficient for unlubricated copper on steel is always 0.35 or 0.85 or 1.5 or, for that matter, any single value. In attempting to estimate or model the frictional behavior of met-als or alloys sliding on themselves or other materials, one should ask the following questions:

- How clean are the surfaces? Clean metals on metals, without lubrication and having normal atmospheric exposure after cleaning, often display sliding friction coefficients in the range $\mu = 0.4–0.8$. Special cleaning methods or high contact pressures will produce values well over 1.0.
- What is the contact pressure? Is the contact pressure so low that the ambient oxides or absorbed layers on the metal surfaces will control friction? If this is so, friction coefficients may range from about $\mu = 0.1$ to 0.3. Very high contact pressures tend to reduce the friction coefficients of many metallic material sliding combinations. The amount of reduction can be more than 50% if the contact pressures are high enough.
- What is the sliding speed? At high sliding speeds, a number of factors can affect friction: surface softening by frictional heating, increased plowing in the softer material, enhanced material transfer, and oxide film formation. The interplay among these factors makes it difficult to predict the friction of a given metallic material combination, and most of what is known about the effects of sliding speed on friction is experimentally based.
- Is the couple residing in a vacuum or reducing atmosphere? Friction coef-ficients of metals rise significantly *in vacuo* or in the absence of protective films. Values of the friction coefficient are usually higher than 1 and can be as high as 10. At some point, the question arises whether the tangential force being measured is really the interfacial friction force or the shear strength of a solid-phase welded joint.

It is interesting to consider that at low contact pressures, the friction of metallic materials may be governed by oxides or tarnish films and consequently be relatively low. At higher contact pressures, film rupture and plowing the subsurface material

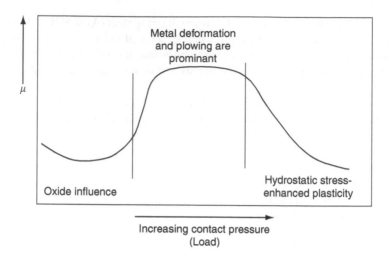

FIGURE 5.1 Effects of contact pressure on friction coefficient.

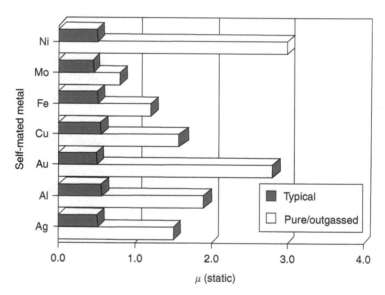

FIGURE 5.2 Static friction coefficients for metals are affected by cleaning method. (Adapted from Rabinowicz, E., *Am. Soc. Lubr. Eng. Trans.*, 71, 198, 1971; Bowden, F.P. and Tabor, D., *The Friction and Lubrication of Solids*, Pergamon Press, Oxford, 1986.)

raise the friction. Transfer is common in metals such as copper, aluminum, titanium, and others.[2] At very high contact pressures, however, the friction tends to decrease again as hydrostatic stress fields produce increased plasticity. This type of behavior is illustrated in Figure 5.1.

The static friction coefficients of metals exposed to the air for some time are generally much lower than those tested immediately after careful cleaning. Figure 5.2 compares the static friction coefficients reported for flat-on-flat tests in 50% relative

TABLE 5.2
Sliding Friction Coefficients of Self-Mated and
Non-Self-Mated Combinations

Metal or Alloy	Counterface	μ
Aluminum alloy 6061	Aluminum alloy 6061	0.34
	Ti–6Al–4V	0.29
	1032 steel	0.25
	Cu	0.23
	PTFE	0.19
	Tempered glass	0.14
AISI 1032 steel	AISI 1032 steel	0.23
	Aluminum alloy 6061	0.25
	PTFE	0.27
	Nickel-plated surface	0.31
	Titanium nitride coating	0.31

Source: Adapted from Gangopadhyay, A., Jahanmir, S., and Peterson, M.B. in *Friction and Wear of Ceramics*, Marcel Dekker, New York, 1993, 163–197.

humidity air[3] with those for the same self-mated metal combinations that were carefully cleaned and outgassed in a vacuum before air was admitted and sliding commenced.[4] The greatest difference was for nickel and gold and the least for molybdenum. Unlike nickel, gold tends not to form oxide films or tarnishes under normal conditions, and the differences in static friction must have been primarily due to the facilitation of adhesion of clean surfaces compared with surfaces that have been exposed to the environment where hydrocarbons and vapors could condense in monolayer thicknesses and reduce adhesive tendencies.

As Rabinowicz[5] pointed out, the friction coefficients for self-mated metals tend to be higher than those for dissimilar metals. Rabinowicz noted that the success of reducing friction by choosing dissimilar couples varies with the degree of compatibility between the materials selected, and he developed charts and guidance for determining the levels of frictional compatibility for pure metal couples.[3] Table 5.2 exemplifies how the self-mated friction coefficient of aluminum alloy 6061-T6 was higher than other sliding combinations, but, as shown by the data for 1032 steel, the self-mated, high-friction tendency is not always true. If, for example, one of the two materials tends to transfer and adhere to the other, then the couple becomes a self-mated situation during sliding. Compatibility should not be the only criterion used to select friction couples, but its principles may be helpful in avoiding bad choices.

The friction coefficient for metals sliding against nonmetallic materials can also be strongly dependent on the surrounding environment. Table 5.3, from the work of Mishna,[6] shows how the sliding friction behavior of several metal pins sliding against a silicon flat plate in a vacuum of 1.5×10^{-9} torr varies from that in air.

The friction of oriented or single-crystal metals is observed to be crystallographically dependent. Early experiments on the effects of crystal structure on the friction

TABLE 5.3
Sliding Friction Coefficients of Several
Metals against Silicon (0.1 N, 0.2 mm/s)

Metal Pin	μ in Vacuum	μ in Air
Silver	0.75–3.1	Up to 0.25
Copper	0.75–3.0	Up to 0.25
Platinum	1.1–3.2	0.27–0.30
Nickel	1.4–5.2	0.37–0.60
Titanium	2.1–7.3	0.41–0.73

TABLE 5.4
Sliding Friction Coefficients of Three Metallic Glasses (Single Stroke against Aluminum Oxide, 0.2 N load, 3.0 mm/min, 3×10^{-8} Pa Vacuum)

Composition	Room Temperature	2000°C	300°C	500°C
$Fe_{67}Co_{18}B_{14}Si_1$	1.4	1.50	1.65	0.25
$Fe_{81}B_{13.5}Si_{3.5}C_2$	1.0	1.10	1.35	0.30
$Fe_{40}Ni_{38}Mo_4B_{18}$	1.5	1.80	2.20	0.35

of pure copper were performed by Bailey and Gwathmey.[7] Crystallographic effects are not usually obvious in polycrystalline engineering alloys. A textured, highly deformed, or work-hardened layer usually develops in metals during sliding and will influence steady-state frictional behavior more than the initial crystallographic orientation.

The friction between metals also depends on the type of interfacial layers that may be developed as wear progresses. For example, when two metals or alloys are initially placed together, each usually has an ambient oxide film on its surface. After some period of rubbing, a mechanically mixed debris layer of oxides and fine metallic particles (a tribolayer) may form. It is this layer that determines frictional behavior—its magnitude and its variability as a function of the time of sliding. Friction process diagrams (FPDs) (see Chapter 8) can help portray the changing attributes of metallic friction as the interfacial structure evolves. The operating point in an FPD changes as the relative proportions of metal and oxide in the instantaneous contact areas change.

An interesting class of metallic friction materials is the so-called metallic glasses. Unlike typical metals and alloys, these multicomponent materials exhibit amorphous microstructures. They are beginning to find application, for example, as brazing foils, as parts for magnetic recording devices, as golf club heads, and as hard, thin coatings for bearings. Miyoshi and Buckley[8] studied the sliding friction and wear characteristics of several Fe-based metallic glasses at various sliding temperatures in high vacuum and in air. Selected data from that study are given in Table 5.4. The large drop in friction above about 500°C was attributed to the segregation of boron to the surfaces of the materials at elevated temperatures.

There have been claims that due to their higher hardness, bulk metallic glasses (BMGs) offer improved tribological characteristics, but this has not been observed by Blau.[9] A sliding friction and wear study of Zr-based, 3.125 mm tip radius, spherical BMG pins against rotating 52100 steel disks was conducted under dry conditions and lubricated by commercial 15W40 engine oil. Under dry conditions at 0.25 m/s and 4.95 N load, the steady-state friction coefficient was 0.74 and under lubricated conditions 0.20. Both wear rate and friction of the BMG under lubricated conditions were greater than that for Type 303 stainless steel and Nickel-200, neither of which are particularly good bearing materials. Probably, the use of BMGs under boundary lubrication conditions will require the development of special lubricants.

5.3   FRICTION OF GLASSES AND CERAMICS

The friction of ceramics against various materials is important to applications engineers involved with the design of ceramic gears, bushings, face seals, slideways, rolling element bearings, tooling and dies, and ceramic components for automotive and aerospace propulsion systems. Ceramics are lightweight, offer high stiffness, and provide resistance to chemical attack.

Just as there are many kinds of steels that are based on one alloy system (Fe–C), there are many forms and grades of ceramics based on the same basic composition. For example, different varieties of zirconia (ZrO_2) differ in composition, grain structure, crystal structure, and properties. Therefore, friction coefficient data for one type of zirconia may not be the same as that for another variety, especially if the contact conditions are such that enough wear occurs to alter the interfacial behavior. Thus, the differences in wear rate (so as to produce different amounts and types of interface debris) can affect frictional behavior. Ceramics often contain additional phases ("sintering aids" or glassy grain-boundary phases) to assist in processing. Furthermore, some silicon-based ceramics, such as silicon carbide and silicon nitride, usually have a few monolayers of SiO_x on the surface. Under conditions of low contact pressure and minimal wear, the friction of such materials may be governed by the silica layer and its reaction products with the surrounding environment (water vapor, in particular).

In the late 1970s, it was believed by some that structural ceramics were so naturally lubricative and wear-resistant that they could be used for in-cylinder components in advanced "adiabatic" engines with no external lubrication. This hypothesis was shown to be false by high-temperature ceramic tribology experiments at Oak Ridge National Laboratory and elsewhere in the early 1980s. Since then, there have been numerous studies of the tribology of ceramics in the United States, England, Japan, and Germany, in particular. Considerable effort was expended in developing solid and liquid lubricants that are compatible with ceramics. Composite microstructures and ceramic coatings containing built-in lubricants are also being studied and developed.[10]

The leading structural ceramics include silicon nitride, SiAlONs (silicon nitride with Al–O additives), silicon carbide, transformation-toughened zirconia, and alumina. Some ceramics are used as wear-resistant or protective coatings for metals. These include such compounds as titanium nitride, titanium carbide, tantalum carbide, hafnium carbide, and chromium carbide. Some ceramics, such as Co-bonded tungsten carbide, are used both as monolithic pieces of material and as coatings. Still others are used as reinforcing or structural matrix phases in composites.

TABLE 5.5
Static Friction Coefficients for Glass on Several Materials

Glass	Counterface Material	μ
Tempered glass	Aluminum alloy 6061	0.17
	1032 steel	0.13
	PTFE	0.10
Clean glass	Clean glass	0.9–1.0
	Metal	0.5–0.7

Source: Blau, P.J., *ASM Handbook, Volume 18: Friction, Lubrication, and Wear Technology*, ASM International, Materials Park, OH, 1992, 70–75.

TABLE 5.6
Sliding Friction Coefficients for Selected Ceramic Materials (Room Temperature in Air)

Ceramic	Counterface	μ	Table Reference Number[a]
Al_2O_3	Al_2O_3	0.33–0.50	1
Al_2O_3	Al_2O_3	0.20–0.9	2
α-Al_2O_3	α-Al_2O_3	0.38–0.42	3
Al_2O_3	Al_2O_3–SiC composite	0.53	1
Al_2O_3–SiC composite	Al_2O_3–SiC composite	0.64–0.84	4
B_4C	B_4C	0.53	1
SiC	SiC	0.52	1
SiC	Si_3N_4	0.53–0.71	1
WC	WC	0.34	1
Si_3N_4	Si_3N_4	0.42–0.82	5

[a] Table references: (1) Ref. 11, pp. 70–75. (2) M. Gee (1992). Results from a U.K. interlaboratory project on dry sliding wear of alumina, in ASTM STP 1167, *Wear Testing of Advanced Materials*, eds. R. Divakar and P. J. Blau, ASTM, Philadelphia, pp. 129–150. (3) Ref. 18. (4) C. DellaCorte (1993). Tribological characteristics of silicon carbide whisker-reinforced alumina at elevated temperatures, in *Friction and Wear of Ceramics*, ed. S. Jahanmir, Marcel Dekker, New York, pp. 225–259. (5) S. Danyluk, M. McNallan, and D. S. Park (1993). Friction and wear of silicon nitride exposed to moisture at high temperatures, in *Friction and Wear of Ceramics,* ed. S. Jahanmir, Marcel Dekker, New York, pp. 61–77.

Published data on the friction of glasses against various materials are scarce compared to that for metals or for structural ceramics. Table 5.5 lists typical friction coefficients for glass from a tabulation in the *ASM Handbook*.[11] Against metals and polymers, the static friction of glass tends to range from 0.1 to 0.7. Clean glass, as one might expect, can exhibit quite high static friction coefficients.

Unlike metals and alloys, rules for compatibility for ceramics, which are largely ionically or covalently bonded materials, are less applicable. Table 5.6 shows that there seems to be little basis for concern over self-mated sliding friction couples. Data should only be used as a rough guide because other factors can affect the friction of ceramics. For example, extensive research has shown that the friction of ceramic materials sliding on other ceramics is sensitive to the relative humidity[12–15] and the chemistry of the surrounding environment.[16]

From a mechanics and materials standpoint, the review of Briscoe and Stolarski[17] points out that the plowing component of the friction of a brittle material (μ_p) is inevitably associated with microcracking and suggests an expression that takes into account the fracture toughness K_{Ic}, the elastic modulus E, the hardness H, and the normal force N:

$$\mu_p = C \frac{K_{Ic}^2}{E\sqrt{HN}} \qquad (5.1)$$

However, mild sliding conditions will not produce significant plowing or fracture (i.e., wear), and friction will then be governed by the shear of interfacial films composed of adsorbed moisture and reaction products.

For the purpose of this discussion, one can consider two basic states of friction for ceramics: one in which there is significant wear or surface fracture and the other in which there is not. In the former case, one would expect sliding friction coefficients to be in the range 0.5–0.8 and in the latter, in the range 0.1–0.3. Friction in the former case is increased by the presence of abrasive third-body layers. Low humidity tends to promote wear in ceramics such as alumina and silicon nitride. Thus, friction coefficients would be expected to increase in low humidity conditions due to the increase in wear debris and surface roughness. Increasing the temperature of sliding (in air) tends to desorb moisture from a ceramic surface and would also tend to increase friction. As temperature increases further, lubricious oxides may begin to form films of sufficient thickness to drive the friction down again. This explanation is consistent with the pin-on-disk test results obtained by Yust,[18] who studied zirconia-toughened alumina and silicon carbide whisker-reinforced alumina, and with those obtained by Jahanmir and Dong,[19] who studied alumina. However, velocity effects on the friction of ceramics occur at both ambient room temperature and imposed elevated temperatures, as demonstrated by the work of Woydt and Habig,[20] but frictional heating at the higher sliding velocities plays a part in that behavior.

Composite ceramics and complex ceramic weaves are being considered for high-temperature seal applications. Dellacorte and Steinmetz[21] tested the friction of ceramic fiber materials for use in seals for hypersonic aircraft. High-temperature pin-on-disk tests were performed in air using fiber bundles (about 6000 fibers of 11 μm diameter in a bundle) fixed to the tip of the pin specimen. The disk was a nickel-chromium-based alloy, Inconel 718. The load was 2.65 N and the speed was 25 mm/s. Selected friction data at three temperatures are shown in Table 5.7. Frictional performance is not a sufficient measure of material suitability. These investigators also conducted durability tests to assess the potential of each candidate material. Alumina exhibited low friction but was not found to be suitable for the seal application because its high elastic modulus and lack of flexibility made it difficult to weave into a braid.

TABLE 5.7

Friction Coefficients of Several Ceramic Fiber Bundles against Inconel 718

Fiber Material	μ (25°C)	μ (500°C)	μ (900°C)
Al_2O_3	0.51 ± 0.04	0.28 ± 0.07	0.32 ± 0.03
Ti–SiC	0.17 ± 0.03	0.34 ± 0.14	0.62 ± 0.15
$62Al_2O_3–24SiO_2–14B_2O_3$	0.59 ± 0.10	0.68 ± 0.09	0.74 ± 0.13
$70AlO_3–28SiO_2–2B_2O_3$	0.49 ± 0.09	0.65 ± 0.13	0.66 ± 0.11

Source: Adapted from Dellacorte, C. and Steinmetz, B.M., *Tribological Comparison and Design Selection of High-Temperature Candidate Ceramic Fiber Seal Materials*, STLE Preprint 93-TC-1E-2, Society of Tribology and Lubrication Engineering, Park Ridge, IL, 1993.

The friction coefficient of couples in which only one member is a ceramic may be governed more by the properties of the counterface material. For example, in metal/ceramic couples, the metal may gradually form a transfer layer on the ceramic, leading to self-mated sliding and raising the friction. Modeling the frictional behavior of such systems must consider the relative transfer and back-transfer tendencies of each member of the couple.

5.4 FRICTION OF POLYMERS

The friction of polymers is important from two standpoints. First, the friction of polymers against such components as guides, spools, and injection molding dies affects their handling during processing. Second, there are applications for polymers, such as bearings, gears, seals, bushings, and gaskets, in which frictional characteristics define a polymer's suitability for that application. The friction of photographic films is an example of both issues. The friction of films is important to the makers of motor-driven cameras because excessive friction in winding and unwinding on the spool can greatly reduce the battery replacement interval in such cameras. Likewise, the friction of film can create great problems in handling the film during its production. ANSI standards have been developed for assessing the degree of lubrication of processed photographic film.[22] One solution to this problem is to eliminate photographic film entirely in favor of electronic imaging technology; another solution is to develop lubricative coatings.

The viscoelastic and hysteresis properties of many polymers, including elastomers, produce interesting but sometimes erratic frictional behavior. However, polytetrafluoroethylene (PTFE) exhibits a low-friction coefficient in many applications, and has been used either by itself or as an additive to composite materials to modify their sliding characteristics.[23] Elastomers, such as the rubber used in tires, have a particular wave-like response to sliding, which is discussed later.

Weaving of yarns and fabrics is affected by the internal friction in the fiber bundles (tows). As described in Chapter 3, special devices have been developed to measure the friction of fibers and fiber bundles. The structure of the fibers and yarns causes friction to vary markedly depending on the direction of rubbing.

TABLE 5.8

Directional Effects of Wool Fiber Friction Sliding on Horn

Condition of the Wool	μ_1 (Root to Tip)	μ_2 (Tip to Root)	μ (Range)
Clean	0.4–0.7	0.8–1.0	0.07–0.43
"Greasy"	0.3–0.4	0.6–0.8	0.20–0.45

Source: Adapted from Bowden, F.P. and Tabor, D., *The Friction and Lubrication of Solids*, Pergamon Press, Oxford, 1986, 170.

Bowden and Tabor[24] defined a *directional coefficient* to describe the anisotropy in the friction of wool rubbing on horn as follows:

$$\mu = \frac{\mu_2 - \mu_1}{\mu_2 + \mu_1} \tag{5.2}$$

where μ_1 is the friction coefficient obtained when sliding from the root to the tip of the fiber and μ_2 is the friction coefficient obtained when rubbing in the opposite direction (against the scales on the fiber). Table 5.8 lists Bowden and Tabor's representative values for cleaned and greasy wool on horn. The authors point out that μ is sensitive to both moisture and the pH of the surrounding environment.

Some polymers, such as polypropylene, can crystallize, and the orientations of the crystallites (spherulites) can affect the magnitude of the friction coefficient and its steadiness during sliding.[25] Vroegop and Bosma[26] proposed that friction-induced subsurface melting and recrystallization of nylon during dry sliding were caused by the dissipation of the energy produced by high-frequency vibrations. In contrast, Yamada and Tanaka[27] found little, if any, effects of crystallinity on the frictional behavior of polyethylene terephthalate lubricated by water. Thus, the effects of crystallinity, or lack thereof, of polymers cannot be generalized and, like other materials, must be addressed for each case.

Frictional instabilities during the sliding of polymers can also be affected by the load dependence of the friction–velocity characteristics. For example, Eiss and McCann[28] found that, for smooth surfaces of acrylonitrile butadiene styrene (ABS) sliding dry at high normal load in a pin-on-disk apparatus, the friction force–velocity curve had a negative slope (favorable for stick-slip) at low velocities and a positive slope at high velocities.

Bartenev and Lavrentev[29] distinguished three types of polymer friction: (a) friction in the glassy and crystalline states, (b) friction in the rubbery state, and (c) friction in the viscous state. In regard to (a), the investigators stated: "in the study of the law of friction for rigid, glassy polymers there is not a consensus on the relation to load of either the friction force or the friction coefficient, nor are there precise values for the friction coefficient." A study of Bartenev and Lavrentev's review of previous work illustrates the complex, nonlinear relationships between friction force of different polymers and the normal load, contact pressure, time of stationary contact before sliding, sliding velocity,

TABLE 5.9

Effects of Temperature on the Friction Coefficient of Several Polymers against Steel

Polymer	μ (T = 20°C)	μ (T = 40°C)	μ (T = 60°C)	μ (T = 80°C)
Polycaprolactam	0.46	0.50	0.74	—
Polyethylene	0.23	0.26	0.29	0.31
PTFE	0.22	0.21	0.19	0.16

TABLE 5.10

Maximum Allowable Temperatures for Friction Pairs of Metals and Polymers

Polymer	Maximum Temperature (°C)
Polyvinylchloride	60–95
Polystyrene	60–95
Polyolefins	70–105
Polyamides	80–110
Epoxy resins	80–135
Polycarbonates	100–135

temperature, and even the modulus of elasticity. Table 5.9, for example,[30] demonstrates that the friction coefficients of some polymers sliding against steel rise with temperature, whereas others fall.

The glass transition temperature in polymers, such as polyetheretherketone (PEEK), can have a pronounced effect on their friction–temperature behavior. Hanchi and Eiss[31] found, for example, that the friction and wear of PEEK rose markedly above the glass transition temperature and that blending PEEK with polyetherimide could reduce some of these effects.

Not only does temperature (both externally imposed and frictional) alter the frictional characteristics of polymers, but it also combines with surface fatigue effects to hasten their deterioration. Bilik[32] listed maximum allowable temperatures for polymer–metal friction pairs as given in Table 5.10.

The tendency of the surfaces of polymers to adsorb other species, to react with them to form other polymers, or to break down leads to a variety of responses of the friction of polymers to various lubricants. Figures 5.3a and 5.3b are plotted from data provided by Bartenev and Lavrentev[29] on the effects of lubrication on the static and kinetic friction coefficients of nylon. Note that the effects of the lubricant on static and kinetic friction are different for the self-mated case, and in contrast, lubrication tends to reduce both static and kinetic friction coefficients for steel counterfaces.

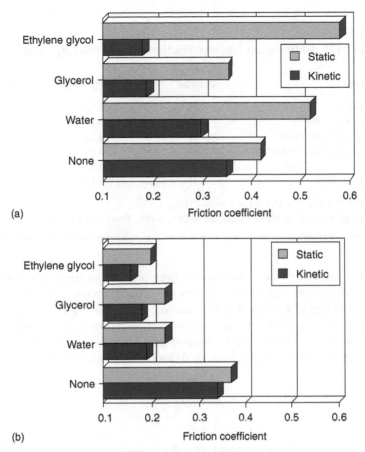

FIGURE 5.3 Effects of lubrication on the static and kinetic friction of nylon: (a) self-mated nylon, (b) steel sliding on nylon. (Adapted from Bartenev, G.M. and Lavrentev, V.V., *Friction and Wear of Polymers*, Elsevier, Amsterdam, 1981.)

Like all tables of friction coefficients for materials, it is especially inadvisable to use values of polymer friction coefficients for final design unless the test conditions were similar in nearly all aspects to those of the intended application. At best, such tables can serve as an initial screening tool to limit candidate materials to a reasonable number. Simulative or in-service testing, therefore, plays an important role in the selection of polymers for friction applications. In conducting such tests, factors such as the combination of linear and rotary motions should be accounted for since they can affect both friction and wear of polymers.[33]

Polymer composites are used to control friction and wear. For example, PTFE is a common additive to other polymers. The effects of PTFE additives on friction and wear were investigated by Mens and de Gee.[34] Table 5.11 lists some of their results for polymer blocks sliding on AISI 52100 steel rings at 0.1 m/s and 500 N load in air.

TABLE 5.11

Effects of PTFE Additives on Polymer Friction against Bearing Steel

Base Material	μ (Base Material)	μ (With PTFE)	Wear Ratio[a]
Polyamide 66	0.57	0.13	0.065
Polyoxymethylene (POM)	0.45	0.21	0.190
Polyether ether ketone (PEEK)	0.49	0.18	0.038
Polyethylene terephthalate (PETP)	0.68	0.14	0.032
Polyphenylene sulfide (PPS)	0.70	0.30	0.074
Polyetherimide (PEI)	0.43	0.21	0.064

[a] Wear rate with PTFE added/wear rate without PTFE.

Source: Based on data from Mens, J.W.M. and de Gee, A.W.J., *Wear*, 149, 255, 1991.

Elastomers, such as the rubber used for tires, are an important class of polymers. Moore[35] reviewed the friction of elastomers in a separate chapter of his 1975 book on tribology. Compared with more rigid materials, the surface of an elastomer loaded against a harder body tends to drape over the hard asperities and assume the surface contour. The actual contact area A is proportional to the ratio of the average contact pressure per asperity (p) to the elastic modulus of the elastomer (E):

$$A = KM \left(\frac{p}{E} \right)^n \tag{5.3}$$

where μ is a constant for the material, M is the number of contacts, and n is about 1.

Based on the groundwork laid by Schallamach,[36] both Briscoe and Stolarski[37] and Moore treated the friction of elastomers in terms of a two-term expression: an adhesive term and a hysteresis term. The adhesive component of friction is significant in the case of elastomer friction. This adhesional component μ_a can be written in terms of the sum of the adhesion strengths j_i of m individual molecular junctions:

$$\mu_a = \sum_{i=1}^{m} \left(\frac{n_i j_i}{P} \right) \tag{5.4}$$

where n_i and P are, respectively, the number of molecular bonds and the contact pressure at a given location i. Moore further defined a hysteresis coefficient of friction μ_H in terms of the number of asperities encountered M, the actual contact area A, the energy dissipated per asperity site E_d, the mean sliding distance between sites λ, and pressure on each asperity p:

$$\mu_H = \left(\frac{M E_d}{A \lambda p} \right) \tag{5.5}$$

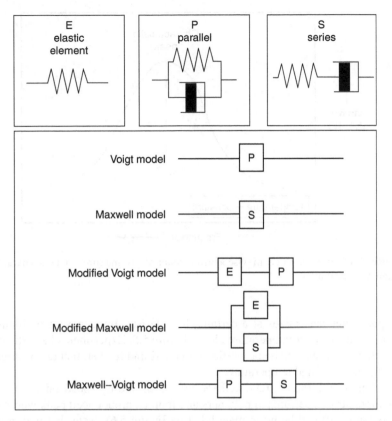

FIGURE 5.4 Models for the viscoelastic behavior of polymers can be represented by various combinations of springs and dashpots in series and in parallel.

Although it is tempting to accept simple equations like Equations 5.4 and 5.5 to explain the frictional behavior of elastomers, the situation is more complex than this, owing to considerations of viscoelasticity. Moore discussed five different models for viscoelastic phenomena, each represented by combinations of springs and dashpots. These are depicted schematically in Figure 5.4. Each of the models expressed a particular relationship between the friction force and the extension. Thus, the stress σ applied to a contact will not, in general, be exactly in phase with the resultant strain ε in the system. As Moore pointed out, this relationship can be expressed as a linear expression with real and imaginary terms:

$$\frac{\sigma}{\varepsilon} = K' + jK'' \tag{5.6}$$

The relationship between the real and imaginary parts K' and K'' can be defined in terms of the tangent:

$$\tan \delta = \frac{K''}{K'} \tag{5.7}$$

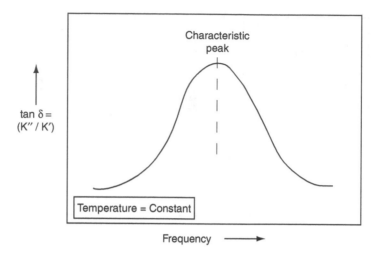

FIGURE 5.5 Friction force in viscoelastic materials is maximized at a characteristic frequency of oscillation.

This parameter maximizes at a certain characteristic frequency of oscillation, as shown by the constant-temperature plot in Figure 5.5. Experiments have shown that the friction force tends to be proportional to $\tan \delta$, and further, that $\tan \delta$ varies with both sliding speed and temperature.

McLaren and Tabor[38] reviewed the effects of sliding speed on the frictional behavior of several self-mated polymers and found a pronounced peak whose speed at maximum differed between materials (see Figure 5.6). Note that within a relatively small range of velocities (say 30–60 cm/s) polypropylene's friction coefficient increases, whereas that for polyethylene decreases. A criterion for stick-slip is that friction coefficient decreases with increasing velocity. Thus, an investigator whose experiments happen to be limited to a small range of velocities might conclude that polyethylene would exhibit stick-slip, but polypropylene would not. Examining the behavior over a larger range of speeds, however, would indicate that both polymers exhibit the same type of frictional behavior, but are just displaced in velocity space.

Various investigators have studied the hysteresis in rubber friction. In 1970s, for example, Yandell[39] applied a new treatment of rubber friction hysteresis to understand the friction of rubber on various roadway surfaces. Validation of Yandell's "mechano-lattice" treatment, which involved different shear stress–shear strain relationships for loading and unloading of frictional elements, was based on an analysis of the measured slopes of bumps on the surfaces of small stone test coupons (tiles) and the determination of the damping factor of the tester, using a pendulum device. The experimental values of friction coefficient of rubber on various stone surfaces in the presence of glycerin at 70°F, given in Table 5.12, agreed to within about 5–10% of the theory.

Similar to treatments of friction as a distributed property of the bulk versus friction as a localized property of asperities, hysteresis treatments of friction can be viewed from the standpoint of macro- and microhysteresis. In macrohysteretic

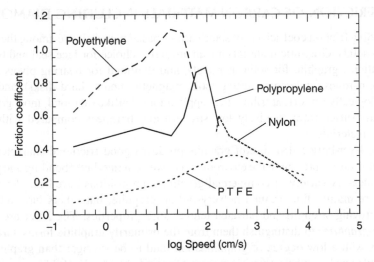

FIGURE 5.6 The maximum friction peak for various polymers occurs at different sliding velocities. (Adapted from Briscoe, J. and Stolarski, T.A., *Characterization of Tribomaterials*, Butterworth-Heinemann, Boston, 1993.)

TABLE 5.12
Friction Coefficients for Rubber on Various Stone Surfaces

Surface Material	μ (Rough Surface)	μ (Polished Surface)
Mica schist	—	0.26
Prospect dolerite	0.43	0.30
Volcanic breccia	0.40	0.34
Siliceous sandstone	0.43	0.38
Gneiss	0.45	0.36

Source: Adapted from McLaren, K.G. and Tabor, D., *Proceedings of Lubrication and Wear Convention*, Institute of Mechanical Engineering, London, 1963, 210–215.

treatments, the influence of fine-scale roughness or individual asperity interactions is minimized, and the phenomenon is treated more as a bulk property. In microhysteresis treatments, asperity interactions are considered, but in both treatments of the problem there is considerable question about when hysteresis considerations give way to adhesional considerations. As Moore[35] stated:

> Indeed, there may be as many as 100 variables to be accounted for in determining tyre-road interactions in rolling, sliding, or cornering, and we must undoubtedly await further developments in research which will quantify as many, of these as possible ... Our lack of complete knowledge in this field may be compensated for by judicious and timely experiments, which guide the theoretical investigations (and vice versa) and provide a link with practical experience.

5.5   FRICTION OF CARBON MATERIALS INCLUDING DIAMOND

The sliding friction coefficients of some carbon-based materials are among the lowest for any solids. Graphitic materials are an attractive choice for face seal and bearing applications, graphite for solid lubricants, and diamond for bearing pivots (watch jewels), hardness indenters, cutting tools, magnetic media (hard disks), and other technologically important tribological applications. Unlike diamond, most graphitic materials suffer from relatively low strength and hardness compared with other bearing materials.

The layer-lattice structure of graphite produces good frictional characteristics, when moisture is present in the environment to weaken interlayer bonding and permit easier shear. A variety of commercial graphite compositions have been developed. Some are mainly fine-grained polycrystalline graphite, but others have additives for improving strength and elevated temperature properties. The latter are called *carbon-graphites* to distinguish them from the primarily graphitic forms. Graphitic carbons with a low degree of graphitization tend to be stronger than graphite, but also tend to be less lubricative. Thus, there appears to be a trade-off between strength and lubricity. In addition to graphitic materials, certain glassy-carbon materials have been developed. They offer some attractive features of being less susceptible to the effects of moisture.[40,41]

Spreadborough[42] investigated the frictional behavior of graphite in 1962. Using a reciprocating, Bowden–Leben apparatus with hemispherical copper sliders, the author conducted friction studies of several types of natural graphite under low contact loads. A summary of the results is given in Table 5.13.

Longley et al.[43] published a review of the frictional behavior of carbon-graphite materials in 1963. In that review, the authors not only described the mechanisms of carbon-graphite friction, but also presented the results of a host of experiments on the effects of gaseous environments and surface texturing on the friction and wear of these materials. Table 5.14 summarizes the effects reported by Longley et al. with annular flat-on-flat ring sliding experiments conducted in a pressurized chamber. Three types of carbon-graphite were used: (a) a hard carbon with low graphite (LG), (b) a more highly graphitized form of the LG material (HG), and (c) a nongraphitic carbon (NG).

TABLE 5.13
Sliding Friction Coefficients for Natural and Compounded Graphite against Cu in Air

Type of Graphite	μ
Natural Madagascar flakes	0.07
Fibrous Ceylon graphite	0.17
Compacted natural flake graphite	0.13
Natural flakes + FeCl$_3$ compound	0.13

Source: Adapted from Spreadborough, J., *Wear*, 5, 18, 1962.

TABLE 5.14
Effects of Environment on Friction Coefficients of Carbon-Graphite Materials

Test Conditions	μ, HG carbon	μ, LG carbon	μ, NG carbon
Self-mated in 50% RH lab air	0.05–0.10	Initial period of 0.08–0.01, then rising to $\geq$0.4	—
Self-mated in CO_2 (5–50 ppm H_2O)	0.05–0.10	Period of $>$0.1 after run-in, then rising to $>$0.5	—
Self-mated in 10^{-4} mm Hg vacuum	High (with high wear)	Up to 0.8	—
Against mild steel in air or CO_2	0.05–0.10	Similar to self-mated LG, but with an extended low-friction period	0.05–0.10

Source: Adapted from Longley, R.I., Midgely, J.W., Strang, A., and Teer, D.G., *Proceedings of Lubrication and Wear Convention*, Institute of Mechanical Engineers, London, 1963, 198–209.

Later studies by Arnell et al.[44] revealed that, like the earlier work on graphitic and nongraphitic carbon, the friction of pyrolytic carbon against mild steel, nitrided steel, and tungsten carbide in air was found not to be strongly dependent on the initial crystallographic orientation of the surface but rather to depend on the counterface surface finish. Smooth surfaces exhibited a sliding friction coefficient of about 0.4, but abraded surfaces produced a friction coefficient as low as 0.1. Long-term, sawtooth-shaped friction variations between about $\mu = 0.2$ and 0.4 occurred with a typical period of hundreds of meters in sliding distance. In experiments with nitrided steel counterfaces, the period of frictional oscillation was found to depend on the crystallographic orientation of the starting carbon surface.

Studies in the author's laboratory indicated that there were considerable differences in the friction and wear behavior of commercially produced carbon-graphites tested in reciprocating ball-on-flat sliding.[45] In unlubricated sliding against Type O1 tool steel and NBD 100 silicon nitride spheres in laboratory air, the friction coefficients tended to lie between about 0.12 and 0.15, but some of the compositions displayed very repeatable frictional transitions from $\mu = 0.12$ to about 0.18 due to the onset of surface fatigue, microcracking, and the subsequent production of abrasive debris after about 4 min of reversed sliding contact (25 N load, 9.53 mm diameter ball slider, 5 Hz, 10 mm stroke). That study also demonstrated that while placing the materials in hot engine oil (150°C) reduced friction, it also greatly accelerated the wear rates of the materials, sometimes by more than 100 times. The possible cause for this behavior is that the lubricant excluded beneficial moisture from the contact. Furthermore, the wear under lubricated conditions produced a more uniform surface than that under ambient air conditions, suggesting that the process of wear changed from a fatigue and adhesion-dominated wear process to one associated with the action of fine debris entrained within the oil film. Thus, strategies used to reduce the friction of carbon-graphites must consider environmental factors.

The tendency of carbon-graphites to fatigue after prolonged sliding contact under certain conditions, forming blisters and delaminations, may promote changes in friction. Clark and Lancaster[46] described this phenomenon in detail. The degradation process was found to involve the progressive breakdown of the near-surface structure until fine regions, of the size of the original carbon-graphite crystallites, are produced. These fine regions can develop a preferred reorientation with respect to the sliding direction. Fatigue can occur through the initiation and growth of subsurface cracks beneath the altered layers.

In addition to primarily mechanical applications, there are electromechanical applications for graphite tribomaterials, notably electric motor brushes, in which the combined effects of electrical current and moisture level can have deleterious effects on motor life and current-carrying capability.[47] In fact, some of the most important, groundbreaking work in understanding asperity contacts, conducted by Ragnar Holm, was prompted by electrical contact problems.[48] Among many other findings, Holm found that the friction of carbon brushes was dependent on the orientation of the basal planes of the graphite and that the friction significantly increased when motors were operated at high altitudes in World War II aircraft where the humidity was low. Longley et al. later demonstrated that a preferred crystallographic texture was developed in LG and HG materials due to directional sliding. They found that if, after a period of unidirectional sliding ($\mu \approx 0.09$), the direction of sliding were reversed, the friction rose quickly (to $\mu \sim 0.11$) and a number of passes over the surface were required to bring the friction down to its former steady-state level. Thus, as Clark and Lancaster also found, the crystallite orientation in the near-surface regions is a significant contributor to the frictional performance of carbon-graphites.

Constriction resistance is the term used to refer to the higher resistance of an electrical contact produced when the current is forced to pass through small, localized regions on the tips of asperities (Holm called these *a-spots*). This increased current density leads to enhanced resistive heating, which, when added to the effects of frictional heating, can cause serious problems in the life and performance of electrical brushes. As the temperature of graphite-laden materials increases, the desorption of moisture occurs, altering their frictional behavior and their wear. These and other effects on the tribology of carbon materials are reviewed in a comprehensive chapter by Badami and Wiggs.[49]

Studies of the friction of diamond have been conducted for many years, and several excellent reviews exist.[50,51] Diamond has many interesting properties: high thermal conductivity, high elastic modulus, high hardness, and low friction if slid in the presence of moisture. In the 1980s and 1990s, developments in the production of diamond thin films on various substrates by a variety of thermochemical methods have reduced the cost of diamond-bearing surfaces and expanded their applications. These films can contain a range of perfection in the crystallography of the diamond structures. Pure diamond exhibits SP^3 type bonds, but forms of films that contain mixtures of diamond and amorphous carbon (diamond-like carbon [DLC] films) have a mixture of SP^3 and SP^2 type bonds. These types of films can be distinguished by Raman spectroscopy.

TABLE 5.15

Types of DLC Films

Film Type	Abbreviation	Comments
Hydrogen-free DLC	a-C	Primarily SP^2 bonding
Hydrogenated DLC	a-C:H	Small amount of SP^3 bonding
Tetrahedral amorphous carbon	ta-C	>70% SP^3 bonded carbon
Hydrogenated tetrahedral amorphous carbon	ta-C:H	Substantial SP^3 bonded carbon
Doped DLC (e.g., silicon, metal, carbides)	Si DLC, Me DLC	

Recent years have witnessed increasing use of DLC films in technological applications such as bearings, tooling, razor blades, and computer hardware, and the term has come to represent not a single structure or composition but a class, as shown in Table 5.15. The characteristics of DLC films, including adhesion and friction, can be adjusted by altering their composition and processing,[52] and their properties begin to degrade at temperatures as low as 250–300°C due to the onset of graphitization.

DLC films can be grown flatter and smoother than diamond films, which tend to have crystallographic pyramidal facets, square plates, or fine-grained cauliflower appearances when viewed in the scanning electron microscope. Blau et al.[53] showed that such faceted films tend to abrade nondiamond counterfaces, leading to friction coefficients considerably higher (e.g., $\mu > 0.5$) than those measured on flat diamond surfaces ($\mu \sim 0.03$–0.15). Similarly, Gupta et al.[54] found that as-deposited, coarse-grained films had friction coefficients as high as 0.42 against alumina sliders, whereas polished films exhibited friction coefficients as low as 0.09 under the same conditions. One would expect this sort of behavior due to the added contribution of the plowing and cutting to the overall friction force. If and when the initial growth facets wear down, as during prolonged sliding, the friction gradually becomes lower. Since smooth films of DLC can be grown, these are finding more rapid commercial application in tooling and magnetic media (hard disk drives) than are the diamond films. Reviews comparing the properties of diamond with those of diamond thin films are available in the literature.[55,56]

In summary, like other carbon-based materials, the friction of diamond is affected by the presence of moisture and oxygen in the environment. Driving off adsorbed water by heating can markedly increase the friction. Likewise, testing in vacuum can produce high-friction levels ($\mu \geq 0.8$). At high temperatures, diamond graphitizes and then oxidizes, so excessive frictional heating at high speeds or use in elevated temperature applications in air may be problematic. The friction and wear behavior of diamond has been convincingly shown, in controlled experiments on single crystals, to be dependent on the crystallographic orientation (frictional anisotropy) as well as on the roughness and lay of the surface material. Diamond often leaves a very thin carbonaceous film on the surface of the material over which it slides, and this film, which is not necessarily graphitic or amorphous, may be critical in producing its low-friction behavior. As discussed in Chapter 4 in the context of superlubricity, the nature of the hydrogen bonding is also important to achieving very low friction in diamond. Clearly, the friction and wear of diamond materials continues to open new avenues of use.

5.6 FRICTION OF ICE

The friction of ice is important for safety (walkways and tires), recreation (skating, skiing, curling, tobogganing, etc.), and international commerce (ice breakers). Structurally, ice is more interesting and complex than its simple formula, H_2O, might seem to imply. Depending on pressure, temperature, and the conditions of formation, ice can take on any of at least eight allotropic forms, the largest number for any known substance. Figure 5.7 shows temperature–pressure relationships between the various ice structures.[57] Owing to experimental difficulties, the boundaries between various structural regions on the diagram are not precisely determined; however, it is known that, under atmospheric pressure, as temperature decreases from 0°C, the dominant structure of ice changes from hexagonal crystals to cubic crystals and then to an amorphous or vitreous structure at temperatures of about −120 to −180°C.

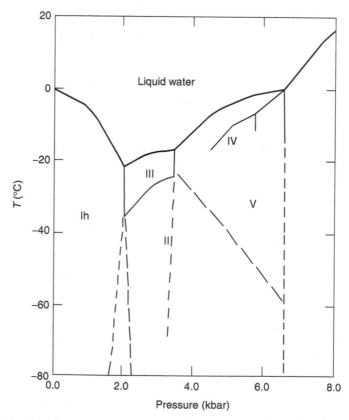

FIGURE 5.7 Pressure–temperature relationships for ice. (Adapted from Hobbs, P.V., *Ice Physics*, Clarendon Press, Oxford, 1974.)

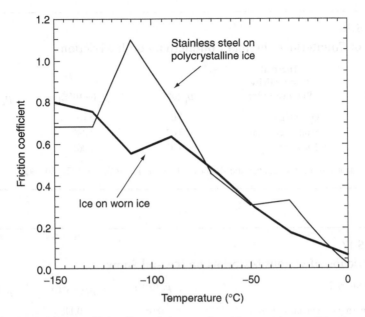

FIGURE 5.8   Friction coefficient for stainless steel and ice on polycrystalline ice as a function of temperature. (Adapted from Bhushan, B. and Gupta, B.K., *Handbook of Tribology: Materials, Coatings, and Surface Treatments*, McGraw Hill, New York, 1991.)

Since the structures of ice have different thermophysical properties at various temperatures, ice frictional characteristics vary considerably.

Schulz and Knappwost[58] studied the effects of temperature on the friction between ice and stainless steel at several sliding speeds and normal loads. Figure 5.8 shows that friction tends to increase as the temperature decreases, but not continuously, and this complex behavior is especially pronounced when sliding against stainless steel. Very near the melting point, there is a marked rise in friction coefficient between ice and stainless steel, then the friction drops slightly.

The subject of ice friction has intrigued investigators for hundreds of years. As early as 1829, George Rennie (see Chapter 1) investigated its effects on skates and sledges. He reported the kinetic friction coefficient for steel on ice as 0.014 and that of ice on ice as 0.028,[59] values in good agreement with more recent studies of ice near its melting point. Until 1939, it was largely believed that the low-friction coefficients between ice and many solids, such as steel ice-skating blades, were primarily caused by pressure-induced melting. However, Bowden and Hughes[60] showed, by a simple calculation of the load per unit area on skis, that the pressures developed would lower the melting point by only a fraction of 1°C, not enough to produce significant melting, especially when the ice temperature was more than a few degrees below zero. Electrical resistance experiments using electrodes embedded in skis confirmed that any surface melting that occurs below −20°C is limited to localized spots and does not form a continuous water film. Thus, Bowden and Hughes advanced the theory that it is frictional heating and not pressure-induced melting, which results in the formation of low-friction interfacial conditions on ice.

TABLE 5.16

Effects of Counterface Thermal Conductivity on Ice Friction

Material	Thermal Conductivity Relative to Ice	μ_k (0°C)	μ_k (−40°C)	μ_k (−80°C)
Brass	Quite high	0.03	0.105	0.160
Ebonite	Moderately higher	0.03	0.080	0.100
Ice	The same	0.02	0.065	0.075

Source: Adapted from Bowden, F.P. and Hughes, T.P., *Proc. Royal Soc. A*, 172, 280, 1939.

TABLE 5.17

Static Friction of Various Materials on Ice and Snow

Surface Material	μ_s (0°C)	μ_s (−5°C)	μ_s (−10°C)
Ski lacquer (pigmented nitrocellulose plasticized with phthalate)	0.05	0.11	0.43
Paraffin wax	0.04	0.27	0.37
Norwegian wax (sulfur-free, bituminous wax)	0.045	0.10	0.20
Swiss wax (highly refined, bituminous, hydrocarbon wax with 1.5% Al powder)	0.05	0.10	0.20
Polytetrafluoroethylene (PTFE)	0.04	0.05	0.55

Source: Adapted from Bowden, F.P., *Proc. Royal Soc. A*, 217, 462, 1953.

Table 5.16, based on their work, shows how the thermal conductivity of the counterface, which affects the degree of frictional heating, affects the friction coefficient of ice at various temperatures. The lower the thermal conductivity of the counterface, the greater was the friction reduction, because more heat was trapped in the interface and could therefore contribute to raising its temperature.

Bowden[61] conducted static friction experiments with skis and small sledges on snow and ice. The static friction coefficient was strongly affected by temperature, as summarized in Table 5.17. For loads between about 2 and 120 kgf, Amontons' laws were obeyed. The material performing the best was the polymer PTFE. Friction rose slightly above 0°C in "slushy" conditions, but this effect was probably due to the need to push semisolid matter aside during sliding.

Kong and Ashby[62] used the concept of frictional heating of ice to compute conditions that would favor the playing of the "Scotsman's favorite sport," curling. Curling stones must be slid such that they fall as close as possible from the "tee," a target marker on the ice, or else possess sufficient kinetic energy to scatter the stones of an opponent. A curling stone, typically made of granite, is about 240 mm in diameter, weighs about 20 kg, and has a handle on top. It rests and slides on a narrow, circular

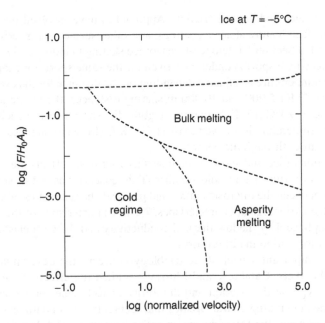

FIGURE 5.9   Transition diagram for ice. (Adapted from Kong, H.S. and Ashby, M.F., *Case Studies in the Application of Temperature Maps for Dry Sliding*, Cambridge University Engineering Department Report, 1991.)

ridge, called the *arridge*. With an estimated contact area of about 10^3 mm^2, Kong and Ashby calculated the nominal contact pressure on the stone during play to be about 0.2 MPa. Using a friction coefficient of 0.5 and typical values for the thermal conductivity, thermal diffusivity, and hardness of the granite and the ice, they plotted transition diagrams for curling at both -5 and $-25°C$ (see Figure 5.9). The axes are normalized force and normalized velocity. Assuming that asperity melting is responsible for effective continued sliding, the stones will continue sliding until the speed reaches 5.0 mm/s when the temperature is $-5°C$ but will stop sliding smoothly at 20 mm/s when the temperature is $-25°C$.

Although frictional heating helps to explain some of the kinetic friction behavior of ice against other solids, it does not adequately explain why ice can remain slippery even for people or objects standing still or traveling at low speeds where frictional heating is minimal. To address that point, more recent fundamental studies of ice friction invoke another phenomenon: the presence of a very thin (~2 nm) layer of quasi-liquid water on the surface of ice.[63] Detailed X-ray photoelectron spectroscopy (XPS) studies by Bluhm et al.[64] suggest that the thickness of the layer, which acts as a boundary lubricating film, can be increased by contaminants, and that helps to explain the role of aromatic molecules in ski waxes and other friction-modifying coatings. It has also been speculated that excess CO_2 could affect these layers, and hence the bonding between ice crystals buried in polar ice formations, leading to more rapid melting.

Spurred by the pressures of international competition, winter sports technology has become increasingly sophisticated in recent years, and new methods are being explored to reduce and control friction. Approaches have involved the use of new materials, surface feature shapes, coatings, and other surface engineering methods. As mentioned in Section 1.1, temperatures for ice skating in indoor rinks are adjusted depending on which sport is under way. Even for the same sport, for example, speed skating, the temperature is adjusted depending on the distance for the event. According to Collins,[65] for 5,000 and 10,000 m skating distances, the temperature is kept slightly cooler −9.4°C (15° F) to maximize glide. For shorter sprints such as 500 and 1000 m, the temperature is allowed to rise to −7.8°C (18° F) to increase edge control for turning harder through the corners.

The friction of ice and snow-covered surfaces seriously affects the safety of aircraft landings on runways in cold weather. This concern has led to studies of the relationship between the microstructures and physical characteristics of ice and snow layers and their tribological characteristics.[66] The characteristics of the tribosystem are also complicated by the low thermal conductivity and the viscoelastic characteristics of the rubber used in aircraft tires.

Another important aspect of ice tribology concerns the development of coatings for icebreakers. Calabrese and Murray[67] studied the effects of surface finish and coating type on the friction and durability of icebreaker hull surfaces. They discussed the demanding characteristics that icebreaker hull coatings require. Icebreakers have to survive abuse during an entire ice season, and therefore they must reduce the drag resistance of the hull while maintaining adequate corrosion protection. For friction testing, they modified a drill press head to load thrust-washer-type ring test coupons against a piece of ice inside a cold box that was cooled by a commercial refrigeration unit. The effects of surface finish and sliding velocity on the kinetic friction coefficient were investigated on bare AISI 1018 steel against ice under a load of 456 N. For an initial finish of 2.5 μm CLA (center-line-average roughness), μ began at about 0.4 and decreased to 0.1 when the sliding velocity increased to 0.22 m/s. It remained at that value up to the maximum test speed of 0.97 m/s. When the roughness was doubled, μ began at 0.5 and decreased monotonically until it reached 0.1 at a speed of 0.75 m/s.

Calabrese and Murray differentiated between breakaway friction and static friction coefficients, the former being that after the surfaces were held under load for a specified period of time, and the latter being the static friction coefficients measured by stopping, starting, and stopping several times after the breakaway value was measured. As might be expected, the breakaway frictional torque increased with holding time. The selected data in Table 5.18 were read from a figure in the paper by Calabrese and Murray and typify values of breakaway (60 s hold), static, and kinetic friction coefficients. Measurements of friction were accompanied by surface observations and wear tests. Overall, a nonsolvented polyurethane and a solvent-free epoxy coating were found to offer the best combination of friction and durability, as demonstrated by several years of successful follow-on testing on icebreaker hulls.

In general, the following can be said about the friction of ice:

1. As temperature drops below the melting point, the general tendency is for the kinetic frictilnon of ice against other materials to increase, but this increase is not continuous over all temperature ranges.

TABLE 5.18
Friction Coefficients (μ) for Candidate Icebreaker Hull Materials

Material	Breakaway μ	Static μ	Kinetic μ
Uncoated 1018 steel	0.65	0.37	0.09
PTFE-filled polyurethane coating	0.28	0.14	0.07
PTFE coating	0.25	0.11	0.02
Epoxy	0.13	0.10	0.05
Polyurethane coating	0.07	0.06	0.03
Polyethylene coating	0.05	0.04	0.03

Source: Adapted from Calabrese, S.J. and Murray, S.F. in *Selection and Use of Wear tests of Coatings*, ASTM STP 769, ASTM, West Conshohocken, PA, 1982, 157–173.

2. A primary mechanism controlling the kinetic friction of ice is frictional heating and the effectiveness of heat dissipation to the surroundings, but the presence of a quasi-fluid molecular layer of water on ice surfaces can also play a role in its frictional behavior. Research on nanometer-thick water films is in its infancy.
3. In the temperature range a few degrees below the melting point, the friction of smooth ice on various materials ranges between about 0.01 and 0.05, but the effects of surface roughness, such as that of heavily used skating rinks, may raise these values.
4. Recent improvements in ice friction control have involved altering contact geometry at macro- and microscales and using tribochemistry to modify the thin layers of water that seem to be present on ice surfaces.

5.7   FRICTION OF TREATED SURFACES

In the present context, treated surfaces include coatings, films, weld overlays, textured finishes, and compositionally modified surfaces. The most common reasons for treating engineering surfaces are as follows:

- Corrosion protection
- Cosmetic appearance
- Modification of optical qualities (mirrors, lenses, filters, gratings, etc.)
- Wear control (reduced abrasion, erosion, etc.)
- Control of thermal characteristics (insulation or enhanced thermal conductivity)
- Control of electrical or magnetic properties (electrical contacts, hard disks, or printed circuits)
- Improved bonding to another surface or coating
- Friction control, including improvement of lubricant flow characteristics

TABLE 5.19

Surface Treatments for Tribological Control

Type of Treatment	Tribological Function(s)
Geometric patterning or grooving	Debris control (trapping, removal)
	Lubricant flow control
	Lubricant retention
	Thermal control
Mechanical working (e.g., shot peening)	Hardening
	Reduction of corrosion-assisted wear, fretting
Diffusion or implantation treatments that change substrate composition without altering roughness or dimensions	Hardness and wear control
	Control surface chemistry; enhance reactions with beneficial lubricant additives
	Maintain surface dimensions but alter the surface physical properties
Surface melting (laser, electron beam, etc.) or weld overlaying	Hardness and wear control (by grain refinement, phase transformation or surface alloying)
	Providing a thick "wear allowance"
Friction stir processing	Refine near-surface microstructure or mix particulates into an existing surface
Coating with thin films (ion plating, rf sputtering, CVD, PVD, flash electroplating)	Solid lubrication
	Protect from mild wear, erosion, or abrasion
Thick coatings (thermal spraying, thick electroplating)	Wear protection
	Thermal barriers (heat control)
	Self-lubricating composite coatings

Usually, some combination of characteristics is required for engineering surfaces. For example, it would be impractical to apply a low-friction coating to surfaces that are exposed to corrosive environments which would quickly attack it. Conversely, the primary reason for applying a coating might be corrosion control, yet some level of acceptable friction is still required. Thus, the *balance of properties* required for each materials system must be established, and constraints may be placed on the friction-control options.

Some methods for engineering the tribological responses of surfaces are given in Table 5.19. This list is not comprehensive but rather suggests that the main approaches to friction and wear control by surface treatment involve changing the composition, morphology, state of stress, and microstructure.

It is important to consider the compatibility of a treated surface with its intended counterface. One surface treatment may work very well with one type of counterface but not with another. Often, a manufacturer may claim that his coating has "excellent frictional properties," but fails to mention the counterface material(s), the substrates involved, or the contact conditions to which that statement applies. As for materials in general, there is no intrinsic friction coefficient or wear rate of a given coating or

surface treatment. Two coatings of identical composition may have quite different
properties when applied to different substrates. This difference can be caused by
factors such as differences in heat flow in the substrate during application, crystallographic mismatch-induced interfacial strains, differential thermal expansion during
heat treatment, chemical compatibility, and residual stresses. Therefore, it is important that surface treatments be considered as part of a total materials system and
used in context of the imposed service conditions.

A comprehensive discussion of the types and uses of surface treatments to control friction and wear is given in the handbook by Bhushan and Gupta[68] and in the
text by Budinski.[69] When selecting such treatments, consider the repeatability of the
performance. Producers of experimental coatings and surface treatments may boast
about wonderful properties based on limited experiments, but this will be proved
only when these properties can be vastly reproduced after a number of surface treatments. Reliability is a vital component in the selection and use of surface treatments
and coatings.

Halling[70] developed an expression for the friction of coated surfaces. Letting τ_1
and τ_2 be the shear strengths of the substrate and film material, respectively, he wrote

$$\mu = \frac{\tau_1 A_1 + \tau_2 A_2}{H_1 A_1 + H_e A_2} \tag{5.8}$$

where A_1 and A_2 are the real areas of contact, H_1 the hardness of the substrate, and
H_e the effective hardness of the substrate covered by the film. This expression served
as a basis to derive the effects of surface roughness and film thickness on friction
coefficient. Experiments using lead films on steel substantiated the applicability of
the equation to soft films on hard substrates. Note that there is an implicit assumption here that the counterface is much harder than the coating-substrate materials
system and does not yield in shear.

Surface engineering is the discipline concerned with the analysis and modification of surfaces to achieve a certain set of performance requirements, and some of
them may involve friction. Hutchings[71] devoted a chapter to this subject in his book
on tribology. In fact, several universities, such as the University of Sheffield, the
University of Hull, the University of Alberta, Case Western Reserve University, and
Northwestern University, have all established surface engineering centers. There is
even a Surface Engineering Association in the United Kingdom. Although the term
surface engineering has been in use for only a few years, its principles are not new. In
a broader context, surface treatments employed for friction and wear protection date
back at least three millennia. Dowson[72] reported that over 3000 years ago copper
nails were pounded into wheel rims to reduce their wear. This was in effect surface
engineering.

The number of commercial surface treatments has greatly increased in recent
years. But with the variety of choices comes increased confusion for designers
and those charged with the responsibility for selecting materials for friction and
wear control. A part of the problem lies in the lack of a comprehensive, searchable
database for standardized tribological data on surface treatment options. Even where
limited compilations do exist, the test methods used may not necessarily correspond

to the application(s) of interest. Therefore, some amount of testing becomes inevitable during the selection of surface treatments for friction control.

Clearly, the technology of surface engineering is expanding at a rapid rate. In the future, expert systems, tribology databases, and various artificial intelligence methods may help reduce the difficulties in narrowing the number of candidates for friction problem solving. In the end, it may be cost and convenience that govern the final selection of a coating or surface treatment. Pressure to find "something that works" may override the more rigorous efforts needed to find an optimal solution.

5.8   FRICTION OF PARTICLE AGGREGATES

The importance of third-body particles in determining the friction of wearing interfaces was discussed in earlier chapters. Papers by Berthier[73] and Alexeyev,[74] for example, described the motion and contributions of particles on friction, and Heshmat et al.[75] studied the fluid-like behavior of powders—a work that eventually led to a quasi-hydrodynamic model for powder lubrication. But there are other instances when powders make up the bulk of the tribosystem's volume, and it is their flow, rather than the motion of the solids on either side, that is of primary interest.

Friction plays a very important role in the flow and handling of particles. The friction of particles is a consideration in such diverse scientific and technical areas as pharmaceuticals (pill production), dry copying, powder metal processing, sedimentology, civil engineering, and mining. Even the whistling (booming) of sand dunes in the desert has been associated with the friction between particles.[76] Factors such as the narrowness of the particle size distribution, particle shape, and grain surface roughness have been associated with the tendency for sand dunes to make noises.

Friction of particle aggregates basically involves two aspects: interparticle friction and friction between the aggregate and the walls of the container. Both aspects must be considered while treating practical problems related to granular materials. Particle mechanics is the subject that deals with the mechanical properties and flow of particles and particle aggregates. Frictional effects are a critical aspect of many particle mechanics problems.

Free-flowing powders in the dry state are pourable and when allowed to form a stable heap assume a characteristic angle of repose, θ_r.[77] If the maximum angle of stability, θ_m, is exceeded, grains begin to flow in a relatively thin boundary layer near the surface of the pile. In the transitional state between these two critical angles, flow characteristics become more complex. The point corresponding to the loosest packing that is still mechanically stable under a given applied force has been termed as the *random loose-packed limit*. More densely packed powders flow by the displacement of blocks (dead zones) within the assemblage. Granular material is inherently heterogeneous, and failure zones between these internal regions permit internal shearing and displacement to occur. To some extent, the operation of such processes depends on the roughness of the wall of the container. Rougher walls tend to promote internal shearing of the powder mass to enable flow.

A Hele–Shaw cell is a device consisting of transparent, closely spaced parallel plates. It was originally developed in 1898 to study the behavior of viscous, immiscible liquids but has also been used to conveniently measure the angle of repose of sand and other kinds of small particles.

Friction between individual particles has been treated as a point-contact problem. For elastic conditions, Adams[82] showed that the friction force at a point contact F_{PC} under a load P can be calculated from

$$F_{PC} = \pi\tau_o \left(\frac{3R}{4E}\right)^{2/3} P^{2/3} + \alpha P \qquad (5.9)$$

where τ_o is the interfacial shear strength, r the particle radius, α the pressure dependence of shear strength, and E the composite elastic modulus, given by

$$E = \left(\frac{1-v_1}{E_1} + \frac{1-v_2}{E_2}\right)^{-1} \qquad (5.10)$$

where the subscripts 1 and 2 denote the elastic moduli (E) and Poisson's ratios (v) for bodies 1 and 2, respectively. When multiple contacts or plasticity is involved, more complex arguments are used. Tanakov et al.[78] developed algebraic expressions for approximating the distributions of hydrostatic stresses in assemblages of small particles. They also used finite element methods to calculate the elastic stresses in cases of 2, 8, and 12 contacting particles.

Despite the more recent work on particle–particle analysis, most of the classical work on frictional characteristics of powders has been treated using distributed properties, or average properties of all particles in an assemblage. Brown and Richards[79] described three principles central to understanding the flow of powders:

1. The principle of dilatancy
2. The principle of mobilization of friction
3. The principle of minimum energy of flowing granules

The first principle states that, as a geometrical consequence of particle assemblages, shearing causes some granules previously in contact to separate and to produce an overall dilation (expansion) of the powder body. The second implies that the exact stress distribution in a powder at rest is indeterminate and that during shear, exceeding a limiting value locally will cause a surface of sliding to be formed within the body. The exact location of this surface cannot generally be predicted due to the heterogeneity of equilibrium states within the body. The third principle implies that when flow occurs, the particles will rearrange themselves progressively so that the mobilization of friction becomes defined by the flow geometry. Thus, discharge rates through orifices and constrictions can be calculated from the shape of the surfaces of sliding at the aperture. Two kinds of relationships between shear stress and the particle packing density are illustrated in Figure 5.10.

An early work by Bagnold[80] attempted to model the stresses and dilations of particles. By assuming grains flowed with a smooth velocity profile, and from kinetic

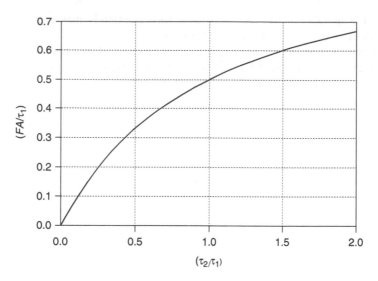

FIGURE 5.10   Effect of packing density on the shear stress of sand particles. (Adapted from Bagnold, R.A., *Proc. Royal Soc. A*, 295, 219, 1966.)

energy arguments, Bagnold developed a law for the average friction force F in a particle mass:

$$F = \frac{mD^2}{2\lambda_e}\left(\frac{dv}{dz}\right)^2 \tag{5.11}$$

where m is the mass of the particle, D the average particle diameter, λ_e the mean distance for the dissipation of excess energy (typically several times D, and a function of packing density), and (dv/dz) the velocity gradient within the zone. The Bagnold law tends to work particularly well for experiments using high shear rates. Jaeger and Nagel[77] extended a friction law for the motion of grams within a pile on the basis of not only kinetic energy, but also potential energy associated with particles rising out of and falling into depressions within the surface of motion. Letting the velocity gradient be designated as γ, accounting for the acceleration due to gravity g, and defining a quantity

$$X = \left(\frac{D\gamma^2}{g}\right) \tag{5.12}$$

Jaeger and Nagel expressed F as follows:

$$F = mg\left(\frac{a}{1+bX} + cX\right) \tag{5.13}$$

where a, b, and c are constants. At zero velocity, Equation 5.13 has a finite value corresponding to the Coulomb friction, but F decreases as velocity increases until a minimum value, after which it starts to rise rapidly.

The mobilization of a frictional resistance to powder flow was expressed in a simpler form by Brown and Richards,[81] expressing the relationship between the compressive stress on a cohesive powder mass N, the shear strength developed at the frictional surface S_f, and tensile strength T_a:

$$S = \mu (N + T_a) \tag{5.14}$$

but for noncohesive granular powder, Equation 5.14 simply becomes

$$S_f = \mu N \tag{5.15}$$

Experimental measurement of the shear properties of power masses is discussed in Chapter 4 of Brown and Richards' book. The most common method is the shear box, which consists of two rectangular boxes filled with powder, similar to a closed-mold sand casting. The upper box is dragged over the lower box under constant normal force and the tangential force is recorded. Using a series of normal loads, a plot of horizontal versus normal load is produced. The slope of the plot is the friction coefficient. The effects of wall friction can be studied in such a device. For example, crushed limestone powder on powder had a higher slope than limestone powder slid against a wall of sandpaper.

A number of review articles and books have addressed the subject of powder friction. For example, Adams[82] has reviewed the friction of granular materials, and comprehensive treatments of this subject may be found in Brown and Richards[79] and in the book by Feda.[83] In addition, there are also several journals devoted to powder technology that contain articles related to powder friction. In Ref. 84, the effects of particle shape and relative humidity on transitions in the frictional behavior of granular layers has been analyzed.

REFERENCES

1. C. H. Scholz (1990). *The Mechanics of Earthquakes and Faulting*, Chapter 2—Rock Friction, Cambridge University Press, Cambridge, pp. 44–96.
2. M. Antler (1964). Processes of metal transfer and wear, in *Mechanisms of Solid Friction*, eds. P. J. Bryant, M. Lavik, and G. Salomon, Elsevier, Amsterdam, pp. 181–203.
3. E. Rabinowicz (1971). Determination of compatibility of metals through static friction tests, *Am. Soc. Lubr. Eng. Trans.*, 71, pp. 198–205.
4. F. P. Bowden and D. Tabor (1986). *The Friction and Lubrication of Solids*, Appendix I, Clarendon Press, Oxford, p. 322.
5. E. Rabinowicz (1965). *Friction and Wear of Materials*, Wiley, New York, p. 30.
6. H. Mishna (1988). Friction and wear of semiconductors in sliding contact with pure metals, *Tribol. Int.*, 21(2), p. 76.
7. J. M. Bailey and A. T. Gwathmey (1962). Friction and surface deformation during sliding on a single crystal of copper, *ASLE Trans.*, 5, pp. 45–56.
8. K. Miyoshi and D. M. Buckley (1983). Friction and wear of some ferrous-base metallic glasses, *ASLE Trans.*, 27(4), pp. 295–304.

9. P. J. Blau (2001). Friction and wear of a Zr-based amorphous metal alloy under dry and lubricated conditions, *Wear*, 250(1), pp. 431–434.

10. A. Gangopadhyay, S. Jahanmir, and M. B. Peterson (1993). Self-lubricating ceramic matrix composites, in *Friction and Wear of Ceramics*, ed. S. Jahanmir, Marcel Dekker, New York, pp. 163–197.

11. P. J. Blau (1992). Appendix: Static and kinetic friction coefficients for selected materials, *ASM Handbook, Volume 18: Friction, Lubrication, and Wear Technology*, 10th ed., ASM International, Materials Park, OH, pp. 70–75.

12. K. Umeda, Y. Enomoto, A. Mitsui, and K. Mannami (1993). Friction and wear of boride ceramics in air and water, *Wear*, 169, pp. 63–38.

13. T. E. Fischer and H. Tomizawa (1985). Interaction of tribochemistry and microfracture in the friction and wear of silicon nitride, *Proceedings of International Conference on Wear of Materials*, ASME, New York, pp. 22–32.

14. H. Ishigaki, R. Nagata, and M. Iwasa (1988). Effect of adsorbed water on friction of hot-pressed silicon nitride and silicon carbide at slow speed sliding, *Wear*, 121, pp. 107–116.

15. K. Demizu, R. Wadabayashi, and H. Ishigaki (1990). Dry friction of oxide ceramics against metals: The effect of humidity, *Tribol. Trans.*, 33(4), pp. 505–510.

16. S. Sasaki (1992). *Effects of Environment on Friction and Wear of Ceramics*, Bull. Mech. Engin. Lab., No. 58, Japan, ISSN 0374–2725, p. 109.

17. J. Briscoe and T. A. Stolarski (1993). Friction, in *Characterization of Tribomaterials*, ed. W. A. Glaeser, Butterworth-Heinemann, Boston, pp. 44–46.

18. C. S. Yust (1992). Tribological behavior of whisker-reinforced ceramic composite materials, in *Friction and Wear of Ceramics*, ed. S. Jahanmir, Marcel Dekker, New York, pp. 199–231.

19. S. Jahanmir and X. Dong (1992). Wear mechanisms of aluminum oxide ceramics, in *Friction and Wear of Ceramics*, ed. S. Jahanmir, Marcel Dekker, New York, pp. 15–49.

20. M. Woydt and K.-H. Habig (1989). High temperature tribology of ceramics, *Tribol. Int.*, 22(2), pp. 75–88.

21. C. Dellacorte and B. M. Steinmetz (1993). Tribological Comparison and Design Selection of High-Temperature Candidate Ceramic Fiber Seal Materials, STLE Preprint 93-TC-1E-2, Society of Tribology and Lubrication Engineering, Park Ridge, IL.

22. ANSI Standard PH 1.47–1972 (1972). Methods for Detecting the Degree of Lubrication on Processes Photographic Film by the Paper-Clip Friction Test, American National Standards Institute, Inc.

23. V. Buck (1986). Self-lubricating polymer cages for space-proofed bearings: Performance and roundness, *Tribol. Int.*, 19(1), pp. 25–28.

24. F. P. Bowden and D. Tabor (1986). *The Friction and Lubrication of Solids*, Pergamon Press, Oxford, p. 170.

25. V. A. Bely, A. I. Sviridenok, M. I. Petrokovets, and V. G. Savkin (1982). *Friction and Wear in Polymer-Based Materials*, Pergamon Press, Oxford, pp. 168–171.

26. P. H. Vroegop and R. Bosma (1985). Subsurface melting of nylon by friction-induced vibrations, *Wear*, 104, pp. 31–47.

27. Y. Yamada and K. Tanaka (1986). Effect of the degree of crystallinity on the friction and wear of poly(ethylene terephthalate) under water lubrication, *Wear*, 11, pp. 63–72.

28. N. S. Eiss and B. P. McCann (1992). Frictional instabilities in polymer-polymer sliding, *Tribol. Trans.*, 36(4), pp. 686–692.

29. G. M. Bartenev and V. V. Lavrentev (1981). *Friction and Wear of Polymers*, Elsevier, Amsterdam.

30. Ref. 28, p. 96.

31. J. Hanchi and N. S. Eiss, Jr. (1993). The Tribological Behavior of Blends of Poly-etheretherketone (PEEK) and Polyetherimide (PEI) and Elevated Temperatures. STLE Preprint 93-TC-1C-1. Society of Tribology and Lubrication Engineering, Park Ridge, IL.
32. M. M. Bilik (1964). *Wear and Friction for Metals and Plastics*, Nauka, Moscow, p. 87.
33. B. J. Briscoe and T. A. Stolarski (1979). Combined rotating and linear motion effects on the wear of polymers, *Nature*, 281(5728), pp. 206–208.
34. J. W. M. Mens and A. W. J. de Gee (1991). Friction and wear behavior of 18 polymers in contact with steel in environments of air and water, *Wear*, 149, pp. 255–268.
35. D. F. Moore (1975). Friction of elastomers, in *Principles and Applications of Tribology*, Pergamon Press, Oxford, pp. 62–85.
36. A. Schallamach (1963). A theory of dynamic rubber friction, *Wear*, 6, pp. 375–382.
37. J. Briscoe and T. A. Stolarski (1993). Friction, in *Characterization of Tribomaterials*, ed. W. A. Glaeser, Butterworth-Heinemann, Boston, pp. 46–48.
38. K. G. McLaren and D. Tabor (1963). Friction of polymers at engineering speeds: influence of speed, temperature, and lubricants, *Proceedings of Lubrication and Wear Convention*, Institute of Mechanical Engineering, London, pp. 210–215.
39. O. Yandell (1970). A new theory of hysteretic sliding friction, *Wear*, 17, pp. 229–244.
40. R. A. Burton and R. Gaines Burton (1988). Ultra-low wear in carbon matrix materials, in *Engineered Materials for Advanced Friction and Wear Applications*, eds. F. A. Smidt and P. J. Blau, ASM International, Materials Park, OH, pp. 95–99.
41. R. A. Burton and R. Gaines Burton (1989). *Friction and Wear of Glassy Carbon in Sliding Contact*, Final report of U.S. DOE Contract DE-AC02-88CE90027, p. 38.
42. J. Spreadborough (1962). The frictional behavior of graphite, *Wear*, 5, pp. 18–30.
43. R. I. Longley, J. W. Midgely, A. Strang, and D. G. Teer (1963). Mechanism of the frictional behaviour of high, low and non-graphitic carbon, *Proceedings of Lubrication and Wear Convention*, Institute of Mechanical Engineers, London, pp. 198–209.
44. R. D. Arnell, J. W. Midgely, and D. G. Teer (1964–1965). Frictional characteristics of pyrolytic carbon, *Proceedings of Lubrication and Wear Convention*, Institute of Mechanical Engineers, London, pp. 115–122.
45. P. J. Blau and R. L. Martin (1994). Friction and wear of carbon-graphite materials against steel and silicon nitride counterfaces, *Tribol. Int.*, 27(6), pp. 413–422.
46. W. T. Clark and J. K. Lancaster (1963). Breakdown and surface fatigue of carbons during repeated sliding, *Wear*, 6, pp. 467–482.
47. R. A. Burton (1992). Friction and wear of electrical contacts, in *ASM Handbook, Volume 18: Friction, Lubrication, and Wear Technology*, ASM International, Materials Park, OH, pp. 682–684.
48. R. Holm (1958). *Electric Contacts Handbook*, Springer-Verlag, Berlin.
49. D. V. Badami and P. K. C. Wiggs (1970). Friction and wear, in *Modern Aspects of Graphite Technology*, ed. L. C. F. Blackman, Academic Press, New York, pp. 224–255 (Chapter VI).
50. D. Tabor (1979). Friction of diamond, in *The Properties of Diamond*, ed. J. E. Field, Academic Press, London, pp. 325–350 (Chapter 10).
51. D. Tabor and J. E. Field (1979). Friction of diamond, in *The Properties of Diamond*, ed. J. E. Field, Academic Press, London, pp. 547–571 (Chapter 14).
52. S. V. Hainsworth and N. J. Uhure (2007). Diamond like carbon coatings for tribology: Production techniques, characterisation methods, and applications, *Intern. Mater. Rev.*, 52(3), pp. 153–174.
53. P. J. Blau, C. S. Yust, L. J. Heatherly, and R. E. Clausing (1989). Morphological aspects of the friction hot-filament-grown diamond thin films, in *Mechanics of Coatings*, eds. D. Downson, C. M. Taylor, and M. Godet, Elsevier, Amsterdam, pp. 399–407.

54. P. K. Gupta, A. Maishe, B. Bhushan, and V. V. Subramanium (1993). Friction and Wear Properties of Chemomechanically Polished Diamond Films, ASME Preprint 93-Trib-20.
55. P. J. Blau (1994). A comparison of the friction behavior of bulk diamond with that of diamond films and diamond-like carbon films, *Diamond Films Technol.*; 4(3), pp. 1–12.
56. M. N. Gardos (1994). Tribology and wear behavior of diamond, in *Synthetic Diamond: Emerging CVD Science and Technology*, eds. K. E. Spear and J. P. Dismukes, Wiley, New York, pp. 419–504.
57. P. V. Hobbs (1974). *Ice Physics*, Clarendon Press, Oxford, p. 61.
58. H. H. Schulz and A. Knappwost (1968). Die Festkorpoerreibung des Eises als Relaxationseffekt, *Wear*, 11, pp. 3–20.
59. Ref. to Rennie in D. Dowson (1979). *History of Tribology*, Longman, London, p. 229.
60. F. P. Bowden and T. P. Hughes (1939). The mechanism of sliding on ice and snow, *Proc. Royal Soc. A*, 172, pp. 280–298.
61. F. P. Bowden (1953). Friction on snow and ice, *Proc. Royal Soc. A*, 217, p. 462.
62. H. S. Kong and M. F. Ashby (1991). *Case Studies in the Application of Temperature Maps for Dry Sliding*, Cambridge University Engineering Department Report, pp. 20–24.
63. N. Nalence and B. C. Rostro (2006). Cold play, *Tribology and Lubr. Tech.*, December, pp. 22–32.
64. H. Bluhm, D. F. Ogletree, C. F. Fadley, Z. Hussain, and M. Salmeron (2002). The premelting of ice studied with photoelectron spectroscopy, *J. Phys.: Condens. Matter*, 14(8), pp. L227–L233.
65. J. Collins (2002). Slip and slide, coast and glide, US Airways *Attaché Magazine*, February, pp. 22–26.
66. N. K. Sinha and A. Norheim (2000). A new retrospect of snow and ice, tribology and aircraft performance, *Snow Engineering—Recent Advances and Developments*, eds. E. Hjorth-Hansen, I. Holand, S. Loset, and H. Norem, A. A. Balkema Pub., Rotterdam, pp. 427–435.
67. S. J. Calabrese and S. F. Murray (1982). Methods of evaluating materials for icebreaker hull coatings, *Selection and Use of Wear tests of Coatings*, ASTM STP 769, ed. R. G. Bayer, ASTM, West Conshohocken, PA, pp. 157–173.
68. B. Bhushan and B. K. Gupta (1991). *Handbook of Tribology: Materials, Coatings, and Surface Treatments*, McGraw Hill, New York.
69. K. Budinski (1988). *Surface Engineering for Wear Resistance*, Prentice Hall, Englewood Cliffs, NJ.
70. J. Halling (1979). Surface coatings, *Tribol. Int.* 12(10), pp. 203–208.
71. I. M. Hutchings (1992). Surface engineering in tribology, in *Tribology: Friction and Wear of Engineering Materials*, CRC Press, Boca Raton, FL, pp. 213–240 (Chapter 8).
72. D. Dowson (1979). *History of Tribology*, Longman, London, p. 43.
73. Y. Berthier (1990). Experimental evidence for friction and wear modelling, *Wear*, 139, pp. 77–92.
74. N. M. Alexeyev (1990). On the motion of materials in the border layer in solid state friction, *Wear*, 139, pp. 33–48.
75. H. Heshmat, O. Pinkus, and M. Godet (1988). On a Common Tribological Mechanism Between Interacting Surfaces, STLE Preprint 88-AM-88-2, STLE, Park Ridge, IL.
76. D. R. Criswell, J. F. Lindsay, and D. L. Reasoner (1975). Seismic and acoustic emissions of a booming dune, *J. Geophys. Res.*, 80, pp. 4963–4974.
77. H. M. Jaeger and S. R. Nagel (1992). Physics of the granular state, *Science*, 225, pp. 1523–1530.

78. M. Yu Tanakov, L. I. Trusov, M. V. Belyi, V. E. Bulgakov, and V. G. Gryaznov (1993). Elastically stressed state in small particles under conditions of Hertzian contacts, *J. Phys. D. Appl. Phys.* 26, pp. 997–1001.
79. R. L. Brown and J. C. Richards (1966). *Principles of Powder Mechanics*, Pergamon Press, Oxford.
80. R. A. Bagnold (1966). The shearing and dilatation of dry sand and the "singing" mechanism, *Proc. Royal Soc. A*, 295, pp. 219–232.
81. Ref. 79, p. 87.
82. M. J. Adams (1992). Friction of granular non-metals, in *Fundamentals of Friction: Macroscopic and Microscopic Processes*, eds. I. L. Singer and H. M. Pollock, Kluwer, Dordrecht, Netherlands, pp. 183–207.
83. J. Feda (1982). *Mechanics of Particulate Materials*, Elsevier, Amsterdam.
84. K. M. Frye and C. Marone (2002). Effect of humidity on granular friction at room temperature, *J. Geophys. Research*, 107(B11), pp. 2309–2326.

6 Lubrication to Control Friction

All things and everything whatsoever however thin it be which is interposed in the middle between objects that rub together lighten the difficulty of their friction.

Leonardo da Vinci's Notebooks

The use of lubricants by far predates da Vinci's observation that interposing materials between sliding bodies can reduce their friction. Lubrication involves not only the selection and formulation of lubricants, but also the design of contact geometry and the methods to deliver, filter, and condition them. Broadly speaking, lubricants can be solids, liquids, or gases. Solid lubricants can be in the form of thin films, constituent phases in composite materials, or powders. The definition of *lubricant*, developed by the Organization for Economic Cooperation and Development (OECD), International Research Group on Wear of Engineering Materials,[1] is as follows:

> *Lubricant*—Any substance interposed between two surfaces in relative motion for the purpose of reducing the friction and wear between them.

Note that the OECD definition for lubricant contains the phrase "interposed … for the purpose of," implying that a substance can be considered a lubricant only if it is applied intentionally. Some materials are considered *lubricious* (the more correct term is *lubricative*), that is, they tend to reduce friction, but by the OECD definition, lubricative substances are not necessarily lubricants unless they are applied intentionally. Lubricative materials include soap, seawater, sputum (spit), blackberry jelly, perspiration, Teflon™, and waxed paper. The oxide that forms on copper in air is lubricative,[2] but it is not strictly a lubricant by some definitions unless it was applied intentionally. It might also be argued that a "traction fluid" used in wet clutches is not a lubricant *per se* even though it may have some lubricative properties.

Various materials, some seemingly unlikely choices, have been used as lubricants. Consider, for example, animal fat, "dolphin head oil," refrigerants, liquid metals, slurries, hand lotion, talcum powder, petroleum jelly, soap, water, and liquid hydrogen. The magnetic field that is used to support a magnetic bearing, although reducing the frictional contact between the surfaces of bearing, is not a lubricant because it is not a *substance*. In addition to friction reduction, lubricants also carry away heat and wear particles, and can serve as the means to distribute corrosion inhibitors and biocides. Sometimes the taste and smell of lubricants also affects their selection, like the lubricants used to draw beverage cans.[3]

Dowson[4] has written the most comprehensive historical account of the development of lubrication technology. The story of lubricants dates back thousands of years and links progress to significant enabling events in the development of civilizations, notably the evolution of transportation technology. Were it not for lubricants, it is unlikely that we would have had a worldwide economy that depends on travel and the transportation of finished goods and bulk commodities over long distances. Many of the most significant twentieth century achievements in lubrication technology have been described in a book edited by Sibley and Kennedy.[5]

In this chapter, we principally consider the frictional characteristics of liquid and solid lubricants and their interaction with bearing materials. A good review of gas lubrication was prepared by Pan,[6] and comprehensive reviews of liquid and solid lubrication are available in texts and handbooks.[7–10] Short courses and practitioner certifications are offered by technical societies such as the Society for Tribologists and Lubrication Engineers (STLE).

6.1 LUBRICATION BY LIQUIDS AND GREASES

6.1.1 Liquid Lubrication

Lubricating films should support the pressure between opposing surfaces, separate them, and reduce the sliding or rolling resistance in the interface. There are several ways to accomplish this. One way is to design a bearing in such a way as to entrain the fluid and create sufficient pressure to separate the opposing surfaces. This method relies on fluid mechanics and the liquid's properties such as viscosity and the dependence of viscosity on temperature and pressure. Another way to reduce friction is to formulate a liquid lubricant in such a way that chemical species within it react with the surface(s) of the bodies to form lubricative films. Surface species need not react with the lubricant but can serve as catalysts to promote the reactions that produce a film.

Klaus and Tewksbury[11] reviewed the following characteristics of liquid lubricants:

- Viscosity and its relationships to temperature and pressure
- Vapor pressure
- Density
- Bulk modulus
- Gas solubility
- Foaming and air entrainment tendencies
- Thermal properties and stability
- Oxidation stability

Of these liquid lubricants, it is especially important to understand the concept of viscosity. Viscosity is expressed as either *absolute viscosity* or *kinematic viscosity*. Sometimes the term *dynamic viscosity* is used in place of absolute viscosity. The absolute viscosity is the ratio of the shear stress causing a flow to the resultant velocity gradient. If one imagines the stack of layers that is free at the top but fixed at the bottom, then a force applied to the top will create the shear situation shown in Figure 6.1. The gradient in this case is the velocity at the top layer divided by the film thickness. The shear stress on the element is the shearing force divided by the

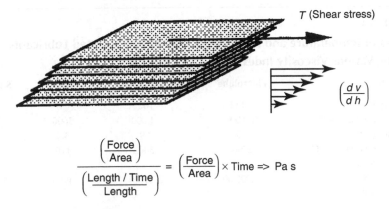

FIGURE 6.1   The basic concept of viscosity.

area over which it is applied. In SI units, the standard unit for absolute viscosity is the Pascal-seconds (Pa · s). Sometimes, the unit *centipoise*, abbreviated cP, is used (1 cP = 0.001 Pa · s). The *kinematic viscosity* is the ratio of the absolute viscosity to the density. The unit of kinematic viscosity works out to be mm²/s. In the old "CGS" system of units, 1 mm²/s was called a *centistoke*, abbreviated cSt.

The viscosity of a fluid usually decreases with temperature and therefore can support less load as temperature rises. The term *viscosity index*, abbreviated VI, is a metric to express this variation. The higher the VI, the lesser the change in viscosity with temperature. One of the types of additives used to reduce the sensitivity of lubricant viscosity to temperature changes is called a *VI improver.* ASTM test method D 2270 explains a procedure used to calculate the VI. This procedure, although fairly awkward, is nevertheless in common use. It is also described step-by-step in the articles by Litt[12] and by Klaus and Tewksbury.[11] The method is based on two test oils (neither of which is still made), uses two different methods of calculation (depending on the magnitude of VI), and relies on charts and tables. Modern methods use computer programs to simplify the tedium of this calculation, but in Litt's opinion the VI is neither a very convenient nor a useful quantity despite its ingrained usage in the oil industry.

ASTM Standard D 341 recommends using the Walther equation to represent the dependence of lubricant viscosity on temperature. Defining Z as the viscosity in cSt plus a constant (typically ranging from 0.6 to 0.8 with ASTM specifying 0.7), T equal to the temperature in Kelvin or Rankin, and A and B being constants for a given oil, then

$$\log_{10}(\log_{10} Z) = A + B \, (\log_{10} T) \tag{6.1}$$

Sanchez-Rubio et al.[13] have suggested an alternative method in which the Walther equation is used. In this case, they define a viscosity number (VN) as follows:

$$VN = \left[1 + \frac{3.55 + B}{3.55} \right] \cdot 100 \tag{6.2}$$

224 Friction Science and Technology: From Concepts to Applications

TABLE 6.1

Effects of Temperature and Pressure on Viscosity of Selected Lubricants Having Various Viscosity Indexes[a]

Quantity	Fluorolube	Hydrocarbon	Ester	Silicone
Viscosity index	−132	100	151	195
Viscosity (cSt) at −40°C	500,000	14,000	3600	150
Viscosity (cSt) at −100°C	2.9	3.9	4.4	9.5
Viscosity (cSt) at −40°C and 138 MPa	2700	340	110	160
Viscosity (cSt) at −40°C and 552 MPa	>1,000,000	270,000	4900	48,000

[a] All these fluids have viscosities of ~20 cSt at 40°C and at 0.1 MPa pressure.

Source: Klaus, E.E. and Tewksbury, E.J. in *The Handbook of Lubrication (Theory and Practice of Tribology)*, CRC Press, Boca Raton, FL, 1984, 229–254.

The value of 3.55 was selected because lubricating oils with a VI of 100 have a value about equal to −3.55. Using this expression implies that VN = 200 would correspond to an idealized oil whose viscosity has no dependence of viscosity on temperature (i.e., $B = 0$).

The pressure to which the oil is subjected can also strongly influence its viscosity. Barus' equation is commonly used to express the relationship between dynamic viscosity and hydrostatic pressure p, and can be represented by

$$\eta = \eta_o \exp(\alpha p) \tag{6.3}$$

where η and α vary with the particular oil. This relationship has been modified for use in special situations such as very high, localized pressures. Coy[14] has reviewed these in the context of internal combustion engines.

Table 6.1, using data from Klaus and Tewksbury, illustrates the wide range of viscosities possible for several liquid lubricants under various temperatures and pressures. The viscosity indices for these oils range from −132 to 195. Viscosity has a significant effect on determining the regime of lubrication and the resultant friction coefficient. Similar to the effect of strain rate on the shear strength of certain metals, such as aluminum, the rate of shear in the fluid can also alter the viscosity of a lubricant.

Ramesh and Clifton[15] used a plate impact device to study the shear strength of lubricants at strain rates up to 900,000 s^{-1} and found significant effects of shear rate on the critical shear stress. In a *Newtonian fluid*, the ratio of shear stress to shear strain does not vary with stress, but there are other cases, such as for greases and solid dispersions in liquids, where the viscosity varies with the rate of shear. Such fluids are termed *non-Newtonian* and the standard methods for measuring viscosity cannot be used.

Table 6.2 lists the viscosities and viscosity index of several Society of Automotive Engineers liquid lubricant grades. The "W" in the designation stands for "winter"

TABLE 6.2

Properties of Several Engine Lubricants Grades

Grade	Viscosity at 40°C (cSt)	Viscosity at 100°C (cSt)	VI
5W-20	38.0	6.92	140
10W-30	66.4	10.2	135
10W-40	77.1	14.4	193
10W-50	117	20.5	194

Source: Klaus, E.E. and Tewksbury, E.J. in *The Handbook of Lubrication (Theory and Practice of Tribology)*, CRC Press, Boca Raton, FL, 1984, 229–254.

TABLE 6.3

Typical Values of (*ZN/p*) for Lubricated Components

Component	Absolute Viscosity (cP)	(*ZN/p*)
Automobile main bearing	7	15
Marine engine main bearing	30	20
Stationary steam engine main bearing	15–60	20
Steam turbine main bearing	2–16	100
Rotary pump shaft bearing	25	200

Source: Hall, A.S., Holowenko, A.R., and Laughlin, H.G. in *Machine Design*, Schaum's Outline Series, McGraw-Hill, New York, 1961, 279.

when the viscosity is higher in the presence of low ambient temperatures. Viscosity indexes in such oils typically range from 100 to 200, and viscosities can vary by more than an order of magnitude.

The concept of *lubrication regimes* is an important one. They determine the effectiveness of fluid film formation and, hence, surface separation and friction. In the last decade of the nineteenth century, Richard Stribeck was a professor of machine engineering at the Technische Hochschule in Dresden, Germany. By 1900, he had moved to Berlin where he began to develop a systematic method to depict regimes of journal bearing lubrication. This approach linked the properties of lubricant viscosity (η), the rotational velocity of a journal (ω), and the contact pressure (p) with the coefficient of friction. Based on the work of Hersey, McKee, and others, the dimensionless group of parameters has evolved into the more notation (*ZN/p*), where Z is viscosity, N is rotational speed, and p is pressure. A detailed discussion of the historical development of this important parameter may be found in a chapter in Hersey's book[16] titled *Dimensions and ZN/p*. Table 6.3 lists several representative values of (*ZN/p*) for operating systems.

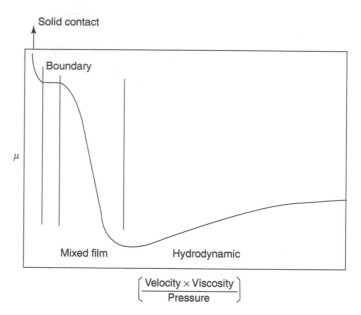

FIGURE 6.2 The general shape of the Stribeck curve.

The so-called *Stribeck curve*, exemplified in Figure 6.2, has been widely used in the design of bearings and to explain various types of behavior in the field of lubrication. At high pressures, or when the lubricant viscosity and speed are very low, surfaces may touch, leading to high friction. In that case, friction coefficients are typically in the range of 0.5–2.0. The level plateau at the left of the curve represents the *boundary lubrication regime* in which friction is lower than that for unlubricated sliding contact ($\mu = 0.05$ to about 0.15). The drop-off in friction is called the *mixed film regime*. The mixed regime refers to a combination of boundary lubrication with *hydrodynamic* or *elastohydrodynamic* lubrication. Beyond the minimum in the curve, hydrodynamic and elastohydrodynamic lubrication regimes are said to occur. Friction coefficients under such conditions can be very low. Typical friction coefficients for various types of rolling element bearings range between 0.0010 and 0.0018.

The conditions under which a journal bearing of length L, diameter D, and radial clearance C (bore radius minus bearing shaft radius) operates in the hydrodynamic regime can be summarized using a dimensionless parameter known as the Sommerfeld number S, defined by

$$S = \frac{\eta NLD}{P}\left(\frac{R}{C}\right)^2 \tag{6.4}$$

where P is the load on the bearing perpendicular to the axis of rotation, N the rotational speed, η the dynamic viscosity of the lubricant, and R the radius of the bore. The more concentrically the bearing operates, the higher the value of S, but as S approaches 0, the lubrication may fail, leading to high friction. Sometimes Stribeck curves are plotted using S instead of (ZN/p) as the abscissa. Raimondi and his coworkers[17] added leakage considerations when they developed design charts in which the logarithm of the Sommerfeld number is plotted against the logarithm of either the friction coefficient or the dimensionless film thickness.

Using small journal bearings, McKee developed the following expression for the coefficient of friction μ on the basis of the journal diameter D, the diametral clearance C, and an experimental variable k, which varies with the length to diameter ratio (L/D) of the bearing:[18]

$$\mu = \left(4.73 \times 10^{-8}\right)\left(\frac{ZN}{P}\right)\left(\frac{D}{C}\right) + k \tag{6.5}$$

Exhibiting a shape similar to that in Figure 6.2, the value of k is about 0.015 at $(L/D) = 0.2$, drops rapidly to a minimum of about 0.0013 at $(L/D) = 1.0$, and rises nearly linearly to about 0.0035 at $(L/D) = 3.0$.

A simpler expression than Equation 6.5, presented by Hutchings,[19] can be used for bearings that have no significant eccentricity:

$$\mu = \frac{2\pi}{S}\frac{h}{R} \tag{6.6}$$

where S is the Sommerfeld number, h the mean film thickness, and R the journal radius. With good hydrodynamic lubrication and good bearing design, μ can be as low as 0.001.

Hydrodynamic lubrication, sometimes called *thick-film lubrication*, depends on the development of a converging wedge of lubricant in the inlet of the tribointerface to generate a pressure profile that forces contact surfaces apart. When the elastic deformation of the solid bodies is similar in extent to the thickness of the lubricant film, then *elastohydrodynamic lubrication* is said to occur. This latter regime is common in rolling element bearings and gears where high Hertzian contact stresses occur. If the contact pressure exceeds the elastic limit of the surfaces, plastic deformation and increasing friction occur.

One parameter for portraying various lubrication regimes is called the *specific film thickness* or *lambda ratio*. It is defined as the ratio of the minimum film thickness in the interface (h) to the composite root-mean-square (rms) surface roughness σ^*:

$$\Lambda = \frac{h}{\sigma^*} \tag{6.7}$$

where the composite surface roughness is calculated from the rms roughness ($\sigma_{1,2}$) of surfaces 1 and 2, respectively:

$$\sigma^* = \sqrt{\sigma_1^2 + \sigma_2^2} \tag{6.8}$$

Note that σ^* is not simply the average roughness of both surfaces but rather involves the sum of both rms values, since the surfaces are stacked upon each other. For the boundary regime, $\Lambda \ll 1$. For the mixed regime $1 < \Lambda < 3$. For the hydrodynamic regime, $\Lambda \gg 6$, and for the elastohydrodynamic regime, $3 < \Lambda < 10$. Thus, it is possible to alter the position of the Stribeck curve for a bearing by changing the surface finish and film thickness. The determination of Λ requires both surface roughness data and film thickness data, the former being relatively easy to obtain from profile measuring instruments. Film thickness is another matter.

Perhaps the best-known mathematical relationships for pressure distribution within a fluid film were derived by Osbourne Reynolds (1842–1912). He is considered by many practitioners to be the founder of the science of lubrication. Reynolds received his early education from his father, the Rev. Osbourne Reynolds, a prominent educator of that time. The junior Reynolds was eventually appointed to the Manchester Chair at Owens College in England, and he pursued rigorous studies of lubrication, which his students reportedly felt "were too severe and his lectures too difficult to follow." Sneck and Vohr[20] and Khonsari and Booser[21] provide good introductions to the *Reynolds equation* in its various forms, a subject that is beyond the scope of this book but that is often the starting point for many studies of lubrication. The general Reynolds equation is a differential equation that relates pressure, velocity, viscosity, and contact dimensions. Fortunately, analytical expressions for the film thickness in different kinds of bearings have been derived, and experimental tribologists have devised clever methods to measure it. The use of experiments to measure EHL film thickness and their comparison with predictions has been reviewed by Spikes.[22]

A calculation for the minimum EHL film thickness in elliptical contacts (h_{min}) was presented in the much cited work of Hamrock and Dowson:[23]

$$h_{min} = 3.63 R_x \frac{U^{0.68} G^{0.49}}{W^{0.073}} (1 - e^{-0.68k}) \tag{6.9a}$$

Symbols are defined as follows:

$G = \alpha E$

$U = \dfrac{\mu_o(u_1 - u_2)}{2ER}$

$W = \dfrac{w}{ERL}$

$\dfrac{1}{E} = \dfrac{1}{2}\left(\dfrac{1-v_1^2}{E_1} + \dfrac{1-v_2^2}{E_2}\right)$

μ_o = viscosity at the inlet temperature
$\mu = \mu_o e^{\alpha p}$, α = pressure viscosity coefficient and p = pressure
w = total load on the cylinder
L = length of the cylinder
R_x = the contact radius is the direction of motion
u_1, u_2 = surface velocities relative to the contact region
k = ratio of contact dimensions perpendicular to and parallel to the direction of motion

Dowson and Higginson[24] had earlier derived a similar expression for line contacts for two rolling elements having radii R_1 and R_2:

$$\left(\frac{h_{min}}{R}\right) = 2.65\frac{G^{0.54}U^{0.7}}{W^{0.13}} \tag{6.9b}$$

where

$$R = \frac{R_1R_2}{R_1 + R_2} \text{ when both surfaces are convex}$$

$$R = \frac{R_1R_2}{R_1 - R_2} \text{ when the surface with a larger radius is concave}$$

The estimation of film thickness is not straightforward when the bearing situation does not allow a constant film thickness to develop. For example, changes in velocity or load during operation can cause the operating point to travel up and down the Stribeck curve. Furthermore, as will be discussed in the next chapter, some surfaces have been engineered with patterns of dimples or pores, and the film thickness changes locally.

In contrast to EHL, boundary lubrication phenomena derive from a combination of liquid properties, thin solid film properties, and substrate characteristics. It produces friction coefficients that are lower than those for unlubricated sliding but higher than those for hydrodynamic lubrication (typically, $0.08 < \mu < 0.15$). In 1969, the American Society of Mechanical Engineers published a comprehensive appraisal of world literature involving boundary lubrication.[25] Fein[26] defined *boundary lubrication* as *a condition of lubrication in which friction and wear between two surfaces in relative motion are determined by the properties of the surfaces, and by the properties of the lubricant other than bulk viscosity.*

Briscoe and Stolarski[27] have reviewed friction under boundary-lubricated conditions. They cited the earlier work of Bowden, who gave the following expression for the friction coefficient (μ) under conditions of boundary lubrication:

$$\mu = \beta\mu_a + (1 - \beta)\mu_1 \tag{6.10}$$

where the adhesive component μ_a and the viscous component of friction μ_1 are given in terms of the shear stress of the adhesive junctions in the solid (metal) τ_m and the shear strength of the boundary film τ_1 under the influence of a contact pressure σ_p:

$$\mu_a = \frac{\tau_m}{\sigma_p} \quad \text{and} \quad \mu_1 = \frac{\tau_1}{\sigma_p} \tag{6.11}$$

The parameter β is called the *fractional film defect*. Briscoe and Stolarski provided the following expression for β:

$$\beta = 1 - \exp\left\{-\left[\frac{(30.9 \times 10^5)T_m^{1/2}}{VM^{1/2}}\right]\exp\left(\frac{-E_c}{RT}\right)\right\} \tag{6.12}$$

where M is the molecular weight of the lubricant, V the sliding velocity, T_m the melt-ing temperature of the lubricant, E_c the energy to desorb the lubricant molecules, R the universal gas constant, and T the absolute temperature.

Moore[28] took a similar approach to modeling boundary lubrication of metals. He assumed that the resultant friction force comprised (a) solid friction at the asper-ity peaks, (b) liquid friction in the voids between asperities, and (c) plowing of hard asperities through the softer surface of the two. Thus, the friction coefficient for boundary lubrication, μ_{BL}, was given by

$$\mu_{BL} = \alpha_W \left(\frac{\tau_s}{p_{ave}} \right) + (1 - \alpha_W) \frac{\tau_{liq}}{p_{ave}} + \mu_{PL} \tag{6.13}$$

where α_W = fraction of the area with solid contact, τ_s = shear strength of the solid, τ_{liq} = shear strength of the lubricant, p_{ave} = the average contact pressure (less than the yield pressure of the solid but more than the hydrostatic pressure of the liquid in the voids), and μ_{PL} = the contribution from plowing. This expression was used to derive the ratio between the friction coefficients for boundary-lubricated sliding and unlubricated (dry) sliding, namely:

$$\frac{\mu_{BL}}{\mu_{dry}} = \alpha_W + (1 - \alpha_W) \frac{\tau_{liq}}{\tau_s} \tag{6.14}$$

Since the rightmost term in Equation 6.14 is very small, boundary-lubricated fric-tion is much less than dry friction, as is the common observation, and the effect of boundary lubrication is proportional to the load supported by solid asperities and trapped liquid.

Graphical methods have been developed to help select boundary lubricants and to help simplify the task of bearing designers. Most are based on the design param-eters of bearing stress (or normal load) and velocity. One method, developed by Glaeser and Dufrane[29] in the mid 1970s, involves the use of design charts for dif-ferent bearing materials, which plot bearing pressure on one axis and velocity on the other. Figure 6.3 illustrates this approach, showing how the lines separating the friction regions vary for two different sleeve bearing materials.

A similar approach was used in developing the so-called "IRG transitions dia-grams" (subsequently abbreviated ITDs).[30] It evolved in the early 1980s, was applied to various bearing steels, and is still being used to define the conditions under which boundary-lubricated tribosystems operate effectively. Instead of pressure, load is plotted on the ordinate. Three regions of ITDs are defined in terms of their fric-tional behavior: region I, in which the friction trace is relatively low and smooth, region II, in which the friction trace begins with a high level then settles down to a lower, smoother level, and region III, in which the friction trace is irregular and remains high. The transitions between regions I and II or between regions I and III are described as the collapse of the liquid lubricating film. The locations of transition boundaries for steels depended more on the surface roughness of the materials and the composition of the lubricants than on the microstructure and composition of the

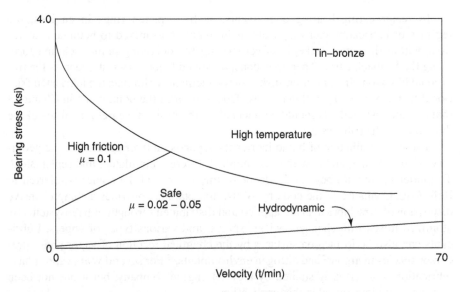

FIGURE 6.3 Example of a design chart developed by Glaeser and Dufrane. (Diagram adapted from Glaeser, W.A. and Dufrane, K.F., *Machine Design*, April 6, 207, 1978.)

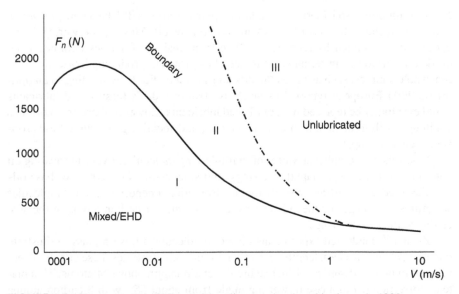

FIGURE 6.4 Example of an IRG transition diagram developed in the early 1980s.

alloys. The following test geometries were used to develop ITDs: four-ball machines, ball-on-cylinder machines, crossed-cylinders machines, and flat-on-flat testing machines (including flat-ended pin-on-disk). Figure 6.4 illustrates an ITD showing its general features. One four-ball testing study by Odi-Owei and Roylance,[31] for example, used the ITD concepts to establish how alumina particle contamination in a base oil altered the transition conditions for steel couples.

In reciprocating bearing components, such as piston rings in engines, the regimes of lubrication can vary from hydrodynamic to mixed to boundary conditions within the same stroke, in effect moving the operating point back and forth along the Stribeck curve. Therefore, constant-speed laboratory tests, such as the traditional block-on-ring or pin-on-disk, cannot adequately simulate the lubricated frictional behavior of many real machines. This reinforces the point made in Chapter 3 that friction test methods should be selected on the basis of a careful analysis of the dynamics of the tribosystem.

The proper function of liquid lubricants depends on their properties, the geometry of the contact, and how they are applied, filtered, circulated, and replenished. The latter factors are beyond the scope of this book, and the reader is referred to Refs 7–10. Chapter 6 of the book by Wills,[7] for example, provides a comprehensive discussion of how lubricants are applied and distributed throughout tribosystems by atomized mists, thin channels, wicks, sprays, and various types of wipers. Lubricants can also be formed on surfaces by the chemical reaction of vapor-phase precursor species in argon[32] and nitrogen environments.[33] For several years vapor-phase lubrication was seriously studied by a diesel engine company, but it has not been commercially introduced in that application.

6.1.2 Composition of Liquid Lubricants

Most commonly used lubricating oils are petroleum-based. The *word petroleum* comes from the Latin *petra* for rock and *oleum* for oil. Most of the world's lubricating oil is obtained by refining distillate from residual fractions obtained from crude oil;[34] however, in recent years interest has grown in producing lubricants from renewable sources, such as vegetable oils and synthetic fluids. According to an optimistic 2004 European report,[35] as much as 90% of today's fossil-based lubricants could eventually be replaced with bio-based lubricants. However, there are enormous challenges still to be met: technical, economic, and social (e.g., growing food crops versus energy crops).

Base oils are complex mixtures of multiple-ring molecules. Every lubricating oil contains aromatic rings, naphthenic rings, and side chains. Consequently, base oils are classified as paraffinic, naphthenic, or aromatic, depending on their molecular structures, the length of the side chains, and the ratio of carbon atoms in the side chains to those in the rings.

Zisman[36] conducted experiments on monomolecular films on glass to illustrate the effect of carbon chain length on friction. Figure 6.5 illustrates these effects. There seemed to be no advantage of increasing the chain length above 14 atoms. This plateau corresponded to a rise in wetting angle from about 55° (with 8 carbon atoms) to a maximum of 70° above chain lengths of 14 carbon atoms. Buckley[37] described similar experiments on the lubrication of tungsten single crystals, which showed a decrease in friction coefficient by about a factor of 2 as the number of carbon atoms in the chain increased from 1 to 10. The effects of increasing molecular weight were also observed for pin-on-disk tests of high (100,000–5,000,000) molecular weight polyethylene oxide polymers.[38] In addition to petroleum oils, Rabinowicz[39] listed the following types of liquid lubricants: polyglycols, silicones, chlorofluorocarbons, polyphenyl ethers, phosphate esters, and dibasic esters.

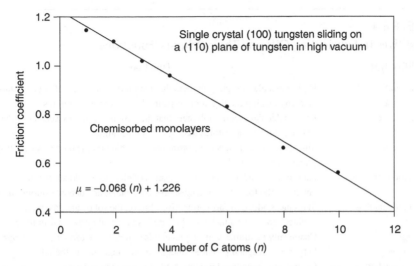

FIGURE 6.5   The effects of hydrocarbon chain length on the average kinetic friction coefficient. (D. H. Buckley's verification of Hardy's results.)

Fresh oil base stocks, like plain mineral oil without additives, are sometimes referred to as *neat oils*. Additives to oil base stocks are used to alter their properties and make them more suitable for certain applications. A specially formulated group of additives for use with a base oil is called an *additive package*. Some companies specialize in supplying additive packages to the oil industry. Some of these additive packages are used to account for variations in the quality of the crude oil used to produce base stocks; therefore, the formulation of additive packages is an ongoing activity. Ten types of additives and their functions are listed in Table 6.4. Comprehensive reviews of oil additives may also be found in publications by Liston,[40] Rizvi,[41] and Rudnick.[42]

Extreme-pressure (EP) additives, such as calcium sulfonate, cause the formation of protective layers on highly loaded bearing surfaces. These compounds commonly contain chlorine, sulfur, and phosphorus, and are designed to react with iron-bearing surfaces. Phosphorus, for example, can react with frictional hot spots on the surface of ferrous bearing surfaces to form low-melting-point phosphide eutectics and thus reduce friction and wear. It is important that additives react very quickly on the bearing surfaces because films removed mechanically during sliding contact must be immediately replenished to maintain stable frictional behavior. Well-functioning films should be stable when formed, adherent to the surface, and easily sheared. These are also the desirable characteristics for solid lubricants, a subject discussed in Section 6.2.

Friction modifiers and antiwear additives to oils have been the subject of extensive proprietary research in oil companies. Only occasionally do articles describing the chemistry and behavior of antifriction additives reach the open literature. One such article by Tung et al.[43] described the screening of various compounds and their combinations using a reciprocating laboratory test. High-chromium (SUJ2) steel was used as the slider, and several other steels, including ion-implanted steels, were used

TABLE 6.4

Additives to Lubricating Oils and Their Primary Functions

Additive Type	Function
Pour point depressors	High-molecular-weight polymers that inhibit the formation of wax crystals, thereby making the liquid more pourable at lower temperatures
VI improvers	High-molecular-weight polymers that increase the relative viscosity of the oil more at high temperatures than at low temperatures
Defoamants	Silicon polymers at low concentrations, which retard the tendency of oils to foam when agitated
Oxidation inhibitors	Substances added to reduce the oxidation of oils exposed to air, thereby reducing the formation of undesirable compounds and deposits during running
Corrosion inhibitors	Substances added to form protective films on the solid surfaces, which reduces corrosive attack by other species in the oil or the environment
Detergents	Chemically neutralize certain precursors to reduce the formation of deposits
Dispersants	Disperse or suspend potential sludge-forming materials in the oil
Antiwear additives	Long-chain, boundary lubrication additives to reduce wear
Antifriction additives	Similar to antiwear additives in that they enhance contact surface lubricity
Extreme-pressure additives	Form oil-insoluble surface films that help to bear high contact pressures and improve wear and friction as well

Source: Adapted from Wills, J.G. in *Lubrication Fundamentals*, Marcel Dekker, New York, 1980.

as the counterface alloy. One percent additions of four different friction modifiers to commercial engine oils of various viscosity grades were used as follows:

FM-1: bis(isoctylphenyl)-dithiophosphates with molybdenum
FM-2: molybdenum disulfide compound dispersed in an organic carrier
FM-3: organic sulfur fatty oil
FM-4: a sulfur-free organomolybdenum compound

FM-1 was claimed to be the most effective of the additives tested in reducing friction. Figure 6.6 shows the effects of various additives on the friction coefficients for reciprocating SUJ2 steel couples (50 Hz, 50–300 N load, Hertzian point contact) in three different oils at 150°C. In contrast to the other two oils, the friction of the SAE 40 oil seemed little changed by the additives. The SAE 30 and SAE 20W-40 oils did not respond equally to the four additives, suggesting that the other constituents in these oils interacted differently with the chemistry of the additives to produce quite different effects on friction.

One of the most popular antiwear additives is zinc dialkyldithiophosphate (sometimes called ZDP or ZDDP). Typically, about 1–1.5% is added to reduce wear by forming films on contact surfaces during use. These sub-micrometer-thick films are not necessarily homogeneous, nor are they evenly distributed on surfaces, and this complicates the task of accurately analyzing their structure, composition, and functionality. In recent years, there has been concern that the sulfur and phosphorous in ZDDP will poison exhaust gas catalytic converters and so there has been

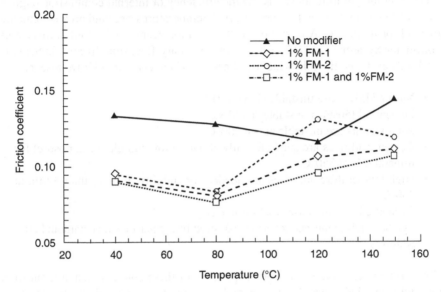

High-carbon chrome steel couples
Reciprocating tests, point contact, 4 h
Base oil SAE 30, API SF/CD
FM-1 = Mo-containing bisdithiophosphates
FM-2 = Organic molybdenum disulfide compound

FIGURE 6.6 Effects of antifriction additives on friction coefficient for reciprocating tests of steel on steel.

an effort to formulate suitable substitutes for ZDDP. A two-article series reviewing the characteristics and formation of ZDDP films was published in 2005 by Spikes' group at Imperial College in London.[44,45]

The extensive use of Fe-based alloys in bearings has prompted extensive work on antiwear additives that react with ferrous surfaces, but interest in using lightweight materials for energy-efficient transportation systems has prompted work on additive formulations that work with aluminum, titanium, and magnesium alloys. In an ideal world, the same oil formulations would work to lubricate and protect surfaces of both traditional and lightweight alloys, but tribochemistry generally does not allow it. However, results of recent piston ring-on-flat experiments at Oak Ridge National Laboratory, using titanium alloys (Ti–6Al–4V) as candidate cylinder liner materials, have shown that ZDDP can be made to form beneficial, lubricating films if the surface of the Ti alloy is first activated by oxidization in air.[46] In fact, nontreated Ti surfaces had as much as six orders of magnitude higher wear rates whereas the friction and wear of the oxidized Ti equaled or bettered that of traditional cast iron against a chrome-plated piston ring.

The complexity of today's additive packages makes it difficult to predict exactly how the oil chemistry changes after exposure to operating conditions, and extensive research has been conducted on this subject, much of it proprietary. There are tribochemical effects from fuel residues, combustion products like soot and ash, and reaction products due to the wear and corrosion of the materials that come in contact with the lubricant. Effects of temperature and oxidative degradation are also of concern.[47] In engines, an oil's residence time in the hottest sections of the engine may be more critical in determining its stability than the mean temperature of the sump or other lower-temperature locations. As the efficiency of internal combustion engines is driven higher by the need to conserve fuel, temperatures rise, and greater demands are made on lubricants. A workshop on "Lubrication Technologies for Future Energy Conversion Systems" held at Northwestern University, Evanston, Illinois, in September 1992 developed the following list of needs and concerns in additive research:

- Most additives are unstable above 180°C.
- Diluent oil (50%) is unstable at 150°C.
- Antioxidants are unstable above 200°C.
- Current additives are stable for only about 3 min (cumulative exposure) for mineral oils at 350°C.
- High-temperature additives are needed for stability of more than 3 min at 450°C.
- Synthetic lubricant base stocks are needed.
- Synthetic lubricants cannot achieve desired future performance without better additives.

More than 15 years after that workshop, many of these concerns remain, but there are additional challenges. Environmental concerns have increasingly limited what can be added to oils, and the trend toward bio-fuels is opening new areas for lubrication research. But the initial burst of enthusiasm for such work was dampened by concerns over the global effects of cultivation and land use on total carbon emissions. Undoubtedly these will continue to influence the direction of research in lubricant formulation.

The unique and demanding environments of space have challenged liquid lubricant technology and led to a number of federal government-funded lubricant research programs over the years. In 1990, NASA published a 23-page report on the use of liquid lubricants in space,[48] and in 1998, a more comprehensive *Space Tribology Handbook* was published by AEA Technology.[49] Major challenges for lubricants include very low ambient pressure, the presence of atomic species other than the normally encountered molecular species (e.g., atomic oxygen), thermal radiation exposure, and zero gravity. The exterior of satellites may operate at pressures less than 10^{-10} Torr and can experience a wide temperature range as the vehicle rotates from darkness to sunlight, or emerges from the shadow of the earth.

Specialized apparatuses that operate at high vacuum and a range of temperatures have been constructed to screen lubricants for space applications, for example, NASA's unusual spiral orbit tribometer (SOT) that was standardized by ASTM.[50] It was designed to evaluate small quantities of liquid lubricants (50–100 µg quantities) as well as solid film lubricants. The test geometry consists of a single bearing ball that is captive between a stationary lower plate and a rotating upper plate that drives the ball in an outward spiral orbit. A guide plate with a force transducer keeps the ball from spinning out and guides it back toward the center on each revolution. Typical

TABLE 6.5
Friction Coefficients of a Metal Pair in Two Refrigerants after 10 min

Conditions	Load (MPa)	CFC-12, Nominal μ	HFC-134a, Nominal μ
Gaseous refrigerant	0.72	0.27 ± 0.05	0.35 ± 0.1
Liquid refrigerant	2.86	0.15 ± 0.02	0.28 ± 0.05
With additive[a]	5.73	0.01	0.05
With additive[a]	10.74	0.03	0.02

[a] Mineral oil in CFC-12, PAG in HFC-134a.
Source: Komatsuzaki, S., Homma, Y., Kawashima, K., and Itoh, Y., *Lubr. Eng.*, 47, 1018, 1991.

friction coefficients of liquid lubricated metals in such applications fall within the boundary lubrication regime ($\mu \sim 0.08-0.15$).

Although solid lubricants are often used for space applications, low-vapor-pressure liquids like those based on silicone are used in some cases. Perfluorinated polyalkyl-ethers (PFPE) have been used in space at temperatures of $-100°C$. Some commercial PFPE compounds, such as Krytox™, have vapor pressures as low as 3×10^{-14} Torr at 20°C, making them suitable for long exposures at low ambient pressure.

Another challenging problem for liquid lubricant additives is in the replacement of chlorofluorocarbons in air-conditioning systems. Chlorinated refrigerants, such as CFC-12, have been replaced with newer, less lubricative refrigerants like HFC-134 (CH_2FCF_3). These will require additives to reduce friction and wear. Polyalkylene glycol (PAG) is one such new lubricating additive.[51] Table 6.5 compares the friction coefficients for Al–15% Si alloy on a grooved plate of chromium–molybdenum steel run under various test chamber pressures at room temperature at 1.0 m/s sliding speed in CFC-12 and HFC-134a.

Adhesive transfer of material from one contacting surface to another can lead to self-mated conditions and high friction. Metals and alloys such as iron and steel, copper and brass, aluminum alloys, and titanium transfer relatively easily during sliding contact. As Heinicke[52] points out, fatty amines, fatty alcohols, and fatty acids have been effective for antiwear and friction-lowering additives, reducing metal transfer by a factor of more than 20,000 times. Figure 6.7 uses the data of Grunberg and Campbell[53] to illustrate the relationship between the friction coefficient and the mass of copper transferred to a steel counterface per unit sliding distance. Trends for white oil and oleic acid are similar.

Water, with or without additives, can also be an effective lubricant. Sometimes materials cannot be lubricated with oils in applications such as in the food-processing industry. In such cases, bearing surfaces might be effectively lubricated with water or with water containing nontoxic additives. For example, Sasaki[54] published a compilation of the effects of water and water with additives on the friction and wear of ceramics. He used a pin-on-disk tribometer at various speeds and normal loads. Several of Sasaki's findings are given in Figures 6.8 through 6.10. In Figure 6.8, the average steady-state friction coefficients are given for four self-mated ceramics sliding in pure water at various speeds. Silicon carbide exhibits the lowest friction coefficient

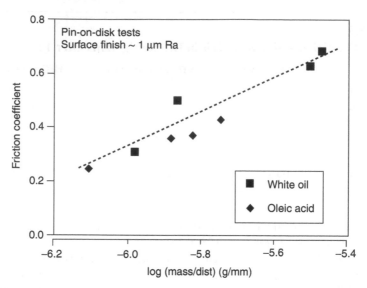

FIGURE 6.7 Effects of lubricating species on the quantities of material transferred during sliding as it affects the friction coefficient. (Adapted from Grunberg, L. and Campbell, R.B., *Proceedings of the Institution of Mech. Eng. Conference on Lubrication and Wear*, London, 1957.)

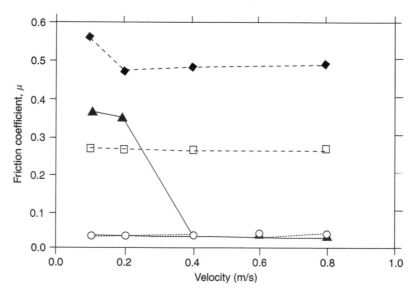

FIGURE 6.8 Effects of water lubrication on the friction of different ceramics, □ Alumina; ◆ PS zirconia; ○ silicon carbide; ▲, silicon nitride. (From Sasaki, S., *Bull. Mech. Eng. Lab. Jpn.*, 58, 32, 1992.)

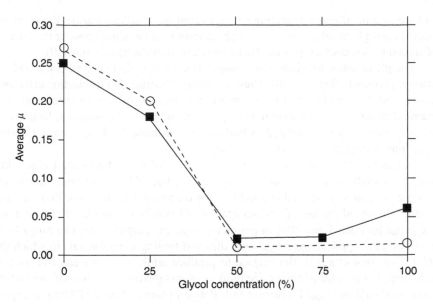

FIGURE 6.9 Effects of glycol additions on the friction of silicon nitride slid in water. ■ Average μ; ○ range in μ. (From Sasaki, S., *Bull. Mech. Eng. Lab. Jpn.*, 58, 32, 1992.)

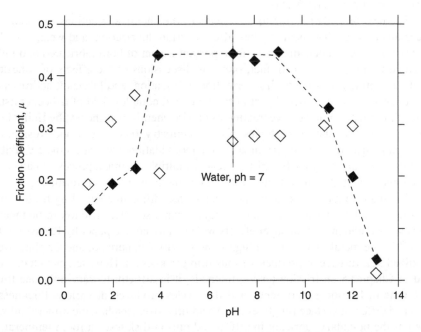

FIGURE 6.10 Effects of pH on the friction of silicon nitride. ■ Average μ; ◊ range in μ. (From Sasaki, S., *Bull. Mech. Eng. Lab. Jpn.*, 58, 3, 1992.)

and seems little affected by sliding speed, in contrast to silicon nitride. Experiments with various glycol additions to water also showed the responsiveness of the friction of silicon nitride couples to water-based lubricant composition (Figure 6.9).

The pH of water solutions also greatly affected the friction of silicon nitride in other experiments (Figure 6.10). These examples illustrate that lubrication effectiveness in reducing friction can be a function not only of the sliding speed and geometrical parameters, but also of the composition and pH of the lubricant, factors that are only indirectly incorporated in traditional, mechanically based bearing design equations through their effects on viscosity.

In Sasaki's tests using water, and in the case of many formulated oils, additives are liquids or species that go into solution, but additives to liquid lubricants and greases can also be used in solid form as dispersants in the fluid. Bhushan and Gupta[55] discussed the use of various graphite dispersions in liquids. Solid contents can range between 1 and 40% in petroleum oils, and particle sizes can range from 0.5 to 60 µm. Applications range from dies and tooling to engine oils in which the solid dispersant clings to the surface to produce additional antiwear and friction reduction. Some solid additives are used to provide extra protection of the sliding surfaces should the liquid lubricant fail. Polytetrafluoroethylene (PTFE) in engine oil additive fluids has become popular to reduce engine friction and improve mileage. It can produce lubricity at lower engine temperatures or during starting when full oil films have not yet been developed on the surfaces. In some cases, the improvement in friction can be noticed immediately and may even require retuning the engine to account for the reduced friction.

The field of study relating to the effects of the chemical reactions between surfaces and their environment, as they affect friction, lubrication, and wear, is called *tribochemistry*. Tribochemistry is an important aspect of both lubricated and unlubricated friction and wear. In fact, previous discussions of the effects of adhesion, relative humidity, oxidation, film formation, lubrication, and lubricant additives on friction can all be considered part of the wide and complex field of tribochemistry. One of the most comprehensive treatments of tribochemistry is the text by Heinicke.[52] He identifies a number of subtopics of tribochemistry, including tribodiffusion, tribosorption, tribodesorption, triboreaction, tribooxidation, tribocatalysis, etc. Tribochemistry is a challenging discipline, because multiple chemical processes can occur simultaneously and aging effects play a major role in lubricant performance.

The important role of oxides and other surface films in controlling friction was introduced in Chapter 3. These surface layers can also affect friction in boundary-lubricated situations. Oxide layer effects were discussed in a paper by Komvopoulos et al.[56] Three metals, oxygen-free high-conductivity Cu, pure Al, and Cr-plate, were oxidized in a furnace to produce various film thicknesses. However, as noted at the end of Chapter 5 on friction-testing methods, it is difficult to vary only one thing at a time in friction experiments, and the oxides produced at various thicknesses also had different surface roughnesses. An additive-free, naphthenic mineral oil was used as the boundary lubricant in self-mated pin-on-disk tests at room temperature in air. A number of those authors' friction coefficient results, obtained from plots of their data for 2 N load and at an angular disk rotation speed of 4.5 rad/s, is summarized in Table 6.6. These data represent only those for thinnest oxides produced in

TABLE 6.6
Effects of Oxide Scales on Boundary-Lubricated Friction

Condition/Parameter	Aluminum	Copper	Chromium
Lubricated but not preoxidized			
Average μ after 0.1 m sliding	0.45	0.18	0.20
Average μ after 50 m sliding	0.20	0.17	0.14
Lubricated and preoxidized			
Oxide layer thicknesses (nm)	4.2	7.0	28.2
Average surface roughness (μm)	0.05	0.1	0.05
μ after 0.1 m sliding	0.12	0.14	0.11
μ after 50 m sliding	0.20	0.14	0.13

Source: After Komvopoulos, K., Saka, N., and Suh, N.P., *J. Tribol.*, 108, 502, 1986.

TABLE 6.7
Effect of Linear Undulations on Boundary-Lubricated Friction of Steel on Titanium (Friction Coefficients at Steady State)

Lubricant	μ, Polished	μ, Undulated
Mineral oil	0.60	0.39
Oleic acid	0.47	0.11
Turbo oil	0.48	0.17
Silicone oil	0.46	0.25
Halocarbon oil	0.17	0.17
Methylene iodide	0.18	0.18

Source: Lancaster, J.K. and Moorhouse, P., *Tribol. Int.*, 18, 39, 1985.

their experiments, tests that typically ran for 50 m sliding distance. Friction in these experiments often exhibited complex behavior associated with the disruption of the oxides and the incorporation of debris into the interface. The thicker oxides tended to be more porous than the thinner ones, making them easier to rupture and producing greater quantities of wear particles. Komvopoulos et al. discussed the surface deformation and wear mechanisms in the interface and suggested several models for the observed behavior on the basis of microscopy of the contact surfaces.

The roughness of boundary-lubricated surfaces can be altered by the presence of oxides whose growth characteristics change as they thicken, but the surface roughness can also be altered intentionally to modify and reduce friction. For example, Tian et al.[57] created linear patterns on titanium surfaces subjected to sliding on 52100 bearing steel to study how those regular features affected the ability of certain boundary lubricants to reduce friction. The testing machine slid the steel pin back and forth 30 mm at an average speed of 1.1 cm/s on the undulated surface at a load of 5 N. The effects of the undulations on the friction of the two metals can be significant, as shown in Table 6.7, but they do not appear to work equally well for all lubricants. Several years earlier, Lancaster and Moorhouse[58] used photolithography to produce pockets in a range of metal substrates for holding solid lubricants (see also Section 6.3). In the case of titanium, a difficult-to-lubricate metal, undulations, coupled with a good choice of lubricant, seem to provide an effective system for lubrication.

Some lubricants are solids over one temperature range, liquids over another, and then become desorbed and cease to function at higher temperatures. Therefore, the conditions of surface contact and the role of the lubricant in separating the surface can change drastically over a range of temperatures. Rabinowicz[59] illustrated this situation using octadecyl alcohol lubricant between copper sliders. Below 40°C when the lubricant was solid, the friction coefficient was about 0.11, but the system experienced a transition between 40 and 60°C to reach $\mu = 0.33$ when the lubricant became liquefied. Friction remained constant until about 120°C, when another transition to a friction coefficient of about 1.0 occurred as the liquid was ultimately desorbed. The wear rate increased correspondingly at each transition temperature because metal transferred to the opposing surfaces with increasing severity as the friction increased.

6.1.2.1 Friction Polymers

Friction polymers (also called *tribopolymers*) are insoluble, strongly adherent resins that are found on contact surfaces in some tribosystems. They can form from hydrocarbon species in the ambient environment. When one or both surfaces are polymeric, friction polymers different in composition than either surface may be formed due to the influence of tribochemistry. First identified on noble metal electric contacts in 1958, friction polymers containing C, H, and O can reduce friction and wear but at the same time may induce electric arcing if they increase the resistivity and separation of the contacts. Although mainly connected with studies of electrical contacts, friction polymers can be found on a variety of tribosurfaces. A review by Lauer and Jones[60] concluded that friction polymers are generally "good" in lubricated metal pairs, that the formation and destruction of friction polymers is a dynamic and repeating process, and that friction polymer behavior is subject to thermal controls—that is, the frictional temperature of contact may rise above the temperature at which the friction polymer degrades, and so it must be controlled to ensure effective lubricating action.

Despite the substantial volume of literature on the effects of all manner of experimental parameters on the behavior of lubricants, exactly how boundary lubricants reduce friction is only partially understood. The complexity of additive interactions makes possible many different tribochemistries. The changing nature of the solid surfaces as the system "ages" and experiences wear complicates the goal of reaching a global understanding of lubricated friction. Tools used in understanding lubricant behavior range from surface forces apparatus to surface chemistry analytical instruments and electron microscopes. Lubricants are collected, filtered, centrifuged, and analyzed. Their nature is investigated as functions of temperature, pressure, and exposure to different solid surfaces. The Holy Grail of this work is a technique to study the formation, tribochemistry, and structure of lubricating films *in situ*.

6.1.2.2 Lubricating Characteristics of Ultrathin Layers

Fundamental studies of thin-layer lubrication have been made possible by the recognition that dipping techniques could be used to produce monolayers of lubricant. A 21-year-old new master's degree recipient working with Irving Langmuir at General Electric Company in 1919, Katherine Burr Blodgett, made the remarkable discovery that very thin—even monomolecular—films could be produced on the surfaces of fluids and transferred to solids by careful control of dipping. This technique led to the possibility of investigating lubrication mechanisms on a very fine scale. A special issue of the journal *Thin Solid Films* was devoted exclusively to Langmuir–Blodgett films.[61]

More recently, there have been advances in molecular-scale measurements of fluid properties. Grannick and others[62,63] have described the molecular-level behavior of lubricating films in terms of shear thinning. Figure 6.11 shows that as the thickness of a film decreases, the friction (shear strength) tends to rise. With the advent of high-speed computers and simulations of interfaces, it has become possible to model the behavior of molecules in narrow frictional interfaces.[64] Such efforts have shown that the structural arrangements of atoms in interfaces change in the vicinity of the solid walls and that the properties of the fluids may be much different adjacent to the boundaries as a result of these changes. Robbins[64] has shown that as the surfaces begin to move, lubricant layers may disorder and then reorder when motion ceases. These fascinating results have implications for understanding the nature of boundary lubrication and stick-slip.

Traditional interpretations of boundary lubrication mechanisms have dealt with the orientation of molecules on surfaces. Polar species tend to align with their "heads"

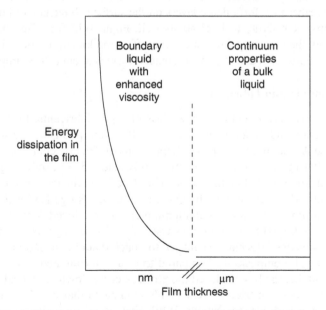

FIGURE 6.11 Schematic depiction of the effects of shear thinning. The shear stress of very thin lubricant layers can rise rapidly as thickness decreases.

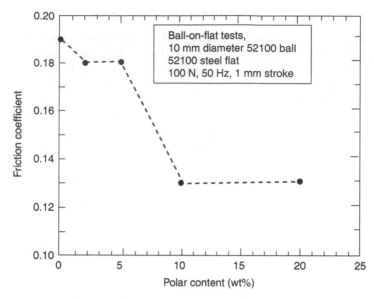

FIGURE 6.12 Effects of percent polar content on the friction coefficient of steel sliding on steel (ratio: 4 parts of saturate to 1 part base stock). (From Benchaita, M.T., Gunsel, S., and Lockwood, F.E., *Tribol. Trans.*, 33, 371, 1990. With permission.)

at the surfaces, their "tails" forming a carpet to provide lubricating action. Long-chain fatty acids exemplify this type of behavior. The shape and side branches of molecules determine whether they form dense layers on the surface. Tendency of liquids to wet surfaces helps their ability to lubricate as well. Figure 6.12 from Benchaita et al.,[65] illustrates how the percent polar content can affect the kinetic friction coefficient of steel on steel. Between 5 and 10% polar content, there is about a 30% drop in friction.

6.1.2.3 Ionic Liquid Lubricants

Ionic liquids (ILs) are a relatively new class of liquid lubricants. In 2001, friction studies of metals and ceramics lubricated by ILs were published by investigators from Chinese Academy of Sciences.[66] Representative friction coefficients, obtained with an oscillating ball-on-disk apparatus (50 N load, 25 Hz oscillating frequency, 1 mm stroke), are provided in Table 6.8a. The IL reduced the friction coefficients by about one-half. More recent work by Qu et al.[67] at Oak Ridge National Laboratory found that certain ILs containing ammonium cations produced friction coefficients in the range of 0.06–0.08 in pin-on-disk tests of 52100 steel against aluminum alloy 6061. The viscosity of ILs can be adjusted to match that of conventional engine oils by altering the IL composition. Compared to a fully formulated 15W40 diesel oil, Qu et al. found that the ILs displaced the Stribeck curve downward, leading to about a one-third reduction in friction. Results also indicated that the ILs could be used either alone or as additives to neat oils. While that work is in early stages at the time of this writing, preliminary findings suggest that traditional additive packages in oils could be replaced by IL additives, a more environmentally friendly alternative.

TABLE 6.8a
Sliding Friction Coefficients (μ) for 52100 Steel Balls against Several Metals and Ceramics Lubricated by an Ionic Liquid and a Fluorinated Lubricant

Disk Material	μ, Ionic Liquid[a]	μ, Reference Lubricant[a]
Type 52100 steel	0.065	0.145
Aluminum alloy 2024	0.040	—
Copper	0.025	0.145
SiO_2	0.060	0.132
SIALON (ceramic)	0.065	0.12

[a] Lubricants: Ionic liquid = 1-methyl-3-hexylimidazolium tetrafluoraborate; reference lubricant = perfluoropolyether (PFPE).

Source: Ye, C., Liu, W., Chen, Y., and Yu, L., *Chem. Commun.*, 2244, 2001.

As fundamental studies of interfacial structure, molecular motions, and tribochemistry are integrated with micro- and macromechanics, improvements in lubrication science and technology will emerge. This interdisciplinary path remains an arduous one, especially in an age where technical specialization and the proliferation of new journals tend to disperse research results. However, powerful computerized search engines are now available to help locate research published in a variety of sources and enable interdisciplinary findings to be integrated.

6.1.3 GREASE LUBRICATION

Grease is a semisolid lubricating material whose main function is to remain in contact with the surfaces and not be forced out of the contact by centrifugal or gravitational forces or by the squeezing action of the applied load. If hydrodynamic films cannot be formed because the relative surface velocity is insufficient to produce them, greases may be used to increase the load-carrying capacity. Greases can also serve as sealants.

A typical grease consists of three parts: an oil or a synthetic fluid, a thickener, and an additive package. The relative amounts of these are typically 80%, 10%, and 10%, respectively. Boehringer,[68] Wills,[7] and Harris[69] have reviewed the nature, composition, and uses of greases. Soaps and calcium or lithium complexes are examples of thickeners. Many of the additives to greases have the same functions as those used for liquid lubricants: friction reducers, antiwear compounds, corrosion inhibitors, and extreme-pressure (EP) modifiers.

One of the most important characteristics of grease is its EP properties. The ASTM four-ball test (ASTM Method D 2596), shown schematically in Figure 6.13, is a common method to determine the EP properties of greases. Three lower balls are locked in a nest and contacted by a loaded upper ball, which is spinning as shown. The test is intended to measure the maximum load-bearing capacity of a given test grease. There is also a five-ball test in which four caged balls are driven by a fifth ball.

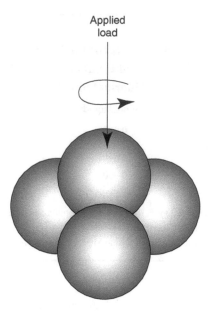

FIGURE 6.13 The four-ball testing geometry.

Another test method, called the Timken test (ASTM D 2509), involves a rotating ring pressed against a block. A grease that sustains an applied load greater than 155–180 N (known as the "OK value") is considered to be an EP grease.

With the drive to remove toxic materials like lead from the environment, new EP additives to greases are being sought. One promising new EP additive is bismuth. Bismuth, among the least toxic of many metal candidates, seems to have many of the desirable characteristics of Pb-based EP additives, such as lead naphthenate. It has the widest temperature range between its melting and boiling points (271–1560°C) and expands when it solidifies. Bismuth is used in lipsticks, nail polishes, eye shadows, and even antacid medicines. In metalworking applications, bismuth naphthenate or bismuth octoate works even better than lead as EP additives to oils and greases. When bismuth reacts with ferrous surfaces in the presence of sulfur, bismuth sulfides and iron sulfides with low friction are produced.[70]

6.1.3.1 Liquid Crystal Lubricants

In the late 1980s, there was interest in *liquid crystals* as additives to greases as well as lubricants in their own right.[71] Lamellar liquid crystals form lamellar, amphiphilic layers organized with the carbon chains back-to-back and separated by thin layers of a polar solvent, like water and glycol. These so-called *lyotropic liquid crystals* were tried as lubricants, reasoning that they might support bearing pressure even when the velocity of relative motion was insufficient to produce a pressurized hydrodynamic film to support the surfaces. A study by Friberg et al.,[72] using oleic acid, triethanolamine, and glycol to form their crystals, determined that the load-carrying capacity of the liquid crystals was limited by the contact pressure. They conducted slow-speed sliding tests of bearing steel on mild steel using liquid crystal lubricants.

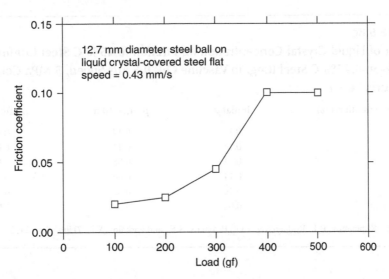

FIGURE 6.14 Frictional effects of liquid crystals in sliding of bearing steel on mild steel. (From Friberg, S.E., Ward, A.J., Gunsel, S., and Lockwood, F.E., Lyotropic liquid crystals in lubrication, in Ref. 51, 101, 1989. With permission.)

TABLE 6.8b
Effect of Liquid Crystal Compositions on Friction (52100 Ball-on-52100 Steel Flat, 0.66 mm/s Speed, 0.15 GPa Contact Pressure)

Liquid Crystal Composition (wt%)			
DBSA[a]	TEA[a]	Glycerol	μ
—	—	100.0	0.205
62.0	33.6	4.0	0.148
59.85	35.15	5.0	0.09
63.24	29.76	7.0	0.12
58.5	31.5	10.0	0.09
59.16	27.84	13.0	0.09
61.0	39.0	—	0.115
65.0	35.0	—	0.086
68.0	32.0	—	0.092

[a] DBSA, dodecylbenzene sulfonic acid; TEA, triethanolamine.

Source: Winoto, S.H., Winer, W.O., Chiu, M., and Friberg, S.E., Film thickness and frictional behavior of some liquid crystals in concentrated point contacts, in Ref. 51, 113, 1989.

Their results are summarized in Figure 6.14. Lee et al.[73] conducted other sliding experiments, which showed how the composition of the liquid crystals affected their friction coefficients with 52100 bearing steel in reciprocating sliding on the same material. Tables 6.8b and 6.8c illustrate the effects of liquid crystal composition and effects of liquid crystal additives to oil, respectively, on friction. Data in Table 6.8b

TABLE 6.8c
Effect of Liquid Crystal Concentration on Friction (0.45% C Steel Conformal Block-on-0.45% C Steel Ring, in Vaseline Oil, 0.5 m/s Speed, 5 MPa Contact Pressure, 76°C)

LC Concentration (%)	Initial μ	μ After 10 h	μ After 20 h
0.0	0.14	0.14	0.14
0.5	0.11	0.11	0.04
1.0	0.11	0.08	0.04
5.0	0.11	0.04	0.03
9.0	0.10	0.02	0.02
100.0	0.03	0.02	0.02

Source: Kupchinov, B.I., Rodnenkov, V.G., Ermakov, S.F., and Parkalov, V.P., *Tribol. Int.*, 24, 25, 1991.

illustrate the relative insensitivity of the friction coefficient to changes in the liquid crystal composition, whereas the data in Table 6.8c illustrate that the concentration of liquid crystal (in Vaseline™; oil) can have a significant effect on friction.[74]

Like liquids, greases must also be dispensed, but their high viscosities require higher pressurized systems. Pumpability of greases is a concern in their selection for various applications, as is the frequency of application to ensure adequate performance of the machinery. Greases tend to behave as non-Newtonian liquids; therefore, the ease of their distribution is affected by the shear rate (pumping speed) to which they are subjected.

6.2 LUBRICATION BY SOLIDS

In the previous section on liquid lubricants, we saw that some additives to liquids can result in the formation of solid films or deposits on contact surfaces to help reduce and control friction. Solid lubricants can be formed in other ways than by reaction within a fluid. For example, Peace[75] identified eight types of solid lubricant systems based on the method of application and the form of the material:

1. Solid lubricant powders
2. Resin-bonded dry film lubricants
3. Dry film lubricants with inorganic binders
4. Dispersions of solids in a nonvolatile carrier
5. Wear-reducing solids (with naturally lubricative surfaces)
6. Soft metal films
7. Plastic lubricants
8. Chemical reaction films (as produced by reactions with lubricant additives)

Bhushan and Gupta's *Handbook of Tribology*[76] contains an extensive discussion of coating and surface modification techniques, including those suitable for use with solid lubricants. Additional reviews of solid lubricants may be found in the literature.[77–79] This section discusses the materials that serve effectively as solid lubricants as well as the factors that affect their performance. In the late 1980s, Dr. Donald H. Buckley, of NASA, Lewis Research Center, is reputed to have said: "In years past, ceramics have been used to lubricate metal surfaces, but in the future metal films will be used to lubricate ceramics."

More and more solid materials are being found to be lubricative, but the fact that they are lubricative does not constitute a sufficient condition for them to be widely used as solid lubricants. Other factors such as ease of application, thermal stability, adequate persistence on the surface (wear resistance), cost, and chemical compatibility with the surfaces and service environment are also factors for selection. In addition, some lubricative metals (such as Pb) can no longer be used due to concerns about their toxicity.

Four primary factors that should be considered when selecting and designing solid lubricating films to reduce friction effectively are as follows:

1. The structure and composition of the solid lubricant species
2. The thickness of the solid lubricating film in the given application
3. The conditions of sliding (contact pressure, velocity, temperature, environment)
4. The manner by which the lubricant is applied and resupplied to the surface as sliding or rolling tends to remove it

The structure and composition of the solid lubricant determine its shear strength, its adhesion characteristics to the substrate, its chemical stability, its durability, and in some cases its tendencies toward anisotropic behavior. The thickness of the film determines its friction coefficient in much the same way that the Stribeck curve, described earlier, determines the lubrication regime. In what are now considered "classic experiments" of indium films deposited on tool steel, as described by Bowden and Tabor,[80] the authors stated:

> ... the friction decreases as thinner films are used because the track width and the area of contact A becomes smaller. There is, however, a limit to this and a minimum of friction is reached when the thickness is the order of 10^{-5} cm. With thicknesses less than this, e.g., 10^{-6} cm (or 30 atomic layers), the film ceases to be effective.

The shape of the friction coefficient versus film thickness curve resembles the "bathtub shape" of the Stribeck curve, but the mechanisms responsible for that behavior are not the same. When films are very thin, asperities can penetrate and disrupt them, so thinner films may work if the surfaces are polished extremely flat. When films become too thick, they behave more like bulk solids. For example, silver is an effective solid lubricant when in thin film form, but the friction coefficient of bulk silver rubbing on the same material can be more than 10 times higher.

The method of surface preparation before the application of solid lubricants must be carefully considered if the full benefits of the solid lubricant are to be achieved. For steels and stainless steels, surface preparation may involve phosphate coating, grit blasting, or grinding. For aluminum alloys, anodization treatments may be applied. Titanium alloys may require grit blasting or chemical etching.

TABLE 6.9

Friction Coefficients for Steel Lubricated by Solid Lubricants

Lubricant	μ_s	μ_k	S-S[a]
None (steel on steel)	0.40–0.80	0.40	
Molybdenum disulfide	0.05–0.11	0.05–0.093	N
Tungsten disulfide	0.098	0.09	N
Selenium disulfide, titanium disulfide, and tellurium disulfide	—	0.25	Y
Mica, talc	—	0.25	N
Graphite	—	0.25	N
Boron nitride (hexagonal)	—	0.25	Y
Vermiculite	0.167	0.160	N
Beeswax (at 60–63°C)	0.055	0.05	N
Paraffin (at 47–77°C)	0.112	0.104	Y
Calcium stearate (157–163°C)	0.113	0.107	N
Carnauba wax (83–86°C)	0.169	0.143	N
Sodium stearate (198–210°C)	0.192	0.164	Y
Lithium 12-hydroxystearate (210–215°C)	0.218	0.211	N

[a] S-S, tendency for stick-slip; Y, yes; N, no.
Source: Sonntag, A., *Electro-Technology*, 66, 108, 1960.

One method of applying solid lubricants is to incorporate them within a synthetic resin binder and to paint or spray them onto a surface. Some of the formulations cure in air, but others require oven drying and curing. According to Lancaster,[78] several common constituents of bonded-film lubricants are MoS_2, WS_2, graphite, PTFE, pthalocyanine, CaF_2/BaF_2, Pb, PbO, PbS, Sb_2O_3, Au, Ag, and In. Sonntag[81] has tabulated data for the static and kinetic friction coefficient of solid lubricants on metals and indicated whether or not they exhibited tendencies for stick-slip. A selection of these data is provided in Table 6.9. Since the conditions of use vary greatly and may differ significantly from Sonntag's testing conditions, these values are only provided as an example of relative differences in the solid lubricating behavior of various materials at room temperature.

Solid lubricants are also used at elevated temperatures or in high vacuum applications where most liquid lubricants would volatilize, oxidize, or otherwise become unstable. Table 6.10 lists the friction coefficients of several solid lubricants obtained under the same high-temperature sliding conditions (7.7 kgf and 7.6 mm/s on steel at 704°C), as reported by Peterson et al.[82] in 1969. A more recent compilation of high-temperature solid lubricant friction coefficients was produced by Allam.[83] From a compilation by Clauss,[84] Table 6.11 provides room temperature friction values of compressed powder pellets sliding on stainless steel.

TABLE 6.10
Kinetic Friction Coefficients for Several Oxides at 704°C

Lubricating Solid	μ
PbO	0.12
B_2O_3	0.14
MoO_3	0.20
Co_2O_3	0.28
Cu_2O	0.44
SnO	0.42
TiO_2	0.50
MnO_2	0.41
K_2MoO_4	0.20
Na_2WO_4	0.17

Source: Peterson, M.B., Murray, S.F., and Florek, J.J., *ASLE Trans.*, 2, 225, 1969.

TABLE 6.11
Properties and Friction Coefficients of Certain Compounds (Pressed 200-Mesh Powders into Compacts, Slid on 440C Stainless Steel at 0.18 m/s and Pressure 0.552 MPa)

Class	Compound	Crystal Structure	μ
Disulfides	MoS_2	Hexagonal	0.21
	WS_2	Hexagonal	0.142
	NbS_2	Hexagonal	0.098
	TaS_2	Hexagonal	0.033
Diselenides	$MoSe_2$	Hexagonal	0.178
	WSe_2	Hexagonal	0.13
	$NbSe_2$	Hexagonal	0.12
	$TaSe_2$	Hexagonal	0.084
Ditellurides	$MoTe_2$	Hexagonal	0.20
	WTe_2	Orthorhombic	0.38
	$NbTe_2$	Trigonal	0.70
	$TaTe_2$	Trigonal	0.53
Graphite	C	Hexagonal	0.14

Source: Clauss, F.J. in *Solid Lubricants and Self-Lubricating Solids*, Academic Press, New York, 1972, 114–115.

6.2.1   ROLE OF LAMELLAR CRYSTAL STRUCTURES

Some of the most important solid lubricating materials exhibit what has been called lamellar behavior. They contain weak shear planes within the crystal structure that can yield preferentially to reduce friction when properly aligned to the sliding direction. Many of the compounds in Table 6.11 form hexagonal crystal structures in which the shear strength is lowest parallel to the basal planes. In the case of molybdenum disulfide, there are weak van der Waals bonds between covalently bonded Mo–S layers. Moisture and air tend to reduce the effectiveness of MoS_2 as a solid lubricant, since they penetrate these layers and enhance their shear strength. However, graphite is observed to be more lubricative in moist environments, since interlayer species reduce its shear strength. Therefore, molybdenum disulfide is very effective in low vacuum (space) applications, but graphite is more effective in moist environments. Table 6.12 illustrates some of these effects, but it should be noted that solid lubricants used in powdered form may exhibit different frictional characteristics than the same compositions applied by other methods, like sputtering or burnishing.

The hexagonal crystal structure and the role of weak van der Waals bonding between the basal planes in graphite and molybdenum disulfide inspired Ali Erdemir of Argonne National Laboratory to seek similar hexagonal compounds that might also serve as solid lubricants. The search eventually led him and his colleagues to boric acid, a triclinic compound with weak shear strength parallel to its basal planes. Friction experiments using boric acid pins (made from compacted H_3BO_3 powders) sliding on steel disks indicated an ability to maintain a low and stable friction coefficients ($\mu \sim 0.1$).[85,86] Since boric acid is soluble in water and it decomposes to form boric oxide at about 170°C, it has some practical limitations. Still, it has been suggested that boric acid could be used in metalworking processes and can be removed by rinsing the part with water. He discussed the use of boric oxide (B_2O_3) and its product with water, boric acid (H_2BO_3), in particular. Boric acid resembles other lubricants with layered structures and produces favorable friction reductions under

---

TABLE 6.12
Effects of Relative Humidity on the Friction Coefficients of Two Solid Lubricants (Collected Slider-on-Disk Data)

Solid Lubricant	μ_k, Dry Air, RH < 6%	μ_k, Moist Air, RH = 85% (After Sliding in Dry Air)	μ_k, Dry Air (After Sliding in Moist Air)
Molybdenum disulfide powder	0.06	0.20	0.06
Molybdenum disulfide bonded film on disk	0.09	0.22	0.09
Molybdenum disulfide bonded film on both slider and disk	0.26	0.34	0.31
Graphite powder	0.06–0.10[a]	0.16	0.19

[a] Initial value before film failure.

Source: Clauss, F.J. in *Solid Lubricants and Self-Lubricating Solids*, Academic Press, New York, 1972.

some circumstances. However, when the temperature rises above about 170°C, boric acid decomposes to boric oxide and loses its layered structure. Furthermore, above about 450°C boric oxide liquefies and tends to react with ceramics corrosively.

6.2.2 SIMPLIFIED MODELS FOR SOLID LUBRICATION

The friction of solid lubricants is sometimes modeled by assuming that the friction force **F** in a given sliding system is determined by the shear strength τ of the interfacial medium:

$$\mathbf{F} = \tau A \tag{6.15}$$

where A is the contact area over which shear force **F** is acting. As discussed earlier in regard to friction modeling, it has been found by Bridgman[87] that τ is a function of the contact pressure p. Thus,

$$\tau = \tau_0 \alpha p \tag{6.16}$$

and the pressure coefficient α determines the change in the saturation shear stress with pressure. Bednar et al.[88] have re-examined the pressure dependence of the yield strength and, using anvil experiments, determined the effective friction coefficient μ_{eff} of metals as a function of applied pressure p and the "saturation yield stress" (represented as τ in Equation 6.13). Thus,

$$\mu_{eff} = \frac{\tau}{p} \tag{6.17}$$

Experimentally determined values for τ_0 and α of several metals are listed in Table 6.13. Of the three metals listed as solid lubricants, silver and tin are used more than indium.

TABLE 6.13
Dependence of Saturation Shear Strength and Friction of Metals on the Applied Pressure

Metal	α	τ_0 (MPa)	μ_{eff} (at $p = 1$ MPa)
Fe	0.075	173.82	0.246
Cu	0.049	107.61	0.160
Ag[a]	0.036	109.04	0.144
Au	0.029	91.82	0.123
Al	0.035	47.56	0.082
Sn[a]	0.012	12.30	0.024
In[a]	0.006	5.65	0.011

[a] Commonly used solid lubricants.

Source: Data of Erdemir, A., Fenske, G.R., Erck, R.A., Nichols, F.A., and Busch, D.E., Tribol. Trans., 47, 179, 1991.

Interestingly, the frictional response of silver is quite suitable for solid lubrication because there is no significant difference between μ_s and μ_k, leading to very smooth sliding with an absence of stick-slip behavior.[89]

6.2.3 GRAPHITE AND MOLYBDENUM DISULFIDE

Graphite and molybdenum disulfide are among the most commonly used solid lubricants, and it is worthwhile to further consider their frictional behavior. In 1967, Winer[90] compiled an extensive review of molybdenum disulfide lubrication, and Fleischauer and Bauer[91] reviewed the chemistry and structure of sputtered MoS_2 films. Later, A. R. Lansdown published an entire book on molybdenum disulfide lubrication.[92] Like other lamellar compounds, graphite and molybdenum disulfide are anisotropic. When the crystals are properly oriented, for example, by running-in the surface to produce easy-shear platelet orientations, friction coefficients for MoS_2 can be as low as 0.02–0.1. Graphite typically exhibits friction coefficients from 0.10 to 0.18 in air.

Fluorination of graphite to produce substoichiometric graphite fluoride CF_x ($0.3 \leq x \leq 1.1$) increases the spacing between the basal planes, making it a good solid lubricant.[93] Another method to spread the basal planes is by inserting additional atoms in a process called *intercalation*. This process can result in significant enhancements of film life as well as frictional performance.

As noted by Peace,[75] the thermal stability of graphite and molybdenum disulfide lubricants depends on factors other than temperature alone. These include relative humidity, oxygen concentration in the environment, and whether the material is in powdered or monolithic form. In furnace oxidation experiments, graphite powder begins to oxidize significantly at about 585°C, compared with 298°C for molybdenum disulfide powder. Fusaro[94] found that oxidation can cause molybdenum disulfide films to blister and fail. This is explained by the tendency of molybdenum disulfide to form several kinds of oxides and sulfides in air. In the case of graphite, the oxidation is highly anisotropic, but the rates of oxidation are slower than that for molybdenum disulfide.[95]

Bisson and Anderson[96] reviewed the properties of solid lubricants, including graphite, molybdenum disulfide, and molybdenum trioxide. Figure 6.15 shows how the friction coefficients of various MoS_2 and MoO_3 films on steel surfaces react to increasing sliding velocities. Clearly, MoO_3 is a poor lubricant. As the temperature increases in air, molybdenum disulfide undergoes changes in color and rate of oxidation. Table 6.14 summarizes these changes, as discussed by Bisson and Anderson.

The manner of application of molybdenum disulfide can influence its frictional behavior. Fine-scale experiments using the friction microprobe described in Chapter 3 were performed on molybdenum disulfide films deposited on smooth, high-purity silicon surfaces by light burnishing and by commercial sputtering.[97] A silicon nitride bearing ball, 1.0 mm diameter, was slid across the surface in air. The normal force was 98.1 mN (10 gf) and the traverse velocity was 10 μm/s. Representative traces of friction versus time are shown in Figure 6.16. The tangential force on the burnished layer was relatively low and steady ($\mu = 0.12$–0.14), but dramatic stick-slip behavior was exhibited by the sputtered surface. Initially, one might suspect that the stick-slip

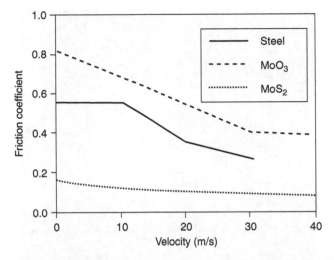

FIGURE 6.15 Effects of sliding velocities on the friction of Mo–S and Mo–O films. (Adapted from Bisson, E.E. and Anderson, W.J., *Advanced Bearing Technology*, NASA SP-38, Washington, DC, 1964. With permission.)

TABLE 6.14
Changes in Molybdenum Disulfide as Temperature Rises

Temperature Range (°F)	Temperature Range (°C)	Behavior
Up to 750	Up to 400	No detectable oxidation rate
750–800	400–427	Thin oxide film forms
800–850	427–454	Slow, but appreciable oxidation
850–900	454–482	Yellowish white MoO forms
Over 900	Over 482	Rapid oxidation

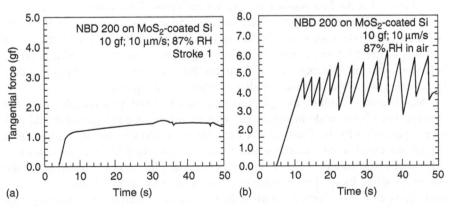

FIGURE 6.16 Stick-slip behavior of molybdenum sulfide films on silicon, measured by the friction microprobe: (a) lightly burnished powder, (b) commercially sputter-coated (1.5 µm thick).

FIGURE 6.17   Optical photomicrograph of an MoS$_2$ coating showing periodic cratering corresponding to frictional stick-slip events as a 440C steel ball was slid along it.

behavior was caused by periodic tearing of the film, but examination of the sliding path showed that the film was still intact after each pass. Rather, the periodic behavior in this case was caused by the slider first creating and then popping out of a series of evenly spaced depressions in the film surface (see Figure 6.17). The relatively low stiffness of the friction microprobe's stage suspension system allowed this elastic behavior to occur. This experiment illustrates that stick-slip can be caused not only by adhesion and bond fracture, but also by other periodic processes.

Molybdenum disulfide and graphite can each be used as solid lubricants, but attempts have been made to determine whether mixing them together would provide synergistic effects. Gardos,[95] for example, reviewed the use of graphite as an oxygen scavenger to help molybdenum disulfide films retain low friction characteristics. Some limited advantages in enhancing the stability and wear resistance of the microcrystalline molybdenum disulfide films in air were reported. About the same time, Bartz et al.[98] in Germany studied the friction of bonded films containing various combinations of graphite, molybdenum disulfide, and antimony thioantimonate

TABLE 6.15
Steady-State Friction Coefficients for Solid Lubricant Combinations (Block-on-Ring, 980 N Load, 1.23 m/s, Steel Substrates)

Film Composition	μ, Steady State
Graphite alone	Unstable μ
MoS$_2$ alone	0.05
Graphite + MoS$_2$ (about 1:2 wt% ratio)	0.01–0.02
Graphite + Sb(SbS$_4$) (about 3:4 wt% ratio)	Unstable μ
MoS$_2$ + Sb(SbS$_4$) (about 4:5:1 wt% ratio)	0.1–0.03
Graphite + MoS$_2$ + Sb(SbS$_4$)	0.04–0.05

Source: After Bartz, W.J., Holinski, R., and Xu, J., *Lubr. Eng.*, 42, 762, 1986.

[Sb(SbS$_4$)]. Using a block-on-ring apparatus, after sandblasting the 100CrMn6 steel ring, they applied bonded films to it, but left the 90MnCrV8 steel block untreated. One way of assessing the effectiveness of the blends was to measure the stable, post-running-in friction coefficient. Table 6.15 lists values of μ for several combinations of lubricants.

6.2.4   SOLID LUBRICATION BY POWDERS

As stated earlier, solid lubricants are used in a variety of forms. One form, familiar to individuals who have attempted to free a stuck car lock by using a tube of dry graphite particles, is powder lubrication. Most of the past work on powder lubrication has been empirical, but Heshmat[99] developed mathematical formalisms by treating powder layers as quasi-hydrodynamic fluid films. This approach only held for a certain range of particle sizes in the specific tribosystem. A powder film system can be visualized as the bounding solids and a powder lubricant layer with intermediate films forming upper and power boundaries on that layer. Heshmat developed relationships for the yield strength and limiting shear stress of powder films and, by applying suitable boundary conditions to establish velocity and pressure profiles in the powder layer, developed relationships for the tapered inlet geometry that permit one to determine the load-carrying capacity of the system and the conditions under which slip will occur within the powder. Several more recent articles by Heshmat address lubricant flow,[100] traction at high temperatures,[101] and the properties of several dry lubricant powders.[102]

Some polymeric materials are naturally lubricative and can be used either as bulk solid lubricants or as additives to less lubricative materials to lower their friction. Notable among these is the compound PTFE, which was discovered by Roy J. Plunkett, a PhD graduate from The Ohio State University. Shortly after joining DuPont, Plunkett was assigned to study refrigerants, and in the course of that work, in 1938, he discovered a slippery powder that clogged a bottle of tetrafluoroethylene gas that he was using. Seizing on the unusual properties of the powder, he continued his studies and eventually patented what is known by trade name Teflon. PTFE has experienced extensive use since its commercial introduction in 1941. In general, PTFE tends to exhibit low and steady friction coefficients against other polymers and some metals (typ. $\mu = 0.04–0.05$) up to 200°C,[103] but its excellent lubricating properties also extend to very low temperatures where it has been used in ball bearings for a liquid hydrogen turbopump application.[104]

PTFE forms the basis for a number of polymers and additive blends that are used for bearings, especially in environments that would ordinarily corrode metals. One PTFE-based family of polymeric bearings is called Rulon®. It was originally developed by Dixon Industries in the United States as a bearing material for the textile machinery but its uses have expanded. There are now as many as 15 commercial varieties of Rulon, and friction data for 7 are shown in Table 6.16. As the data indicate, most of the friction coefficients for these varieties fall within the range of 0.1–0.3 so it is other factors, such as environmental stability, counterface material compatibility, and certification for use with food and pharmaceuticals, that prompt the selection of these varieties for particular applications.

TABLE 6.16
Recommended Uses for PTFE-Based Rulon Compounds[a]

Product Name	Description or Recommended Use	Range of μ (Static and Kinetic)
Rulon LR	Sleeve bearings for use against carbon steel shafting	0.15–0.25
Rulon J	For softer materials, such as stainless steel, aluminum, and other polymers	0.12–0.20
Rulon 641	A white colored material certified by the U.S. Food and Drug Administration for use with mild steel, stainless steel, and harder alloys	0.10–0.30
Rulon W2	For use in fresh water pumps and food processing, not for very soft surfaces or when electrical insulation is required	0.15–0.30
Rulon 123	Nonabrasive for use on softer materials and when good thermal and electrostatic dissipation is required	0.10–0.30
Rulon 488	Wear resistant for use in dry or gas environments, such as hydrogen or natural gas compressors	0.10–0.30
Rulon XL	Designed for compatibility with titanium alloys and anodized surfaces	0.10–0.25

[a] High Performance Fluoropolymer Materials: Rulon, Saint-Gobain Performance Plastics, product literature.

Briscoe[105] has reviewed the mechanisms of organic polymer friction, stating that two noninteracting contributions, adhesion and plowing, can be used to model behavior. In this treatment, frictional energy is assumed to be dissipated in an interface zone (adhesive) and a subsurface zone (deformation or plowing). In the latter zone, behavior in polymers may be viscoelastic, plastic, or brittle. In Section 4.3.1 it was shown that the friction coefficient could be derived from geometric arguments to produce the form

$$\mu = \frac{2}{\pi} \tan \theta \qquad (6.18)$$

where the angle θ was associated with the roughness of the surface. Briscoe found that if PTFE became brittle, as it did after irradiation by gamma rays, the same expression could be written as

$$\mu = \frac{x}{\pi} \tan \theta \qquad (6.19)$$

where the value of x, the slope of the dependence of friction on $\tan \theta$, varied from 0 to 2 depending on the degree of embrittlement (i.e., the extent of plastic flow). When $\tan \theta$ exceeded about 2.3, irrespective of x, the PTFE began to exhibit chip-forming characteristics rather than flow. Like other materials, the shear stress of PTFE was seen to vary with contact pressure, following Equation 6.9.

Since PTFE has a relatively low hardness, various additives are mixed with it to improve its wear resistance. Table 6.17 shows the effects of certain additives on the friction coefficient and wear of PTFE sliding on steel at 0.01 m/s.[106,107] Comparing the first and last row of data shows how it is possible to increase wear resistance by more than three orders of magnitude while raising the sliding friction coefficient of the material by at most about 0.03.

TABLE 6.17
Effect of Additives on the Friction of PTFE Blends

Material Composition	Wear Rate Improvement[a]	μ_k
Unfilled PTFE	1	0.10
15 wt% Graphite	588	0.12
15 wt% Glass fiber	2857	0.09
12.5 wt% Glass fiber and 12.5 wt% MoS$_2$	3333	0.09
55 wt% Bronze and 5 wt% MoS$_2$	4000	0.13

[a] Ratio of the wear rate of unfilled PTFE to that of the given material. The larger the value, the greater the effect of the additive in reducing wear of the PTFE.

Note: Collected data of Bhushan, B. and Gupta, B.K. in *Handbook of Tribology*, McGraw-Hill, New York, 1991, 5.57.

Erdemir[107] reviewed some of the key mechanisms responsible for the lubricating action of solid films on ceramics such as silicon carbide, silicon nitride, and aluminum oxide. He stated that solid lubrication may be the only option available to help lubricate ceramics in severe environments, but noted that like other types of lubricants, solid lubricants suffer from finite lifetimes. One clever approach to addressing the resupply problem in orbiting satellites uses *in situ* physical vapor deposition of MoS_2 from tiny heating elements designed into the bearing housing itself. This was described in a plenary lecture by Koshi Adachi at the 2007 International Conference on Wear of Materials.[108]

Although some materials can be used as solid lubricants in the unadulterated state, the benefits of combining additives with liquids, greases, and solid lubricating materials have also been discussed. In the following section it is shown how liquid and lubricants can be added to solid structural materials to help make them "self-lubricating." As exemplified by the Rulon family of bearing materials, PTFE and other lubricants provide the basis for the self-lubricating qualities of many types of commercial metal and polymer composites.[109]

6.3   ENGINEERED SELF-LUBRICATING MATERIALS

Supplying a lubricant to a bearing surface can be done by incorporating certain design features into the component that induce the transport of the lubricant as the device operates, or by using feed systems. In some applications, it is inconvenient or unfeasible to provide an external lubrication supply for the contact. In confined spaces, remote locations, or permanently sealed components, lubrication must be "built-in" (in the case of liquid lubricants a common term is "fill for life"). In airplanes and satellites, the added weight of a mechanical lubricant supply system may not be acceptable. Critical tribosystems, such as helicopter transmissions, may require backup lubrication in case there is a sudden loss of the primary fluid. In such situations, designers must ensure that the contact will be satisfactorily lubricated over the entire lifetime of the system, or at least for known intervals between inspections or maintenance. Clearly, the use of materials that tend to lubricate themselves can be a distinct advantage. If the structural materials in the bearing are not inherently lubricative like PTFE or diamond, then constituent phases can be added to the surfaces (with coatings or surface modifications) or to the bulk materials during processing. In doing so it is important that the strength, corrosion resistance, stiffness, and other desirable features of the material are not unduly compromised. Therefore, in formulating self-lubricating materials, there may be a need to trade off optimal lubrication characteristics to achieve other needed properties.

According to Peterson,[110] oil-soaked wood was the first documented self-lubricating composite material. In the 1920s, bronze was prepared with graphite plugs inserted into small holes. In 1922, General Electric introduced Genelite™, a material containing graphite dispersed in a bronze matrix. Porous metal bearings were prepared in the 1920s by pressing and sintering copper and tin powders to produce a microstructure in which the pores were interconnected. These "sponges" could then be vacuum impregnated with lubricating oil. By the 1930s plastics with various lubricating fillers were in use.

In porous bearings, the fraction of porosity f can be expressed in terms of the ratio of the apparent density of the material ρ and the density of the material in the nonporous form ρ_0

$$f = 1 - \frac{\rho}{\rho_0} \tag{6.20}$$

The oil content OC is given as the ratio of the volume of oil V_0 to that of the entire part V:

$$OC = \frac{V_0}{V} \tag{6.21}$$

Another useful figure for the assessment of porous metal bearings is called the impregnation factor, IF:

$$IF = \frac{OC}{f} \tag{6.22}$$

The size of the pores, the quantity of porosity, and the viscosity of the permeating fluid determine the porosity of metal bearings. According to Morgan,[111] it is possible in practice to impregnate 85–95% of the available porosity.

Pore closure is a problem with porous metal bearings. If the contact pressure is too high, the pores will begin to close up, permanently reducing the permeability of the bearing surface. Pores may also become clogged with wear particles (wear debris). Morgan states that there is a limit to the amount of pore closure that can occur during running-in without causing one or more of the following undesirable effects:

- High initial running temperature
- High initial oil oxidation
- High initial oil loss
- Seizure from differential expansion
- Debris from wear products
- Seizure from "pickup" (transfer)

The friction coefficient of a porous bearing is typically dependent on the applied contact pressure, as discussed by Youssef and Eudier.[112] At high pressure, the lubricating film is no longer effective, leading to a rise in bearing friction, temperature, and wear. Modifications of the atmospheric pressure applied to the bearing to control lubricant flow in the pores can result in extending the life of the bearing, and the introduction of intermediate foil layers with controlled pore sizes can raise the critical pressure at which the friction rises.

Ceramics have also been used to form porous bearings. In 1987, a patent was granted to Dobbs et al.[113] for a porous silicon nitride material (with 2 wt% aluminum oxide and 6 wt% yttrium oxide) in which lubricant was contained in pores occupying from 3 to 10.5 vol% of the body. In addition to light oils, the authors suggested that molybdenum

TABLE 6.18
Midlife Friction Coefficients for Etched-Pocket Materials

Substrate	μ PTFE + 26 vol% Pb Filler	μ MoS$_2$ + Polyimide
Phosphor bronze	0.13	0.08
Stellite	0.14–0.20	0.06–0.08
EN 47 low-alloy steel	0.11–0.13	0.06–0.07
Hardened steel	0.12	0.08

Source: Lancaster, J. K. and Moorhouse, P., *Tribol. Int.*, 18, 139, 1985.

disulfide could be impregnated by pressure infiltration and that boron nitride could be introduced by mixing it with the powder prior to pressing and sintering.

Other methods besides sintering have been used to produce self-lubricating materials. Lancaster and Moorhouse[114] described so-called "etched-pocket materials" in which a pattern of circular pores is photoetched on the surface of a bearing material and filled with several materials such as lead, PTFE, and molybdenum disulfide. Table 6.18 illustrates the effects of substrate material on the midlife friction coefficient of etched-pocket materials. Rotating 25.0 mm diameter ring-on-flat sliding tests were performed using hardened tool steel rings. At the same time as the ring was rotating, the flat specimen was oscillated back and forth. The applied load was 250 N and the average sliding speed was 0.67 m/s. Under the same conditions, the friction coefficient for a porous bronze bearing was between 0.1 and 0.13. Therefore, the friction of the etched-pocket materials was quite satisfactory. However, the length of time that the low friction persisted depended on the time to wear the surface down below the depth of the pockets.

Several high-temperature, self-lubricating composite materials were developed by NASA. These materials consist of a complex combination of CaF$_2$ and BaF$_2$, sometimes with added Ag, in a chrome carbide matrix.[115] The silver was designed to help lubricate at low temperatures and the eutectic at high temperatures. Initially, the material was created by plasma spraying onto substrates (hence the prefix "PS" in the material designation), but later work produced sintered powder compacts and hot isostatically pressed materials.[116]

Gangopadhyay et al.[117] conducted a series of fundamental experiments in which a pin-on-ring tribometer was modified to hold two 6.25 mm diameter pins in tandem against a rotating 52100 steel ring. One of the pins was ceramic, either silicon nitride or alumina, and the other was made from compressed nickel chloride-intercalated graphite powder. Ring rotation was directed so that the ceramic pin was first contacted then the graphite pin. Tests were conducted in air (10–30% RH) with a normal load of 33 N and sliding speed of 0.14 m/s. Table 6.19 compares the friction of the various pin combinations and not only demonstrates that solid lubricants can be applied effectively by applying a rub-shoe of the lubricant against one of the sliding surfaces, but also illustrates the potential for developing self-lubricating materials with built-in pockets of solid lubricants. Solid lubricants are sometimes distributed within bearings by adding lubricating species to the retainers that separate the rolling elements. This approach is called transfer lubrication.

TABLE 6.19
Effects of Graphite in Tandem with Ceramics in Dual Pin-on-Ring Tests

Sliding Material	μ
Al_2O_3 only	0.48
Si_3N_4 only	0.45
Intercalated graphite only	0.16
Al_2O_3 + graphite	0.19
Si_3N_4 + graphite	0.14

Source: Gangopadhyay, A., Janahmir, S., and Peterson, M.B. in *Friction and Wear of Ceramics*, Marcel Dekker, New York, 1994, 163–197.

The key aspects of designing and selecting self-lubricating materials to reduce or control friction in a given application involve the following:

1. Confirming that the materials can withstand the pressure, speed, and the operating environment
2. Demonstrating that the self-lubricating material can be applied effectively and has an adequate lifetime
3. Verifying that the given material is cost-effective for that application

Despite the number of self-lubricating materials now available, the requirements for some extreme-condition applications, like high-temperature engines,[118] continue to drive research and development efforts in this area of friction technology. Polymeric and ceramic-based self-lubricating materials in particular are being developed. The use of combined approaches, like micro-texturing surfaces plus the introduction of solid and liquid lubricants, offer still more possibilities.

REFERENCES

1. Glossary of terms and definitions in the field of friction, wear, and lubrication (1980), in Appendix to the *ASME Wear Control Handbook*, eds. M. B. Peterson and W. O. Winer, ASME, New York, pp. 1143–1303.
2. E. Rabinowicz and S. B. Narayan (1984). Thickness effects using cupric oxide as a solid lubricant for copper, *Proceedings of the Third International Conference on Solid Lubrication*, ASLE.
3. W. A. Hardwick (1983). Ultimate flavor influence on canned beer caused by lubricants employed in two-piece can manufacture, *Lub. Eng.*, 40(10), pp. 605–607.
4. D. Dowson (1979). *History of Tribology*, Longman, London.
5. L. B. Sibley and F. E. Kennedy (1990). *Achievements in Tribology*, ASME, New York.
6. C. H. T. Pan (1990). Gas lubrication (1915–1990), in *Achievements in Tribology*, eds. L. B. Sibley and F. E. Kennedy, ASME, New York, pp. 31–56.
7. J. G. Wills (1980). *Lubrication Fundamentals*, Marcel Dekker, New York.

8. *Handbook of Lubrication (Theory and Practice of Tribology)*, Vols. I (1983) and II (1984), ed. E. R. Booser, CRC Press, Boca Raton, FL.
9. Various authors (1992). Lubricants and lubrication, in *ASM Handbook, Volume 18: Friction, Lubrication, and Wear Technology*, 10th ed., ASM International, Materials Park, OH.
10. G. E. Totten, ed. (2006). *The Handbook of Lubrication and Tribology*, CRC Press, Boca Raton, FL, p. 1224.
11. E. E. Klaus and E. J. Tewksbury (1984). Liquid lubricants, in *The Handbook of Lubrication (Theory and Practice of Tribology)*, Vol. II, ed. E. R. Booser, CRC Press, Boca Raton, FL, pp. 229–254.
12. F. A. Litt. Viscosity index calculations, in *Starting From Scratch—Tribology Basics*, STLE, Park Ridge, IL, pp. 9–10.
13. M. Sanchez-Rubio, A. Heredia-Veloz, J. E. Puig, and S. Gonzalez-Lozano (1992). A better viscosity-temperature relationship for petroleum products, *Lubr. Eng.*, 48(10), pp. 821–826.
14. R. C. Coy (1997). Practical application of lubrication models to engines, in *New Directions in Tribology*, ed. I. M. Hutchings, Mech. Engr. Press Ltd., UK, pp. 197–210.
15. K. T. Ramesh and R. J. Clifton (1987). A pressure-shear plate impact experiment for studying the rheology of lubricants at high pressures and high shear rates, *J. Tribol.*, 109, pp. 215–222.
16. M. D. Hersey (1966). *Theory and Research in Lubrication*, Wiley, New York, pp. 123–158.
17. A. A. Raimondi (1968). Analysis and design of sliding bearings, in *Standard Handbook of Lubrication Engineering*, McGraw-Hill, New York (Chapter 5).
18. A. S. Hall, A. R. Holowenko, and H. G. Laughlin (1961). Lubrication and bearing design, in *Machine Design*, Schaum's Outline Series, McGraw-Hill, New York, p. 279.
19. I. M. Hutchings (1992). *Tribology—Friction and Wear of Engineering Materials*, CRC Press, Boca Raton, FL, p. 65.
20. H. J. Sneck and J. H. Vohr (1983). Hydrodynamic Lubrication, in *Handbook of Lubrication (Theory and Practice of Tribology)*, Vol. II, ed. E. R. Booser, CRC Press, Boca Raton, FL, pp. 69–91.
21. M. M. Khonsari and E. R. Booser (2001). *Applied Tribology: Bearing Design and Lubrication*, Wiley, New York.
22. H. A. Spikes (1999). Thin films in elastohydrodynamic lubrication: The contribution of experiment, *Proc. Inst. Mech. Eng., Part J, J. Eng. Tribol.*, 213(5), pp. 335–352.
23. B. J. Hamrock and D. Dowson (1981). *Ball Bearing Lubrication—the Elastohydrodynamics of Elliptical Contacts*, Wiley-Interscience, New York, NY.
24. D. Dowson and G. R. Higginson (1959). A numerical solution to the elastohydrodynamic problem, *J. Mech. Eng. Sci.*, 1, pp. 6–15.
25. F. F. Ling, E. E. Klaus, and R. S. Fein (1969). *Boundary Lubrication—an Appraisal of World Literature*, ASME, New York.
26. R. S. Fein (1991). Boundary lubrication, *Lubr. Eng.*, 47(12), pp. 1005–1008.
27. B. J. Briscoe and T. A. Stolarski (1993). Friction, in *Characterization of Tribological Materials*, ed. W. A. Glaeser, Butterworth Heinemann, Boston, pp. 48–51 (Chapter 3).
28. D. F. Moore (1975). *Principles and Applications of Tribology*, Pergamon Press, Oxford, UK, pp. 136–138.
29. W. A. Glaeser and K. F. Dufrane (1978). New design methods for boundary lubricated sleeve bearings, *Machine Design*, April 6, pp. 207–213.
30. A. W. J. deGee, A. Begelinger, and G. Salomon (1984). *Mixed Lubrication and Lubricated Wear*, Butterworth, London.
31. S. Odi-Owei and B. J. Roylance (1986). The effect of solid contaminants on the wear and critical failure load in a sliding lubricated contact, *Wear*, 112, p. 239.

32. J. J. Lauer and S. R. Dwyer (1990). Tribochemical lubrication of ceramics by carbonaceous vapors, *ASME/STLE Tribol. Conf.*, Toronto, October 7–10.

33. N. deGouvea Pinto, J. L. Duda, E. E. Graham, and E. E. Klaus (1984). In Situ Formation of Solid Lubricating Films From Conventional Mineral Oil and Ester Base Lubricants, *ASLE SP-14*, pp. 98–104.

34. Ref. 7, p. 9.

35. Anon. (2004). Biolubricants Market Data Sheet, Interactive European Network for Industrial Crops and Applications (IENICA), August, http://www.ienica.net/.

36. W. A. Zisman (1959). Durability and wettability properties of monomolecular films on solids, in *Friction and Wear*, ed. R. Davies, Elsevier, Amsterdam, pp. 110–148.

37. D. H. Buckley (1981). *Surface Effects in Friction and Wear*, Elsevier, Amsterdam, p. 515.

38. Ref. 30, p. 378.

39. E. Rabinowicz (1965). *Friction and Wear of Materials*, Wiley, New York, p. 219.

40. T. V. Liston (1992). Engine lubricant additives—what they are and how they function, *Lubr. Eng.*, 48(5), pp. 389–397.

41. S. Q. A. Rizvi (1992). Lubricant additives and their functions, in *ASM Handbook, Volume 18: Friction, Lubrication, and Wear Technology*, 10th ed., ASM International, Materials Park, OH, pp. 98–112.

42. L. R. Rudnick (2003). *Lubricant Additives: Chemistry and Applications*, Marcel Dekker, New York, NY, p. 760.

43. C.-Y. Tung, S. K. Hsieh, G. S. Huang, and L. Kuo (1988). Determination of friction-reducing and antiwear characteristics of lubricating engine oils compounded with friction modifiers, *Lubr. Eng.*, 44(10), pp. 856–865.

44. H. Fujita, R. P. Glovnea, and H. A. Spikes (2005). Study of zinc dialkydithiophosphate antiwear film formation and removal processes, Part I: Experimental, *Tribol. Trans.*, 48(4), pp. 558–566.

45. H. Fujita and H. A. Spikes (2005). Study of zinc dialkyldithiophosphate antiwear film formation and removal processes, Part II: Kinetic Model, *Tribol. Trans.*, 48(4), pp. 567–575.

46. J. Qu and P. J. Blau (2005–2007). Experiments on thermally-oxidized titanium alloy surfaces to improve their friction and wear performance, research at Oak Ridge National Laboratory, Tennessee, supported by the U.S. Department of Energy, Office of FreedomCAR and Vehicle Technologies.

47. S. M. Hsu, C. S. Ku, and P. T. Pei (1986). Oxidative degradation mechanisms of lubricants, in *Aspects of Lubrication*, eds. W. H. Stadtmiller and A. N. Smith, ASTM Special Technical Pub. 916, ASTM, Philadelphia, pp. 27–48.

48. E. V. Zaretsky (1990). *Liquid Lubrication in Space*, NASA RP-1240, p. 23.

49. Various authors (1998). *Space Tribology Handbook*, European Space Tribology Laboratory, AEA Technologies, Warrington, UK.

50. W. R. Jones, S. V. Pepper, M. J. Jansen, Q. G. N. Nguyen, E. P. Kingsbury, S. H. Lowenthal, and R. E. Predmore (2000). A New Apparatus to Evaluate Lubricants for Space Applications—The Spiral Orbit Tribometer (SOT), NASA Tech. memo., NASA/TM-2000–029935, 7 pp. (Also see: ASTM F2661-07 "Standard Test Method for Determining the Tribological Behavior and the Relative Lifetime of a Fluid Lubricant using the Spiral Orbit Tribometer.")

51. S. Komatsuzaki, Y. Homma, K. Kawashima, and Y. Itoh (1991). Anti-seizure and anti-wear properties of lubricating oils under refrigerant gas environments, *Lubr. Eng.*, 47(12), pp. 1018–1025.

52. G. Heinicke (1984). *Tribochemistry*, Carl Hanser Verlag, Munich, p. 446.

53. L. Grunberg and R. B. Campbell (1957). Metal transfer in boundary lubrication and the effect of sliding velocity and surface roughness, *Proceedings of the Institution of Mech. Eng. Conference on Lubrication and Wear*, London, pp. 291–301.

54. S. Sasaki (1992). Effects of water-soluble additives on friction and wear of ceramics under lubrication with water, in *Effects of Environment on the Friction and Wear of Ceramics, Bull. Mech. Eng. Lab. Jpn.*, No. 58, pp. 32–53.

55. B. Bhushan and B. K. Gupta (1991). *Handbook of Tribology*, McGraw-Hill, New York, pp. 5–11 and 5–12.

56. K. Komvopoulos, N. Saka, and N. P. Suh (1986). The significance of oxide layers in boundary lubrication, *J. Tribol.*, 108, pp. 502–513.

57. H. Tian, N. Saka, and N. P. Suh (1989). Boundary lubrication studies on undulated titanium surfaces, *Tribol. Trans.*, 32(3), pp. 289–296.

58. J. K. Lancaster and P. Moorhouse (1985). Etched pocket bearing materials, *Tribol. Int.*, 18(3), pp. 139–149.

59. E. Rabinowicz (1965). *Friction and Wear of Materials*, Wiley, New York, p. 211.

60. J. J. Lauer and W. R. Jones (1986). Friction polymers, in *Tribology and Mechanics of Magnetic Storage Systems*, Vol. III, eds. B. Bhushan and N. S. Eiss, ASME, New York, pp. 14–24.

61. Langmuir-Blodgett Films (1980). A special collection of papers in *Thin Solid Films*, 68(1).

62. S. Grannick (1992). Molecular tribology of fluids, in *Fundamentals of Friction*, eds. I. L. Singer and H. M. Pollock, Kluwer, Dordrecht, pp. 387–396.

63. G. Carson, H.-W. Hu, and S. Grannick (1992). Molecular tribology of fluid lubrication: Shear thinning, *Tribol. Trans.*, 35(3), pp. 405–410.

64. M. O. Robbins, P. A. Thompson, and G. S. Grest (1993). Simulations of nanometer-thick lubricating films, *Mater. Res. Soc. Bull.*, May, pp. 45–49.

65. M. T. Benchaita, S. Gunsel, and F. E. Lockwood (1990). Wear behavior of base oil fractions and their mixtures, *Tribol. Trans.*, 33(3), pp. 371–383.

66. C. Ye, W. Liu, Y. Chen, and L. Yu (2001). Room-temperature ionic liquids: A bovel versatile lubricant, *Chem. Commun.*, pp. 2244–2245.

67. J. Qu, J. J. Truhan, S. Dai, H. Luo, and P. J. Blau (2006). Ionic liquids with ammonium cations as lubricants or additives, *Tribol. Lett.*, 22, pp. 207–214.

68. R. H. Boehringer (1992). Grease, in *ASM Handbook, Volume 18: Friction, Lubrication, and Wear Technology*, 10th ed., ASM International, Materials Park, OH, pp. 123–131.

69. J. H. Harris (1967). Lubricating greases, in *Lubricants and Lubrication*, ed. E. R. Braithwaite, Elsevier, Amsterdam, pp. 197–268.

70. V. Villena-Denton (1993). Bismuth found best lead substitute, *Lubr. World*, September, p. A15.

71. G. Biresaw, ed. (1989). *Tribology and the Liquid-Crystalline State*, American Chemical Society, Washington, DC.

72. S. E. Friberg, A. J. Ward, S. Gunsel, and F. E. Lockwood (1990). Lyotropic liquid crystals in lubrication, in *Tribology and The Liquid Crystalline State*, Amer. Chem. Soc., Washington, DC, pp. 101–111.

73. H. S. Lee, S. H. Winoto, W. O. Winer, M. Chiu, and S. E. Friberg (1990). Film thickness and frictional behavior of some liquid crystals in concentrated point contacts, in *Tribology and The Liquid Crystalline State*, Amer. Chem. Soc., Washington, DC, pp. 113–121.

74. B. I. Kupchinov, V. G. Rodnenkov, S. F. Ermakov, and V. P. Parkalov (1991). A study of lubrication by liquid crystals, *Tribol. Int.*, 24(1), p. 25.

75. J. B. Peace (1967). Solid lubricants, in *Lubrication and Lubricants*, ed. E. R. Braithwaite, Elsevier, Amsterdam, pp. 67–118 (Chapter 2).

76. B. Bhushan and B. K. Gupta (1991). *Handbook of Tribology*, McGraw-Hill, New York.

77. F. J. Clauss (1972). *Solid Lubricants and Self-Lubricating Solids*, Academic Press, New York.

78. J. K. Lancaster (1984). Solid lubricants, in *CRC Handbook of Lubrication*, ed. E. R. Booser, CRC Press, Boca Raton, FL, pp. 269–290.

79. H. E. Sliney (1992). Solid lubricants, in *ASM Handbook, Volume 18: Friction, Lubrication, and Wear Technology*, 10th ed., in ASM International, Materials Park, OH, pp. 113–122.

80. F. P. Bowden and D. Tabor (1986). *The Friction and Lubrication of Solids*, 2nd ed., Oxford Science Press, Oxford, pp. 114–115.

81. A. Sonntag (1960). Lubrication by solids as a design parameter, *Electro-Technology*, 66, pp. 108–115.

82. M. B. Peterson, S. F. Murray, and J. J. Florek (1969). Consideration of lubricants for temperatures above 1000°F, *ASLE Trans.*, 2, pp. 225–234.

83. I. M. Allam (1991). Solid lubricants for applications at elevated temperatures, *J. Mater. Sci.*, 26, pp. 3977–3984.

84. Ref. 77, pp. 114–115.

85. A. Erdemir (1991). Tribological properties of boric acid and boric-acid-forming surfaces. Part I: Crystal chemistry and mechanism of self-lubrication of boric acid, *Tribol. Trans.*, 47(3), pp. 168–173.

86. A. Erdemir, G. R. Fenske, R. A. Erck, F. A. Nichols, and D. E. Busch (1991). Tribological properties of boric acid and boric-acid-forming surfaces. Part II: Mechanisms of formation and self-lubrication of boric acid films on Coron- and Boric acid-containing surfaces, *Tribol. Trans.*, 47(3), pp. 179–184.

87. P. W. Bridgman (1935). Effects of high shearing stress combined with high hydrostatic pressure, *Phys. Rev.*, 48, pp. 825–847.

88. M. S. Bednar, B. C. Cai, and D. Kuhlmann-Wilsdorf (1993). Pressure and structure dependence of solid lubrication, *Lub. Eng.*, 49(10), pp. 741–749.

89. F. P. Bowden and D. Tabor (1986). *The Friction and Lubrication of Solids*, 2nd ed., Clarendon Press, Oxford, pp. 110–111.

90. W. O. Winer (1967). Molybdenum disulfide as a lubricant: A review of the fundamental knowledge, *Wear*, 10, pp. 422–252.

91. P. D. Fleischauer and R. Bauer (1987). Chemical and structural effects on the lubrication properties of sputtered MoS_2 films, *Tribol. Trans.*, 31(2), pp. 239–250.

92. A. R. Lansdown (1999). *Molybdenum Disulphide Lubrication*, Elsevier, UK, 406 pp.

93. P. Sutor (1991). Solid lubricants: Overview and recent developments, *Mater. Res. Bull.*, May, pp. 24–30.

94. R. L. Fusaro (1978). Lubrication and Failure Mechanisms of Molybdenum Disulfide Films, I—Effect of Atmosphere, National Aeronautical and Space Administration special pub., NASA TP-1343.

95. M. N. Gardos (1987). The synergistic effects of graphite on the friction and wear of MoS_2 films in air, *Tribol. Trans.*, 31(2), pp. 214–227.

96. E. E. Bisson and W. J. Anderson (1964). *Advanced Bearing Technology*, NASA SP-38, Washington, DC.

97. P. J. Blau and C. S. Yust (1993). Friction microprobe studies of model self-lubricating surfaces, *Surf. Coat. Tech.*, 62, pp. 380–387.

98. W. J. Bartz, R. Holinski, and J. Xu (1986). Wear life and frictional behavior of bonded solid lubricants, *Lubr. Eng.*, 42(12), pp. 762–769.

99. H. Heshmat (1991). The rheology and hydrodynamics of dry powder lubrication, *Tribol. Trans.*, 34(3), pp. 433–439.

100. H. Heshmat (1991). The Quasi-Hydrodynamic Mechanism of Powder Lubrication Part I: Lubricant Flow Visualization, STLE Preprint 91-AM-4D-1.

101. H. Heshmat (1992). Traction characteristics of high-temperature powder-lubricated ceramics ($Si_3N_4/\alpha SiC$), *Tribol. Trans.*, 35(2), pp. 360–366.

102. H. Heshmat (1992). On some experimental theological aspects of triboparticulates, in *Wear Particles: From the Cradle to the Grave*, M. Godet, D. Dowson, and C. Taylor (eds.), Elsevier, Amsterdam, pp. 357–368.

103. F. P. Bowden and D. Tabor (1986). *The Friction and Lubrication of Solids*, 2nd ed., Oxford Science Press, Oxford, pp. 164–166.

104. M. Nosaka, M. Dike, M. Kikuchi, K. Kamijo, and M. Tajiri (1993). Tribo-characteristics of self-lubricating ball bearings for the LE-7 liquid hydrogen rocket turbopump, *Tribol. Trans.*, 36(3), pp. 432–442.

105. B. J. Briscoe (1992). Friction of organic polymers, in *Fundamentals of Friction: Macroscopic and Microscopic Processes*, eds. I. L. Singer and H. M. Pollock, Kluwer, Dordrecht, pp. 167–182.

106. B. Bhushan and B. K. Gupta (1991). *Handbook of Tribology*, McGraw-Hill, New York, pp. 5–57.

107. A. Erdemir (1994). A review of the lubrication of ceramics with thin solid films, in *Friction and Wear of Ceramics*, ed. S. Jahanmir, Marcel Dekker, New York, pp. 119–162.

108. K. Adachi (2007). Tribology for space applications: In-situ and on-demand tribo-coatings, Plenary Lecture, 16th International Conference on Wear of Materials, Montreal, Canada, April 15–19.

109. G. C. Pratt (1967). Plastic-based bearings, in *Lubrication and Lubricants*, ed. E. R. Braithwaite, Elsevier, Amsterdam, pp. 412–422 (Chapter 8).

110. M. B. Peterson (1990). Advances in tribomaterials, in *Achievements in Tribology*, eds. L. B. Sibley and F. E. Kennedy, Amer. Soc. of Mechanical Engineers, New York, NY, pp. 91–109.

111. V. T. Morgan (1970). Porous metal bearings, in *Friction and Antifriction Materials*, Plenum Press, New York, pp. 187–210.

112. H. Youssef and M. Eudier (1970). Introduction and properties of a new porous bearing, in *Friction and Antifriction Materials*, Plenum Press, New York, pp. 291–301.

113. R. J. Dobbs, D. E. Thomas, and D. E. Wittmer (1987). Self Lubricating Silicon Nitride Article, U.S. Patent Number 4,650,592, March 17.

114. J. K. Lancaster and P. Moorhouse (1985). Etched-pocket, dry-bearing materials, *Tribol. Int.*, 18(3), pp. 139–148.

115. C. DellaCorte and H. E. Sliney (1987). Composition optimization of self-lubricating chromium carbide-based composite coatings for use to 760°C, *ASLE Trans.*, 30(1), pp. 77–83.

116. M. S. Bogdanski, H. E. Sliney, and C. DellaCorte. (1992). Tribological and Microstructural Comparison of HIPped PM212 and PM212/Au Self-Lubricating Composites, NASA Tech. Memo, TM-105615.

117. A. Gangopadhyay, S. Janahmir, and M.B. Peterson (1994). Self-lubricating ceramic matric composites, in *Friction and Wear of Ceramics*, ed. S. Jahanmir, Marcel Dekker, New York, pp. 163–197.

118. R. Kamo and W. Bryzik (1992). Solid lubricants for an adiabatic engine, *Lubr. Eng.*, 48(10), pp. 809–815.

7 Effects of Tribosystem Variables on Friction

The friction coefficient (μ) is a convenient way to characterize the resistance to relative motion between surfaces, but it is not a material property, nor is it a physical constant like the speed of light in a vacuum or Avogadro's number or the elementary charge on the electron. For a given pair of materials, μ may, on average, remain relatively constant over a range of contact conditions, but transitions in friction sometimes occur and one would be ill-advised to extrapolate (or even sometimes to interpolate) the value of μ far beyond the conditions under which it was originally measured. Therefore, it is of both fundamental and practical importance to understand the effects of variables such as sliding speed, normal force (contact pressure), temperature, environment, and material properties on frictional behavior.

Examples of the effects of tribosystem variables on frictional response can be found throughout this book, but in this chapter, we will address the effects of several specific variables. While reviewing the following examples, recognize that it is difficult to vary only one parameter at a time in a friction experiment. For example, if one raises the normal force, the wear rate may increase, in turn altering the surface roughness and introducing debris particles into the interface. When studying the effects of sliding speed over a wide range, the temperature in the interface can vary as well. Isolating the effects of only one variable on tribological transitions is a challenge in light of such synergistic factors.

Table 7.1 lists some of the things that can affect friction and be affected by friction. Their relative importance depends on the specifics of the tribosystem. Sections 7.1–7.7 describe the effects of the most-studied frictional variables.

7.1 EFFECTS OF SURFACE FINISH

Historically, the origins of friction were first linked with the imagined structure of surface features, and in fact surface geometry-related models for friction are still being developed. In some tribosystems there is a direct relationship between friction and the initial contact surface texture, which includes roughness, waviness, and lay. In other cases, however, wear quickly alters the surface(s) and any attempt to correlate post wear-in frictional behavior with the starting surface finish is obviously inappropriate. Thus, at the risk of oversimplifying

> If the contact pressure is low, if the contacting materials are relatively hard, or if a thick lubricating film is present, then initial surface features can be preserved for an extended period of sliding time and can continue to affect friction. If, however, contact pressures are high or if one or both contacting materials is relatively soft, the initial surface features will quickly be obliterated. Instead, the surface morphology will be determined by the wear properties of the materials and other factors, like transfer and debris layer formation.

TABLE 7.1
Factors That Can Affect Friction Depending on Contact Circumstances

General Category	Factor
Mechanical	Contact geometry: macro, micro, nano
	Load and contact pressure distribution at various scales
	Loading history
	System dynamics: vibrations, stiffness, damping, hysteresis
	Type of motion and velocity profile
Materials	Pairing of materials
	Composition and purity of materials
	Adhesive characteristics
	Microstructure and the sizes of microstructural features relative to the size of the tribocontact
	Elastic and plastic mechanical properties
	Property gradients in the near-surface regions
	Thermophysical properties: thermal conductivity, thermal expansion, etc.
	Method of creating the surface (finishing, machining artifacts)
	Residual stress state in the near-surface regions
Thermal effects	Frictional heating
	External heat sources
	Thermoelastic instability
	Thermally induced phase transformations: softening, melting
	Tribochemical activation
	Thermal shock during cycling
Lubrication	Quantity and means of supply
	Regime of lubrication
	Properties of the lubricant
	Lubricant chemistry (tribochemistry)
	Lubricant "aging"
	Filtration and cleanliness
Tribochemistry	Relative humidity
	Surface reactivity/catalysis
	Cleanliness
	Composition of the surrounding environment
	Tribopolymerization
	Friction polymer formation
	Oxides and tarnish films
Third bodies	Transfer particle formation
	Mechanics and lubricity of triboformed layers (tribomaterials)
	Wear particle concentration and agglomeration
	Sizes, shapes, and morphology of particles
	External contaminants
	Flow of third bodies in and out of the contact

TABLE 7.2

Friction Coefficient (μ) and Wear of Metals for Different Contact Durations

Pin Material	Disk Material	μ After 5 Rotations	μ After 100 Rotations	Appearance After 100 Rotations Pin	Disk
Cu	Ni	0.28	1.00	Wide abraded area	Wide abraded area
Ni	Fe	0.82	1.10	Wide abraded area	Wide abraded area
Ti	Zr	0.69	0.64	Wide area, some abrasion	Wide area, some abrasion
W	Fe	0.19	0.54	Narrow, no abrasion	Narrow, significant abrasion
Ag	Fe	0.15	0.18	Medium area, smeared	No abrasion observed
Pb	Fe	0.38	0.93	Large area, some abrasion	Wide area, smeared

Source: Coffin, L.F. in *Friction and Wear*, Elsevier, New York, NY, 1959, 42. With permission.

High-friction sliding couples have more energy available to alter the structure of the materials in the contact zone, including the promotion of wear. As a result, the magnitude of friction can affect the progression of wear, which in turn can alter surface characteristics and affect friction. In such cases, the cause and effect relationship between friction and wear during continued sliding contact becomes a recursive one.

In the mid-1950s, Coffin[1] related the changes in friction of metal pairs in a pin-on-disk apparatus to the appearance of the sliding surfaces of both specimens (see Table 7.2). Usually the friction coefficient increased with time (but not in every instance), and the changes in friction between 5 and 100 rotations were not equal among the difference pairs. Such observations are useful, but they lack the detailed information needed to derive quantitative relationships between the friction and the texture of surfaces produced by wear. Results of Coffin's work also broach the important subject of running-in, which is discussed in more detail in Chapter 8.

Experiments on unlubricated, ground and polished ceramics indicate that the initial surface finish can be retained for substantial periods of time yet fail to correlate directly with the level of friction.[2] The R_a, R_t, and R_q of ground or ground and polished ceramic disks were measured in several directions and averaged. Pin-on-disk friction and wear tests were then performed in air on dry surfaces using 9.525 mm diameter bearing balls as sliders with a normal force of 10 N and sliding speed of 0.1 m/s. The relationship between roughness parameters and steady-state friction coefficients for SiC spheres sliding on Si_3N_4 disks is shown in Figure 7.1. Clearly, changes in roughness could not adequately explain the frictional behavior of these materials. Initial surface finish did, however, affect the length of the running-in period. The smoother the surface, the longer it took to reach the steady-state friction.

Early friction models used simple geometrical shapes such as sawteeth or rigid cones, but these inadequately described surfaces that were produced by industrial

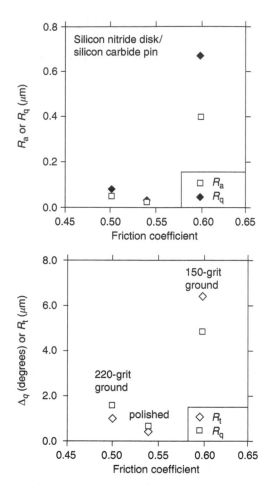

FIGURE 7.1 The relationship between roughness parameters and steady-state friction coefficients for 9.53 mm diameter SiC spheres sliding on Si_3N_4 disks.

machining and finishing operations. Realizing this shortcoming, engineers relied on field trials or laboratory tests to design bearings. In 1940, a conference on "Friction and Surface Finish" was held at the Massachusetts Institute of Technology.[3] The proceedings from that conference provide a historical insight into the approaches adopted by the engineers at the time to control friction using surface finish. Most of the papers focused on lubricated surfaces, such as the bores of cylinders in internal combustion engines, but a few papers discussed fundamental work on unlubricated sliding. For lubricated sliding, it is known that the smoothest surfaces may not provide the best frictional behavior. In 1940, Swigert[4] of Chrysler Corporation remarked:[4]

> Down through the ages but particularly during the last thirty years has the striving for smoother bearing surfaces been prevalent .... it seems strange to anyone who has

followed this progress of smoother surfaces to find that a few concerns and businesses [that] have been founded upon the development of smoother surfaces should now suddenly discover that smoother surfaces were not desirable and that a measurable degree of surface roughness was necessary.

One important example of the optimization of roughness is in the control of stiction in computer hard disks, but there are many other examples, particularly in the field of lubrication engineering where roughness, embodied in the "Λ ratio," affects the regime of lubrication.

The tightening and loosening of bolts is one illustration of how surface finish affects friction. Karamis and Selcuk[5] found that the torsional moment (i.e., friction force acting at the diameter of the threads times the distance to the center of rotation) for tightening steel bolts decreased approximately linearly with increasing surface roughness. An increase in R_a by a factor of about 7.5 (e.g., raised from 0.2 to 1.5 µm) produced a fourfold decrease in torsional moment.

Not only surface roughness but also lay (i.e., the directionality of finishing marks or scratches) affects friction. For example, some computer hard disks, instead of having circumferential grooves, are given a lay at 10–20° from the traveling direction of the flying head to reduce friction. In addition, automotive internal combustion engine cylinders are given special honing treatments to produce crosshatched grinding marks at an angle of about ±24° to the direction of sliding. The author developed a simple method to simulate this pattern to make laboratory-scale friction tests of piston rings against materials for cylinder liners more simulative.[6]

In 1940, Gowger[7] discussed results of tests on the effects of both initial surface roughness and orientation of the finishing scratches on the friction of several metals on steel. Tests were conducted using a 200 mm diameter finished steel disk rotating slowly at about 0.05 rpm with a flat slider loaded against it. Two experiments from that series have been plotted in Figure 7.2 to illustrate that different pairs of metals can react differently to the orientation of finishing scratches. The copper-on-steel couple exhibited a less pronounced effect of surface lay on friction than did the gold-on-steel couple. Although Gowger provided no interpretation, it is likely that the difference in the effects had to do with two differences between the Cu and Au: (a) Cu tends to form copper oxide layers, which under the right conditions can serve as a solid lubricant, but Au is a noble metal and does not form such lubricative films, and (b) the Au, being much softer than Cu, is more sensitive to the abrasion of the surface ridges, which can dig into the surface and shave it off, especially when they lie perpendicular to the sliding direction. This shaving action can also promote the transfer of Au to the disk surface. It would have been instructive had Gowger also published photographs of the disk wear track surfaces at various angles of lay.

Eiss and Smyth[8] reported friction coefficient data for sliding polymers parallel and perpendicular to the lay of ground steel surfaces. Single passes were performed using a normal load of 1.96 N (200 g) and sliding speed of 9 mm/s. Selected data from that study are plotted in Figure 7.3, where the ratios of the sliding friction coefficient perpendicular to the lay to that obtained parallel to the lay for various roughnesses are shown for polyvinyl chloride (PVC) and low-density polyethylene (LDPE). Friction coefficients sliding transversely to the lay were more than 50% higher than those along the lay, and there seemed to be a larger effect of lay at the

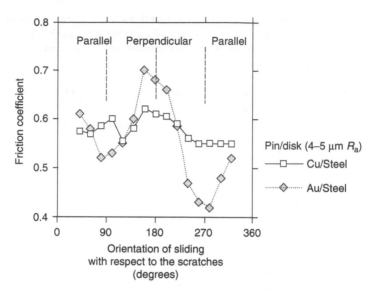

FIGURE 7.2 Selected data from Ref. 7 illustrating that different materials couples can exhibit differing responses to the orientation of the finishing scratches.

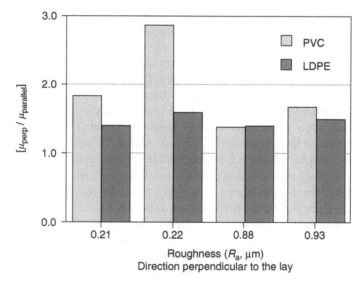

FIGURE 7.3 Ratio of the sliding friction coefficient perpendicular to the lay to that obtained parallel to the lay for various roughnesses on polyvinyl chloride (PVC) and low-density polyethylene (LDPE). (Adapted from Eiss, N.S. Jr. and Smyth, K.A., *Trans. ASME, J. Lubr. Tech.*, 103, 266, 1981.)

lower surface roughness values. The latter finding is in contrast to the work of Jeng[9] on lubricated friction, who constructed model surfaces using plastic films containing microgrooves. Jeng found that lower roughness height yielded lower friction and that transverse roughness produced lower friction than longitudinal roughness.

Unlike the unlubricated tests of Eiss and Smyth, Jeng's results were influenced by the flow of the lubricant within the channels of the surface and produced quite different effects on friction of both roughness and direction of the lay.

A similar effect to that of lay was observed for the role of microstructural orientation in the friction of polymers sliding on oriented composite materials. In a study by Roberts,[10] Lexan (polycarbonate) pins, 3 mm diameter, were slid over the surface of a graphite fiber–epoxy composite material in which the fibers were oriented in different directions to the sliding direction. Three surface roughnesses were used for the composite specimen. A load of 8.9 N at 0.1 m/s produced the results plotted in Figure 7.4. Sliding at 45° to the fiber orientation minimized the effects of surface roughness, and only for the roughest of the three surfaces, did the transverse orientation of the fibers offer a frictional advantage. Thus, control of both roughness and fiber orientation on the surface of composite materials can provide the optimal friction response. It should be noted, however, that it is very difficult to polish composite materials with constituents of differing hardness, and that the recession of the softer phase during abrasive polishing may produce a lay parallel to the fiber orientation, thereby enhancing roughness effects.

The roughness of surfaces changes during the running-in process. Machined surfaces tend to become smoother during running-in as the tallest peaks of the ground or turned surface profiles are worn off. Attempts have been made to calculate the "equilibrium roughness" of surfaces after running-in. For example, Kragelskii and Kombalov[11] developed a calculation on the basis of a contact of a rigid rough surface and a deformable, smooth one. Later, an entire chapter of the book by Kragelskii et al.[12] was devoted to the subject.

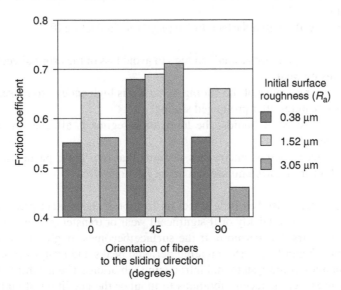

FIGURE 7.4 Effects of fiber orientation in tests involving (polycarbonate) pins, 3 mm diameter, sliding over surfaces of graphite fiber–epoxy composite material in which the fibers were oriented in different directions with respect to the sliding direction. (Adapted from Roberts, J.C., *ASLE Trans.*, 28, 503, 1985.)

Blau[13] has described a sequence of periodic wear stages for metallic surfaces during quasi-steady-state sliding under nominally dry or lubricant-starved conditions:

1. The surface experiences plastic deformation, and fatigue cracks grow.
2. Spalls are produced by the linking of fatigue cracks, which leave pits in the surface.
3. Plateaus remain to bear the load as the surrounding material wears away.
4. The plateaus support heavier and heavier loads per unit area.
5. The weakened plateaus are sheared away and the process repeats.

The evolutionary nature of surface morphology changes is one of the causes for periodic changes in friction as a function of sliding time.

Increasingly realistic portrayals of the effects of surface finish on friction have been enabled by the advances in computer-assisted modeling. It is no longer necessary to assume that surfaces are composed of simple shapes like spheres or cones. Elastic distortions and changes in the contact zone can be addressed on an individual asperity level to generate time sequences. An example from the work of Wang's group at Northwestern University[14,15] is shown in Figures 7.5a through 7.5d. Figure 7.5a is the three-dimensional representation of a surface with an arithmetic average roughness of 0.33 μm, and Figure 7.5b is a map of the corresponding von-Mises stress (σ_{VM}) normalized by the yield strength (Y). The elastic stresses can be calculated by using a convolution between surface tractions and influence coefficients. Then, using the von-Mises criterion, the onset of plastic deformation is determined, which allows calculation of strains (Λ), as depicted in Figure 7.5c. It is also possible to calculate the real area of contact (37.1% of the apparent area of contact in the example of Figure 7.5d).

In summary, the *initial* surface finish plays role in friction if:

1. The contact pressure is small enough to avoid loss of the original geometric features by wear.
2. The wear resistance of the sliding materials is high enough to preserve the surface features after prolonged sliding.
3. In the presence of liquids, the lubrication regime is affected by surface roughness.
4. The surface microgeometry has features that trap loose particles, which would otherwise accumulate and affect friction.

Initial surface finish has little effect on friction if the contact pressure is high enough to obliterate it quickly or if significant wear or transfer of material from the counterface occurs. Furthermore, if the surface becomes rough on a macroscale, the slider may begin to bounce, creating periodic peaks and troughs in load, altering the dynamics of the system and introducing vibrations. The machining industry commonly uses wear-induced vibrations to monitor the condition of cutting tools, and vibration monitoring is also used to assess the health of turbo-machinery bearings. (See Section 7.7 for a discussion of friction–vibration effects.)

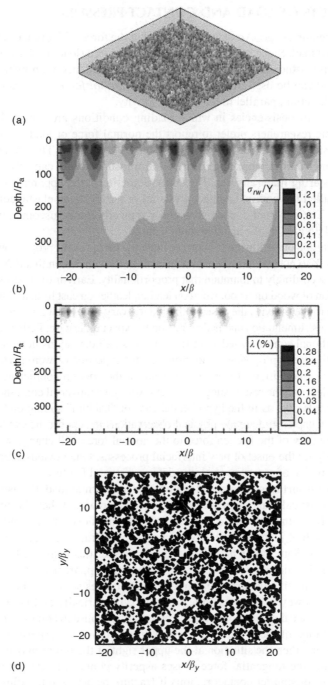

FIGURE 7.5 Example of contact modeling on an asperity level. (a) Initial geometry of a rough surface, (b) the von-Mises stress normalized by the yield strength in a vertical cross-sectional plane, (c) effective plastic strain in a vertical cross-sectional plane, and (d) real contact area (black zones). (Reprinted with permission of Q. Wang, Northwestern University.)

7.2   EFFECTS OF LOAD AND CONTACT PRESSURE

Over three centuries ago, Amontons asserted that the force of friction is proportional to the applied load. Since then, the effect of normal force (load) on friction has been one of the most-studied tribological variables. In accordance with metric practice, the term *load* can be used interchangeably with *normal force* if it is applied in the downward direction, parallel to the force of gravity.

There are inconsistencies in which loading conditions are reported in the literature. Some researchers prefer to report the normal force or load, yet others prefer contact stress or contact pressure–velocity (*PV*) products. The *PV* product is a convenient metric for comparing the maximum load-carrying capacity of different polymeric bearing materials (i.e., *PV* limit). Sometimes, the apparent contact pressure (also called the nominal contact pressure) or the Hertzian elastic contact stress is reported. The apparent contact pressure is typically a macroscopic quantity: the applied load divided by the apparent contact area.

The definition of the kinetic friction coefficient μ (or μ_k) as the proportionality of the friction force **F** to normal force **N** naturally leads to the conclusion that if **N** changes, **F** will change accordingly to maintain that proportionality. Early investigators who studied the friction of wood on wood, wood on leather, leather on cast iron, and so on, found that over the range of loads they used μ did indeed vary little. However, they used relatively crude instrumentation and had no reason to expect otherwise. Perhaps, the expectation of frictional constancy led some to overlook small deviations in their results. As experimental science progressed, and more careful experiments were performed over a wide range of conditions, the effects of load on the friction coefficient were more clearly revealed, and the results suggested that μ was not a universal constant for a given pair of materials. Just as in the hypothetical case of "tribium" sliding on "frictionite" discussed in Chapter 4, there are physical observations to support the expectation that the proportionality of the friction force to the normal force will change when contact conditions trigger the onset of new interfacial processes. One obvious change is from elastic deformation of asperities to plastic deformation and fracture.

When fresh surfaces are placed together under a normal load, the asperities will deform first elastically and then plastically as load increases. When the applied pressure is exactly balanced by the counterforce, distributed over the expanding contact area, the growth of junctions will stop. Figure 7.6 illustrates several representations of this concept. Some wear models assume that asperities begin as hemispherical lumps. Then the distribution of contacts within the apparent contact area might appear something like the arrangement of contact circles shown at the upper left of the figure. If, however, contacts do not deform symmetrically and homogeneously, as in the case when deformation is anisotropic (i.e., response to compression depends on the orientation of the crystal slip systems of the individual asperities to the axis of compression), then the situation at the upper right of the figure may occur. When sliding occurs, the tangential force causes asperity contacts to stretch or to break up into chains of smaller contact regions if fracture occurs (see the illustrations on the lower side in Figure 7.6). Comparing the lower right of Figure 7.6 (reproduced from the first edition of this book) with Figure 7.5d, it is gratifying that the author's portrayal of the physical arrangement of contact spots was verified 10 years later by more rigorous analytical methods.

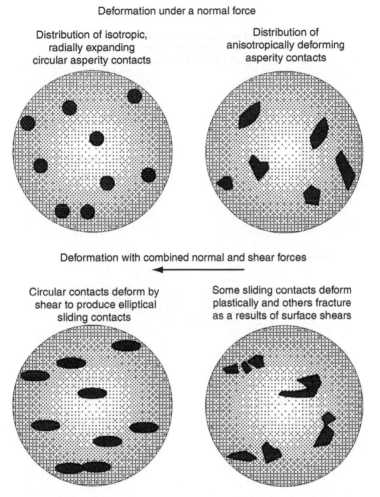

FIGURE 7.6 Illustration of the development of contact junctions when surfaces are placed together under load.

Ground surfaces usually have a "lay" or a directionality that reflects the abrasion of the finishing step. Therefore, rather than beginning with a uniform or nondirectional feature arrangement, bearing surfaces can begin with other than a random texture. Figure 7.7 illustrates how the same lay might respond differently to sliding if it were oriented perpendicular or parallel to the sliding direction.

Simple models for friction have included the assumption that the real area A of contact supporting the normal load on a plastically deforming surface can be calculated directly from the hardness H and normal force $\mathbf{P}$, as $A = \mathbf{P}/H$. This has prompted numerous attempts to measure the true contact area. Such studies have involved techniques such as electrical contact resistance measurements, cross-sectioning of surfaces, and viewing loaded surfaces through transparent plates. Since the "real area of contact" is such a central concept in tribology that extended discussions of it are contained in the classic texts by Bowden and Tabor[16] and Kragelskii.[17]

Deformation under a normal force

The lay of machining grooves
produce elongated and
oriented asperity contacts

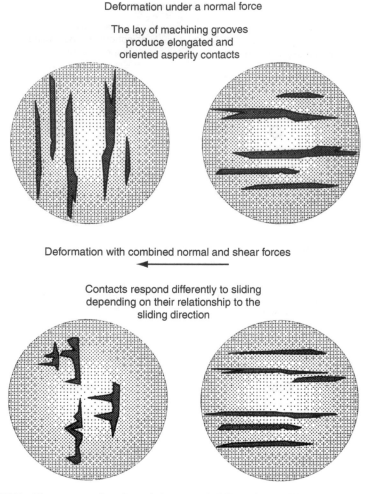

Deformation with combined normal and shear forces

Contacts respond differently to sliding
depending on their relationship to the
sliding direction

FIGURE 7.7   The same surface lay might respond differently to sliding if oriented perpendicular or parallel to the sliding direction.

Bely et al.[18] listed 13 different methods for determining the real area of contact as well as their limitations. More recent interest in this subject was prompted by the need to understand the behavior of momentary contacts with magnetic disk drives in computers.[19] Other work has used fractals to characterize surface contact.[20,21] Depending on the conditions, the ratio of the true area of contact to the apparent (nominal) area of contact can range from 1.0 to 1/100,000 or less. Table 7.3 summarizes the data from Bowden and Tabor[22] in which steel plates were pressed together under a range of loads and the fraction of true contact area to apparent contact area was estimated from electrical resistivity measurements. The data in Table 7.3 indicate a linear relationship between load (P) and area fraction (A_f). Namely, $A_f = (5 \times 10^{-6})P + (7.49 \times 10^{-12}) \approx (5 \times 10^{-6})P$.

TABLE 7.3

Contact Resistance-Based Estimates of the Area Fraction (A_f) of Contact on Polished Flat Steel Plates

P (kg)	Contact Resistance (Ω)	A_f (dimensionless)
2	50.0	0.01×10^{-3}
5	25.0	0.025×10^{-3}
20	9.0	0.1×10^{-3}
100	2.5	0.5×10^{-3}
500	0.9	2.5×10^{-3}

Source: Adapted from Bowden, F.P. and Tabor, D. in *The Friction and Lubrication of Solids*, Oxford Science Press, UK, 1986, 31.

One can divide the numerator and denominator in the definition for the friction coefficient by the true area of contact A to get

$$\mu = \frac{F/A}{P/A} \tag{7.1}$$

Since interfacial shear stress at yielding τ is the friction force per unit area, defining the hardness H as the pressure required to resist penetration gives:

$$\mu = \frac{\tau}{H} \tag{7.2}$$

From this argument by Bowden and Tabor,[23] friction coefficients for metals can be estimated from the shear strength of their asperity junctions and their hardness. Holm[24] pointed out that the hardness of metals is about two to three times the shear strength; therefore, Equation 7.2 predicts friction coefficients between about 0.33 and 0.5. This range does not agree with the value obtained in friction experiments on clean metals. For example, the measured kinetic friction coefficients for dry steel on steel is typically 0.6–0.7, for copper or bronze on steel in air it is typically over 0.8, and for unlubricated self-mated aluminum alloys it commonly exceeds 1. Furthermore, temperature and strain rate (sliding speed) can affect both τ and H. And we also know from the work of Bridgman[25] that τ is a function of contact pressure, especially at high pressures.

The hardness number for a surface is conventionally determined with the load applied vertically. However, that may not reflect the hardness of the surface when there is a significant lateral component (shear component) deriving from the force distribution beneath a slider. Using a series of high-purity metals of varying hardness, Blau[26] showed that the impression diameter (D) obtained with a spherically tipped indenter, applied vertically, is not equal to the scratch width (w) when the same indenter, a spherically tipped diamond, is slid along the surface under the same load (see Table 7.4). A spherical impression of diameter D gives a projected area A_P

$$A_P = \frac{\pi}{4}D^2 \tag{7.3}$$

Similarly, the projected area of the front half of a sphere traveling along a surface, as in a scratch test with a rounded indenter tip, is

$$A_S = \frac{1}{2}\pi\left(\frac{w}{2}\right)^2 = \frac{\pi}{8}w^2 \tag{7.4}$$

Both quasi-static hardness (H) and scratch hardness (H_S) can be defined as the applied load per unit of projected area, and their ratio at equal load is

$$\frac{H_S}{H} = \frac{P/A_S}{P/A_P} = \frac{2D^2}{w^2} = 2\left(\frac{D}{w}\right)^2 \tag{7.5}$$

It follows from Equation 7.5 that the (D/w) ratio, if H_S equaled H, would be 0.707; however, this was not true for the metals listed in Table 7.4. In fact, the ratio of scratch hardness to quasi-static hardness from these simple experiments is shown in the right-hand column of the table, and it is always greater than 1. This departure makes sense in light of the observation that metals often display higher strength when the imposed strain rate is high, which in turn leads to the notion that the dynamic properties of materials may be more useful in modeling friction (and wear) than do quasi-static characteristics, like traditional hardness numbers. However, the author's experience with low-speed scratch testing of metals and alloys using a 0.2 mm radius spherical diamond tip (as in the scratch hardness standard ASTM G 181)

TABLE 7.4
Ratio of Static Indentation Diameter to Scratch Width in High-Purity Metals

Metal	PAH (kg/mm²)[a]	D/w (From Ref. 26)[b]	H_S/H (From Equation 7.5)
Tin	7.1	1.11	2.46
Cadmium	23.1	1.54	4.74
Nickel	98.2	0.93	1.73
Iron	198.0	1.08	2.33
Molybdenum	224.0	1.16	2.69
Cobalt	250.0	1.13	2.55

[a] The projected area hardness (PAH) number was determined using both long and short impression diagonals of a Knoop microindentation at 0.98 N load. This approach accounts for elastic recovery and plastic anisotropy in the residual impression.
[b] Indentations were made using the same diamond stylus with a 76.2 μm tip radius at 0.98 N load. In one case the tip was vertically applied, and in the other case it was sliding at ≈1.2 mm/s.

indicates that the difference between the scratch hardness number and Vickers or Knoop microindentation hardness numbers is not as large as might be expected from Table 7.4. This suggests that more research is needed on the effects of stain rate on hardness under scratching and quasi-static indentation conditions.

Considering elastic and plastic conditions, different dependencies of friction force on normal force can be demonstrated. If we accept the relationship $\mathbf{F} = \tau A$, then for μ to be independent of normal force, shear stress τ must be independent of load and the true contact area A becomes proportional to $\mathbf{P}$. In contrast to this linear dependency of friction force on $\mathbf{P}$, the elastic contact equations of Hertz show that for a sphere pressed against a flat body, the area of contact is proportional to $\mathbf{P}^{2/3}$. In that case, the friction coefficient is expected to decrease as a function of contact pressure. Figure 7.8 illustrates that trend, with all the other factors remaining constant. However, the fact that friction coefficient decreases with increasing load in some cases does not automatically guarantee that Hertzian elastic behavior is the cause. Other possible causes include the effects of increasing load on wear, which causes more surface roughening or generates a layer of triboformed material.

It is convenient to consider load effects on the friction of solids from three perspectives:

1. *Persistence of films.* Surface adsorbates, oxide films, and other reaction products can affect friction if the tangential force or the wear rate is insufficient to rub them off.
2. *Contact pressure effects on microstructure.* Mechanical properties of the near-surface microstructural constituents are influenced by contact pressure. The shear strength of solids may be changed by increasing pressure (e.g., work-hardening in the case of metals or effects of hydrostatic pressure on shear stress). The deformation of materials during sliding can alter the

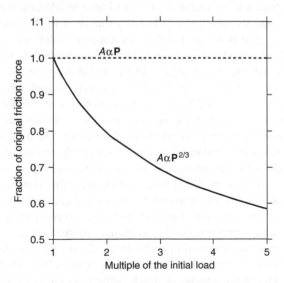

FIGURE 7.8 Decrease in friction coefficient owing to the elastic contact area's proportionality to $\mathbf{P}^{2/3}$ and the proportionality of $\mathbf{F}$ to the elastic contact area.

near-surface crystallographic orientation to promote easier shear or to induce phase transformations.

3. *Wear transitions*. Load-induced wear transitions affect friction. Wear processes alter the state of the surface, which in turn affects friction.

The first set of effects is different from the second two because the latter tend to be strongly dominated by properties of the bulk materials and of the deformed layers that form during sliding. The first type of effect originates from tribochemistry— reactions of exposed surfaces with gases in the surrounding environment and the tendency of species to be adsorbed. Some of these effects are discussed in the section on static friction and stick-slip behavior.

Figure 3.34, the last figure in Chapter 3, summarizes the author's results for friction experiments in which the same two materials (440C stainless steel and Ni_3Al alloy) were slid against each other at three loads on each of three different test rigs.[27] Overall, a load range of four orders of magnitude was explored (98 mN to 100 N). For each set of tests, the friction coefficient decreased with increasing normal load, but the reasons for this behavior were different in each case. In the lowest load case, Hertzian contact arguments apply. That is, the contact area increases as the load to the (2/3) power and the friction force times the area equals the friction coefficient. Under higher loads, the shear strength in the interface was affected both by the progressive roughening of the contact surfaces and by the accumulation of debris from increased wear at higher and higher loads. Therefore, the slopes of the friction–load dependency differ between the three cases.

The role of thin surface films in friction has been described earlier in regard to static friction. The main point here is that the depth below the surface at which the maximum shear stress occurs during sliding affects the degree of importance that a film will have on friction. If the sliding-induced shear stress is concentrated within the film, then a low-shear-stress film will tend to lubricate the surface. If the film is stronger than its substrate and bonds strongly to it, then the substrate may fail instead. If the film is brittle and the film–substrate interface is weak, the film might delaminate and expose the substrate. The various frictional scenarios that film-covered surfaces may experience suggest that material-dependent transitions in frictional response as a function of load would be observed, and this phenomenon is indeed a common one. Table 7.5, adapted from Bowden and Tabor's 1956 book,[28] shows that under controlled sliding conditions, there were certain loads for transitions from oxide-on-oxide behavior to metal-on-metal behavior. The limiting friction coefficients for the clean metal and oxidized case, respectively, are listed to either side of the transition load. Some transitions involved higher friction and others lower friction. This behavior results from the differences in shear strength, brittleness, and hardness between the metals and their corresponding oxides. For example, zinc oxide is about five times harder than zinc metal, but aluminum oxide is more than 10 times harder than aluminum metal. Furthermore, the structure of oxide layers as they form on surfaces can affect their shear strength parallel to the shear directions that are imposed during sliding. Some oxides tend to flake off while others adhere strongly.

Kato et al.[29] conducted a series of simple, sphere-on-flat laboratory experiments to demonstrate how the load affected friction in the presence of thin metallic films.

TABLE 7.5
Friction Transition Loads for Oxidized to "Bare Metal" Behavior

Metal	μ (Oxide on Oxide)	Transition Load (g)	μ (Metal on Metal)
Silver	0.8	0.003	1.0
Aluminum	0.8	0.2	1.2
Zinc	1.2	0.5	0.8
Copper	0.8	1.0	1.6
Iron	1.0	10.0	0.6
Chromium	0.4	>1000	—

Source: Adapted from Bowden, F.P. and Tabor, D. in *Friction and Lubrication*, Wiley, 1956, 37.

For a variety of coatings, they proposed that the general decrease in friction with increasing load **P** could be expressed in terms of the friction extrapolated in zero load (μ_1), the friction at high loads μ_2, and a function $f(\mathbf{P})$ as follows:

$$\mu = \mu_2 + (\mu_1 - \mu_2)e^{-f(\mathbf{P})} \tag{7.6}$$

The load-dependence exponent $f(\mathbf{P})$ was given as follows:

$$f(\mathbf{P}) = \gamma\left(\frac{\mathbf{P}}{\pi R t \tau_0}\right)\delta \tag{7.7}$$

where the denominator is the load-carrying capacity of the film, as a product of the slider tip radius R, the film thickness t, and the shear stress of the film τ_0. The constants γ (range 0.5–0.25) and δ (≈ 1.1) were determined by experiment. Classic experiments by Bowden and Tabor on indium showed that the friction of thin film-covered surfaces can go through a minimum at a certain value of film thickness.

Bowden and Tabor's book (Ref. 28, p. 150) illustrated how electrolytically polished copper on copper had low friction ($\mu < 0.5$) at low loads (<1.0 gf) when an oxide film was present, and reached high friction coefficients ($\mu = 1.6$) at higher loads (a few kgf) when the oxide was penetrated. These kinds of experiments were done with relatively low loads that allowed the effects of persistent oxide films to be observed. Low-load frictional response is important for applications such as electrical contacts and where large contact areas enable the applied load to be distributed over many asperity contacts with a relatively large area. In fact, the signal work of Holm[24] in the 1950s was motivated by studies of electrical contacts.

At the other extreme, the friction of metals at high loads and high localized pressures is important in metalworking. Peterson and Ling[30] investigated the frictional properties of seven metals under pressure ranges typical of metalworking operations: 8,000–200,000 psi (55.2–1380 MPa). A metal foil (about 1.0 mm thick) was squeezed in a rectangular area of contact between an anvil and a flat plate of M-2 tool steel (HRC 62), and the instrumented anvil was moved relative to the plate.

TABLE 7.6
Effect of High Pressure on the Friction Coefficient of Annealed Pure Metals

Metal Foil	μ (At 57.5 MPa)	μ (At 1380.0 MPa)
Indium	0.34	0.05
Tin	0.30	0.10
Cadmium	0.21	0.09
Aluminum	0.71	0.18
Silver	0.40	0.20
Zinc	0.66	0.22
Nickel	0.52	0.32

Note: Pressure data converted from ksi.

Source: Adapted from Peterson, M.B. and Ling, F.F. in *Friction and Lubrication in Metal Processing*, American Society of Mechanical Engineers, New York, NY, 1966, 39–67.

During the experiment, the foil thickness was reduced by as much as a factor of 4, but tearing did not occur. The friction coefficients of all seven metals fell on increasing applied pressure, as shown in Table 7.6, read from a plot in the paper by Peterson and Ling. By increasing the pressure about 24 times, the friction coefficients of the metals could be reduced from one-half to about one-seventh of the former value.

The effects of contact pressure and load on friction are important in assessing the tendency of materials to gall. Budinski[31] reported the behavior of metals and alloys subjected to the standard ASTM galling test method G 98 in which the annular end of a 28 mm diameter cylinder (apparent contact area about 1.28×10^{-5} m^2) is rotated slowly on a flat counterface and the normal force is increased stepwise until galling occurs. At low loads, the friction coefficient is relatively low, about 0.12–0.15, but at higher loads, say in the range of 2000–4000 N, the friction coefficient tends toward a maximum in the range 0.3–0.6 (N.B. the argument of Bowden and Tabor for Equation 7.2). Still higher loads produce a drop in friction coefficient, and by the time the load exceeds about 18,000–20,000 N, it returns to about 0.1–0.15.

Increasing the load can have opposite effects on the friction of different solid lubricants. The classic case of graphite and molybdenum disulfide friction, as discussed by Buckley,[32] illustrates this phenomenon. With graphite, the friction coefficient was observed to increase with increasing load, whereas with molybdenum disulfide, the opposite behavior was observed. The causes for this difference lie in the response of the crystal structures and shear stresses of each material to increasing hydrostatic pressure.

In summary, the dependency of friction coefficient on contact pressure and load cannot be generalized. When the load is increased and the friction force changes correspondingly, the amount of energy available to do work or to generate heat is also changed. However, another way to alter the available energy in a sliding interface is to change the sliding velocity.

7.3   EFFECTS OF SLIDING VELOCITY

The friction force **F** acting over a distance x generates frictional work equal to **F**x. The rate of energy input depends on the distance slid per unit time, that is, the velocity v. The product of the friction force and the velocity **F**v provides the means to determine such things as the temperature rise due to sliding (see Section 4.4). Figure 7.9 schematically shows that the energy generated during sliding can be dissipated in many ways. From early work in cannon boring, studied by Benjamin Thompson (a.k.a. Count Rumford), it was determined that most of the energy (>90–95%) produced by frictional contact is transformed into heat (4.186 J = 1 cal); nevertheless, a portion of the remaining energy is used to deform the material while creating new surface area (i.e., wear) and some is stored as defects in the contacting materials. Czichos[33] described the energy position process by identifying two "planes": the work plane and the thermal plane.

Early interest on the effects of sliding velocity on friction were conducted in the late 1850s by Bochet,[34] who was interested in the action of brakes and railway wagon wheels. He expressed the friction coefficient semiempirically in terms of the sliding velocity as follows:

$$\mu = \gamma + \left(\frac{k - \gamma}{1 + cv} \right) \tag{7.8}$$

where $\gamma = 0$ and $c = 0.23$. For dry rails $k = 0.45$, and for wet rails $k = 025$. This expression suited the work at the time but should not be applied to the general case.

Fundamental studies of friction are commonly conducted at low sliding velocities, say <0.1 m/s, to minimize the complicating effects of frictional heating, but there have been a number of studies using higher speeds. Such studies are of interest to those involved with such applications as brakes, clutches, high-speed metal

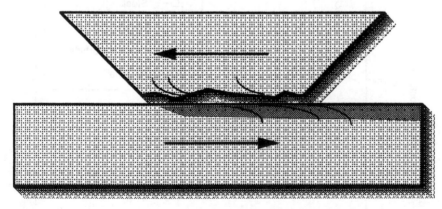

FIGURE 7.9   The energy generated during sliding can be dissipated in many ways: temperature rises in response to energy input, alterations in wear rate and surface roughness, alterations in the microstructure of materials in or near the interface, alterations in materials behavior (strain rate effects on properties), emission of visible light (triboluminescence), or exo-electrons.

fabrication processes, turbine engines, compressor seals, gun barrels, and projectiles. Figures 7.10 and 7.11 show data for high-speed sliding friction in weapons and for laboratory tests using an ultracentrifuge, respectively.[35] In all cases, the tendency is for friction to decrease as the velocity increases. The reasons for this general trend can be explained by one or more of the following: lubricating oxides form at elevated temperatures, the shear strength of most materials decreases at high frictional

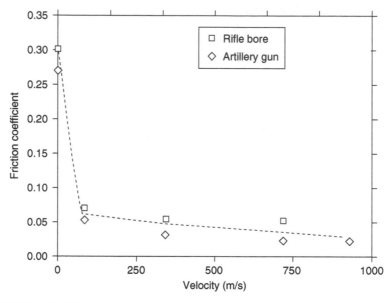

FIGURE 7.10   Published data for high-speed sliding friction in weapons.

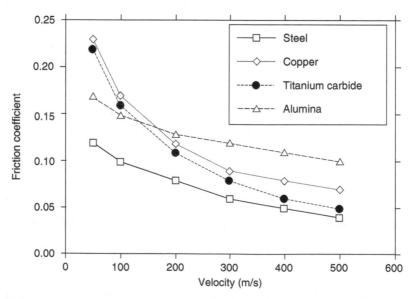

FIGURE 7.11   Data for high-speed laboratory friction tests using an ultracentrifuge.

temperatures, and if the surface frictionally melts, the molten liquid can lubricate the asperity contacts.

In the previous examples, friction was shown to decrease as a function of velocity for a range of materials and a wide range of sliding speeds, but over smaller velocity ranges and under some conditions, other types of the friction–velocity relationship can be obtained. Under some conditions, increasing the velocity increases the wear rate, which in turn roughens the surface and raises the plowing and cutting contributions to friction. Continuing to increase velocity produces a surface softening and the friction decreases again. Therefore, before assuming that the friction always decreases as velocity increases, it is wise to consider the specific materials and systems involved. Figure 7.12 shows the six possible relationships between friction force

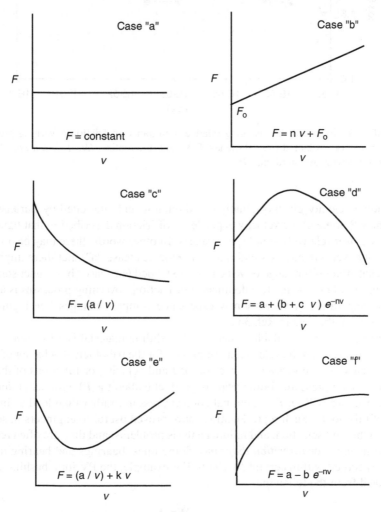

FIGURE 7.12 Six possible relationships between friction force and sliding velocity as originally developed by Ahkmatov. (Adapted from Bartenev, G.M. and Lavrentev, V.V., *Friction and Wear of Polymers*, Elsevier, Amsterdam, 1981, p.53.)

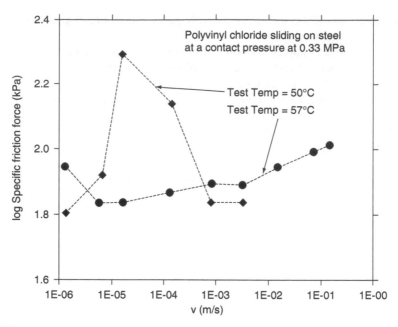

FIGURE 7.13   Effect of the changing relative dominance of friction-governing processes on μ–v behavior. (Adapted from Bartenev, G.M. and Lavrentev, V.V., *Friction and Wear of Polymers*, Elsevier, Amsterdam, 1981.)

and sliding velocity given originally by Akhmatov and later cited by Bartenev and Lavrentev.[36] Some of the velocity dependence of friction described in that figure can take place over relatively small speed ranges. In other words, there may be no effect of velocity over a relatively small range of velocities (case "a"), but there might be a significant effect if the range of velocities is extended (e.g. case "b"). Under some circumstances, where the relative dominance of friction-governing processes is a function of temperature, μ versus v may experience complex changes (see Figure 7.13, replotted from the data in Ref. 36).

   Since the energy available to heat and the sliding material is a function of both the friction force and the velocity, it is common for the tribological behavior of materials systems to be portrayed on axes of load and velocity, or functions of the two, like the contact pressure-sliding velocity product (called the *PV* product). (Note that in this case, we will let P = nominal contact pressure, rather than load as in other parts of this book.) Such representations have proven useful to engineers in solving design, lubricant selection, and failure analysis problems, and they are often referred to as *wear maps* or *transition diagrams*. Sometimes, bearings and bearing materials are specified within certain *PV limits*. For example, the PV for a bushing can be calculated from the following:

$$PV = \frac{W\pi\Omega}{L} \qquad (7.9)$$

where W is the load, L the length of the building, and Ω the rpm of the shaft. (The diameter of the bushing "cancels out" in the derivation of this expression.) In terms of English units

$$P = \frac{W}{Ld} \qquad (7.10)$$

where P is pressure (psi), L bearing length (in.), d shaft diameter (in.), and

$$V = 0.262 \; \Omega d \qquad (7.11)$$

where V is velocity (fpm). Therefore, an alternate form of Equation 7.9 that uses traditional units is

$$PV = 0.262 \frac{W\Omega}{L} \qquad (7.12)$$

An extensive listing of PV limits for various bearing materials can be found in the ASME *Wear Control Handbook*[37] and in plastic bearing manufacturers' literature. Some examples are given in Table 7.7.

Wear mechanism maps, developed by Lim and Ashby,[38] are another useful way to present load and velocity relationships as they affect the thermally activated interfacial sliding processes in simple sliding systems. The axes of the map are dimensionless, normalized load F and dimensionless normalized velocity V:

$$F = \frac{P}{A_n H_o} \qquad (7.13)$$

TABLE 7.7
Typical *PV* Limits for Bushing Materials

Material	PV (ksi ft/min)	PV (Pa m/s)
Tin bronze (C90500)	50	1.75
Beryllium copper	30	1.05
Porous bronze	50	1.75
Tool steel	200	7.01
Aluminum bronze (C95400)	50	1.75
Tribaloy	60	2.10
Carbon graphite	10–50	0.35–1.75
Filled PTFE	10	0.35

Source: Adapted from Glaeser, W.A. in *Wear Control Handbook*, American Society of Mechanical Engineers, New York, NY, 1980, 590–592.

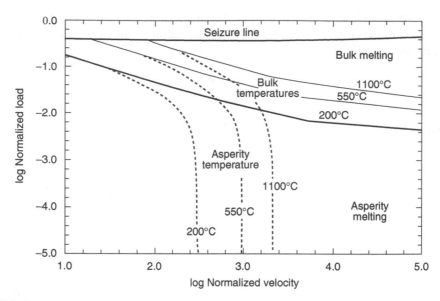

FIGURE 7.14   The Lim–Ashby wear map construction for the case of unlubricated steel sliding on steel.

and

$$V = \frac{vr_0}{a} \tag{7.14}$$

where **P** is the applied load, A_n the nominal area of contact, H_0 the hardness of the softer surface, v the sliding velocity, a the thermal diffusivity of the two surfaces, and

$$r_0 = \sqrt{\frac{A_n}{\pi}} \tag{7.15}$$

Figure 7.14 exemplifies this construction for unlubricated steel sliding on steel. Dashed lines indicate progressively increasing asperity heating (*flash temperatures*) and bulk heating, and regions dominated by various processes are identified. Wear mechanism maps have been applied to a variety of situations. For example, Figure 7.15 shows how the steel-on-steel map can be superimposed on conditions for typical applications. A software package, originally called "T-Maps," was developed and made available to personal computer users in the late 1980s. Table 7.8 lists some of the applications of wear mechanism maps and the associated references.

The effects of velocity on frictional behavior are usually studied by using constant velocity or by gradually increasing or decreasing the velocity. In many engineering applications, however, velocity is neither steady nor periodic. Rapid accelerations and decelerations are common in such applications as reciprocating shafts, brakes, clutches, and magnetic disk drives. It therefore becomes practically important to understand not only how friction and velocity are related at steady state, but also how

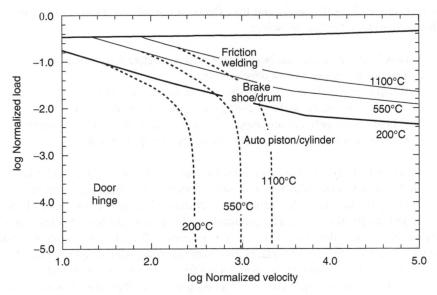

FIGURE 7.15 The Lim–Ashby construction of the steel-on-steel wear map can be superimposed on conditions for typical applications.

TABLE 7.8

Publications on Applications for Lim–Ashby Wear Mechanism Maps

Application or Materials Combination	Reference
Bronze/steel, steel/steel, sapphire/steel, silicon nitride/sapphire, magnesium oxide/steel	39
Various applications such as brakes, gears, friction welding	40
Diamond/glass, steel/rosin, alumina/alumina	41
Friction welding, friction cutting, striking of matches, curling (the sport), head crashes on magnetic disks	42

transients in friction are related to velocity changes. Examples include stick-slip, but there are other situations where such effects come into play. These are discussed in Section 7.4.

7.4 EFFECTS OF TYPE OF SLIDING MOTION

Velocity is a vector quantity, possessing the attributes of magnitude, direction, and sense. The previous section largely considered the direction of the sliding velocity to be parallel to the contact surfaces, constant in sense, and therefore concentrated on the effects of changes in magnitude. Now we will consider the effects of changes in direction and sense. Such situations commonly occur in oscillating sliding (fretting or reversed sliding). To understand the effects of complex motions on friction, we should consider all of the following:

1. How long the contact has been at rest before acceleration (contact aging)
2. The rate of acceleration or deceleration
3. Whether or not steady-state friction conditions are achieved
4. Whether the contact surfaces change (e.g., due to running-in)
5. Whether wear particles are trapped in the interface or allowed to escape
6. Whether other parameters, such as load, are changing as well

Changes in the type of motion create numerous possibilities for frictional behavior. The following examples illustrate what sort of responses in friction can occur.

Fretting is low-amplitude sliding that is common in applications such as bolted joints and electrical contacts. The amplitude of the motion can be only a few tens of micrometers, the order of the size of a few asperities. Particles of wear debris may become trapped between the surfaces and form tiny rolls, which tend to reduce friction.[43] This is one reason why friction during fretting may begin high and decrease as the duration of fretting increases. Waterhouse[44] described this phenomenon in detail. In the case of metals such as iron and copper, the fretting debris consists of oxide (for steel, this is something called red mud). Table 7.9 lists the steady-state friction coefficients of various fretting material combinations, including oxides studied by Godfrey and Bailey.[45] Note that the friction does not decrease much when starting out with oxidized surfaces.

Various factors can affect friction during fretting. For example, Endo et al.[46] discovered that as they decreased the oscillating frequency from 50 to 2.8 Hz, the friction force for steel-on-steel fretting increased, and Mason and White[47] showed how the frictional response of fretting surfaces of nickel silver wire on phenolic resin became more erratic as the amplitude of fretting exceeded about 10 μm due to the

TABLE 7.9
Friction Coefficients for Fretting

Ball Material	Flat Material	$\mu_{initial}$	$\mu_{steady-state}$
Cu	Cu	1.2	0.60
Cu	Glass	1.2	0.65
Fe	Fe	0.7	0.65
Fe	Glass	0.8	0.85
Steel	Steel	0.6	0.55
Steel	Glass	1.3	0.62
CuO	CuO	0.7	0.65
Cu_2O	Cu_2O	0.65	0.62
Fe_2O_3	Fe_2O_3	0.62	0.60
Fe_3O_4	Fe_3O_4	0.3	No steady state

Note: 150–175 gf load, 5 cycles/min, 12.5 mm diameter sphere on flat, amplitude 150 μm.
Source: Adapted from Godfrey, D. and Bailey, J.M., Lubr. Eng., 10, 155, 1954.

TABLE 7.10

Fretting Friction Coefficients at Various Humidity Levels

Metal	$\mu_{\text{steady-state}}$		
	0–2% RH	10–12% RH	49–50% RH
Fe	0.45	0.4	0.25
Ni	0.42	—	0.19
Ti	0.35	0.4	0.28
Cr	0.25	—	0.22

Note: 300 gf load, 60 Hz, 9.52 mm diameter sphere on flat, amplitude 80 μm.
Source: Adapted from Goto, H. and Buckley, D.H., *Int. Tribol.*, 18(4), 237–244, 1985.

onset of stick-slip behavior. Contact surfaces and debris particles can react with the moisture in the environment to alter friction in the fretting interface. Goto and Buckley[48] demonstrated how relative humidity affects friction during fretting. Table 7.10 summarizes representative data for metals. There was a tendency for higher humidity to reduce friction but that was not always the case.

When the amplitude of oscillatory motion exceeds that for fretting, debris entrapment conditions change and changes in frictional behavior follow. For example, Abarou et al.[49] studied wear and friction transitions in polymeric composite materials by varying the stroke length in oscillating sliding of a pin on a disk (i.e., producing an arc-shaped wear scar). To help quantify the wear processes, they defined the *mutual overlap coefficient* (MOC) as the ratio of the nominal contact area of the slider (rectangular-ended, 1.5 × 5.0 mm pin) to the nominal wear area on the counterface (disk). For low-amplitude sliding, the MOC is large, but for long path lengths, the MOC is small. Not only was the MOC varied, but the effects of having the long or short direction of the rectangular pin parallel to the sliding direction were also studied. Figure 7.16 summarizes those results.

Ball-on-flat experiments by the author have demonstrated the effects of motion direction on the friction and wear of unlubricated, self-mated silicon nitride (Table 7.11). The normal force was 10 N, the ball diameter was 9.525 mm, and the environment was laboratory air at room temperature. One set of tests was conducted on a pin-on-disk tester (unidirectional sliding) and the other set on a sinusoidally reciprocating ball-on-flat machine using a stroke length of 10 mm. A key to understanding this behavior lies with two factors: (a) differences in the entrapment of debris and (b) the effects of unidirectional versus reciprocating shear-induced stress fields on the fatigue wear process. It should be noted that an increase in sliding speed reduced the friction slightly in the unidirectional case but had no obvious effect in the other case.

Changes in the sliding direction with respect to the lay of the surface can have an effect on friction. For example, Briscoe and Stolarski[50] constructed a rather novel pin-on-disk machine in which pin rotation was separately controlled from disk rotation. Studies were conducted with five polymers sliding on mild steel. The

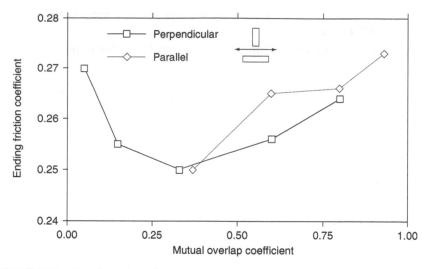

FIGURE 7.16 Experiments on the mutual overlap coefficient wherein the effects of having the long or short directions of the rectangular pin parallel to the sliding direction were evaluated.

TABLE 7.11

Effects of Motion Direction on the Steady-State Friction Coefficient of Silicon Nitride[a]

Sliding Velocity (m/s)	μ, Unidirectional Motion	μ, Reciprocating Motion[b]
0.2	0.70	0.87
0.5	0.65	0.87

[a] NBD 100 silicon nitride ball; polished hot-pressed Si_3N_4 flat specimen, average of at least two tests.
[b] The average velocity was based on a 10 mm stroke length; 10 Hz = 0.2 m/s and 25 Hz = 0.5 m/s.

polymers were as follows: polytetrafluoroethylene (PTFE), high-density polyethylene (HDPE), polymethylmethacrylate (PMMA), ultrahigh-molecular-weight polyethylene (UHMPE), and LDPE. Two conclusions of these studies were as follows:

1. Intrinsic rotation of the contact area in pin-on-disk machines can be responsible for the special wear behavior of PTFE and polyethylenes due to the sensitivity of wear to polymer chain orientation in the contact region.
2. Rotating the pin with an angular velocity opposite to but equal to that of the disk can counteract the effect of the intrinsic rotation.

Interestingly, more recent work on the testing of UHMPE for medical implants confirmed the important role of cross-path motion on correctly simulating wear (and friction) in those applications.[51]

The importance of the direction of motion should be taken into account when selecting friction tests. When screening materials for specific engineering applications, using the appropriate type of motion is particularly important.

7.5 EFFECTS OF TEMPERATURE

Thermal energy in a sliding interface affects the properties of the materials in and near the interface. There are two main sources of thermal energy in tribological interfaces: external (from the temperature of the surroundings) and internal (from frictional heating). For the sake of completeness, one might consider heat sources from chemical reactions in the interface, but the latter is minimal in most practical situations.

The main effects of temperature on friction arise from the following:

- The shear strength of interfacial materials is dependent on temperature.
- The viscosity of liquid and solid lubricants is dependent on temperature.
- The tendency of the surfaces of materials to react with the surrounding environment to form films or tarnishes is dependent on temperature.
- The tendency of formulated liquid lubricants to change chemical composition (e.g., oxidize or change molecular weight) is dependent on temperature and time of exposure.
- Wear processes, which affect surface roughness and traction, change with temperature.
- The ability of a metal or alloy to work-harden or otherwise alter its structure and properties depends on temperature.
- The tendency of materials to adhere and transfer to the rubbing partner may depend on temperature.
- The ability of a surface to adsorb or desorb contaminants is affected by temperature.

In view of the foregoing, the dependence of the friction of a given system of contacting materials on temperature may not be straightforward. If only one factor, such as the reduction of shear strength of metallic materials at high temperatures, dominated, then friction–temperature relationships could be modeled more easily. However, which factor primarily controls frictional behavior depends on the temperature range. It is not unusual to observe irregular behavior of friction as a function of temperature over a wide range of temperatures because different processes operate. For example, Yust[52] at Oak Ridge National Laboratory observed that friction coefficients for ceramics were similar at room temperature ($\mu \sim 0.3$–0.5) and at $>800°C$ ($\mu \sim 0.4$–0.5) but were higher ($\mu \sim 0.65$–0.8) at $400°C$. The interpretation was that adsorbed moisture lowered friction at room temperature and that low-shear-stress, glassy debris layers controlled friction at high temperature. Moisture was desorbed at intermediate temperatures that were too low to promote the formation of protective glassy tribolayers. Therefore, both friction and wear increased at the intermediate temperatures.

The friction of ice on steel and other materials as a function of temperature is an excellent example of complex behavior that depends on its crystal structure and properties (see Chapter 5). Near room temperature, it is quite low ($\mu \approx 0.02$), but it rises quickly as temperature decreases. The increases and decreases in ice friction with falling temperature are a matter of great importance to the builders of icebreakers (ships) as well as to downhill skiers and competition skating rink operators.

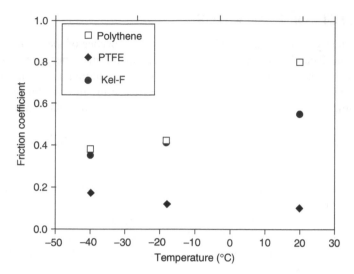

FIGURE 7.17 Data of Bowden and Tabor[53] that show the relatively smooth dependence of the friction of four polymers on temperature.

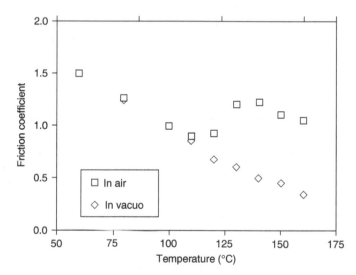

FIGURE 7.18 Data of Bartenev and Lavrentev[54] showing the effects of the rise in moderate temperature on the friction of vulcanized natural rubber on steel.

Several examples of the effects of temperature on friction follow to illustrate the need to understand the characteristics of the tribosystems. Figure 7.17, adapted from the data of Bowden and Tabor,[53] shows the relative smooth dependence of the friction of four polymers on temperature. However, Figure 7.18, adapted from Bartenev and Lavrentev,[54] shows the effects of the rise in moderate temperature on the friction of vulcanized natural rubber on steel. The bifurcation of the two curves, one obtained in air and the other at a reduced atmospheric pressure of about 10^{-3} Torr, shows how oxidation of the sliding material(s) can alter the effects of temperature on friction.

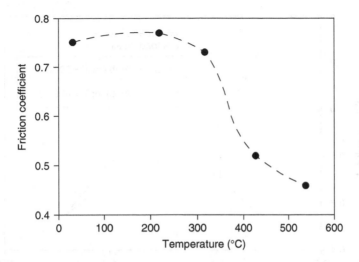

FIGURE 7.19 Data of Johnson and Bisson[55] for a cast Inconel alloy pin sliding unlubricated on an M-10 tool steel disk.

The sliding friction coefficients of many engineering alloys tend to decrease as temperature increases. This type of behavior, which tends to reflect a reduction in shear strength of the materials, is illustrated in Figure 7.19 for the case of a cast Inconel alloy pin sliding unlubricated on an M-10 tool steel disk.[55] There is a relatively temperature-insensitive period, and then the friction coefficient drops off more rapidly.

The importance of considering both internal and external sources of temperature on the friction is illustrated by studies of the friction and wear of ceramics by Woydt and Habig.[56] These investigators used curve-on-flat sliding geometry in a high-temperature friction and wear testing apparatus capable of attaining 1000°C in air. Using a normal force of 10 N, they measured the steady-state friction coefficient of several ceramics after 1000 m of sliding and at various velocities. Figure 7.20 summarizes their results on zirconia (ZrO_2) at room temperature and 400°C. At room temperature, the major heating was from friction, but at 400°C, external and internal frictional heating from high sliding speeds altered the behavior of the materials in the interface. Zirconia undergoes several crystallographic changes depending on the temperature. The low-temperature monoclinic phase is less wear-resistant than the higher-temperature tetragonal phase. The change in wear characteristics induced by combined frictional heating and external heating may have produced the complex friction–temperature relationship.

Various soft metals and oxides have been used as lubricants to reduce the friction of machines operating at elevated temperatures. Table 7.12 lists sliding friction coefficients for a number of candidate lubricating materials at 704°C.[57] The materials were tested in loose, powdered form.

To achieve higher operating efficiencies and greater power density, the turbine engines are being forced to operate at high speeds, and hence the bearings and seals within them require improvements in lubrication and durability. Thermally sprayed coatings have been used for some of these demanding applications. One coating system

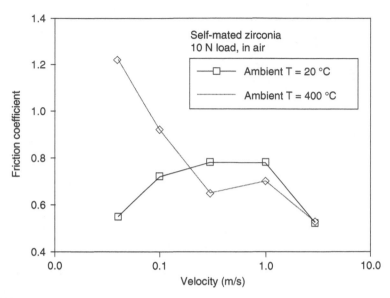

FIGURE 7.20   The steady-state friction coefficients for zirconia (ZrO_2) at room temperature and 400°C after 1000 m of sliding at various velocities. (Adapted from Woydt, M. and Habig, K.-H., *Tribol. Int.*, 22, 75, 1989.)

TABLE 7.12
Friction Coefficients of Various Lubricating Powders at 704°C

Oxide	Melting Point (°C)	μ
PbO	888	0.12
MoO_3	795	0.20
Co_2O_2	900[a]	0.28
WO_3	2130	0.55
Cu_2O	1235	0.44
ZnO	1800	0.33
CdO	900–1000[a]	0.48
$PbMoO_4$	1065	0.32
$NaWO_3$	700	0.17

[a] Decomposes.

Source: Adapted from Peterson, M.B., Florek, J.J., and Murray, S.F., *ASLE Trans.*,
            2, 225, 1960.

developed by researchers at the National Aeronautics and Space Administration's Lewis Research Center consists of chromium carbide (Cr_3C_2), silver (Ag), and a eutectic of calcium fluoride (CaF_2) and barium fluoride (BaF_2). Designed to operate over a range of engine bearing temperatures, this coating was subjected to a variety of friction and wear tests since the mid-1980s. It is currently referred to as PS-200. A later composition was designated as PS-212. Start–stop conditions were particularly

TABLE 7.13

Effects of Lubricant Additives on Starting Friction Coefficients of a Plasma-Sprayed Coating for Foil Bearings

Temperature (°C)	μ_i (No Additive)	μ_i (With Additives)
25	0.64	0.44
200	0.55	0.32
425	0.42	0.31
650	0.36	0.29

Source: Wagner, R.C. and Sliney, H.E., *Lubr. Eng.*, 42, 594, 1986. With permission.

of interest in the early tests, since engine lubrication conditions change markedly during such events. A special journal bearing testing apparatus was used by Wagner and Sliney[58] to study friction during starting and while the bearing was running at the high surface speed of 28 m/s, corresponding to 13,800 rpm of the bearing. The unit bearing pressure on these foil bearing tests was 14 kPa. Starting friction data are given in Table 7.13 for a Ni–Al bonded Cr_2C_3 material with and without additions of 10% Ag and 10% eutectic. The journal surface was coated, and the foil surface was a preoxidized Ni-base superalloy. Friction was reduced slightly at room temperature and 200°C (to $\mu < 0.05$) using the additives, but there was little effect at higher temperatures (all compositions exhibited μ (28 m/s) of 0.05–0.08). The benefits of the additives seemed to be primarily in reducing the starting friction.

Temperature effects on friction are more complicated in the case of polyphase or composite materials, like those used in brake shoes or clutch plates. The effects of temperature on the properties vary from material to material. Therefore, each of the materials in a composite may respond differently to temperature, and its area fraction on the contact surface will govern its particular influence on the effects of temperature on the friction of the tribosystem. One example of this is in so-called self-lubricating materials, such as polymer blends containing PTFE and other additives, where one phase is normally a low-shear-strength phase that spreads over the load-bearing matrix phase to lubricate it. If rising temperature reduces its shear strength too much, then it may be squeezed out of the interface or wiped away too quickly to function effectively. Alternately, some solid lubricants, such as MoS_2, oxidize at elevated temperatures to form compounds that lubricate less effectively. The oxides of other types of materials, such as BN, may actually lubricate better than the compound, which is stable at lower temperatures. Each tribosystem requires separate analysis.

Studies by Gamulya et al.[59] indicated that the friction of several sprayed solid lubricant films containing compositions based on MoS_2, graphite, or PTFE, sliding on steel in a vacuum of 10^{-5} Pa, changed little between 20 and -153°C. But the wear lives of those coatings were, in most cases, greatly increased at low temperature. The effect of temperature on wear, not friction, was the prime consideration in that case.

Understanding of the effects of temperature on friction for materials in specific applications must be supplemented by a clear understanding of the temperature fluctuations. It may not be sufficient to know the temperature due to the surroundings or due to the effects of frictional heating at steady state alone. The changes in the surface that occurred during one high-temperature excursion may persist as the temperature is reduced. For example, a lubricative oxide that forms at high temperature and helps reduce friction in that regime may form a brittle crust when the system is cooled. That crust may produce wear particles or serve to raise the friction of the surfaces at lower temperatures. A discussion of cyclic or transient temperature effects on frictional behavior is a topic beyond the scope of this book, and it is recommended that readers with specific applications involving cyclic temperature effects conduct simulative experiments.

7.6   EFFECTS OF SURFACE FILMS AND CHEMICAL ENVIRONMENTS

Any substance interposed between sliding surfaces has the potential to affect friction. Under some circumstances, the contact conditions may cause that substance to be penetrated or wiped away quickly, and its effects will then be suppressed. However, surface films and chemical environments that cause such films to be formed and replenished can have a major effect on frictional behavior. Surface oxides and tarnishes can significantly affect both the static and kinetic friction in solids. Oxides and other reaction products, however, are not the only source of surface films. Table 7.14 lists several sources of friction-modifying films other than intentionally introduced liquid and solid lubricant films.

Relative humidity has been shown to influence the frictional behavior of materials, but its influence varies from one material combination to another. In addition to producing an adsorbed layer of friction-modifying water molecules, water can penetrate into the intermolecular layers of films to modify their shear properties. Some materials such as carbon exhibit lower friction coefficients in the presence of water vapor, while others such as molybdenum disulfide do not.

Figure 7.21, from Demizu et al.,[60] illustrates the effects of relative humidity on the average kinetic friction of several high-purity metal pins (1.5 mm hemispherical

TABLE 7.14
Eight Friction-Modifying Films

Type of Film	Sources or Conditions for Occurrence
Adsorbates	Moisture from the air; airborne contaminants like hydrocarbon molecules
Residues	Remaining from solvent or other cleaning procedures
Friction polymers	Polymeric films that form on contact surfaces aided by the catalytic action of the surface material
Oxides	Common in metallic materials exposed to the ambient environment, but also present on semimetals like Si and ceramics like SiC and Si_3N_4
Sulfides	Common tarnishes on metals like silver and copper
Melted layers	Thin molten films due to intense frictional heating
Transfer films	Thin, shear-thinned layers transferred to a surface from the sliding partner
Surface segregates	Solute species that diffuse out of the underlying material to form deposits on the surface (e.g., dezincification: Zn surface segregation on Cu–Zn alloys)

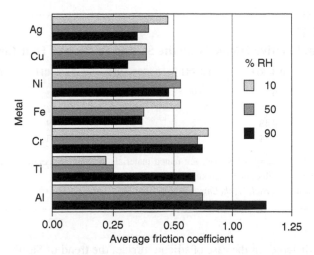

FIGURE 7.21 Effects of relative humidity on the average kinetic friction of several high-purity metal pins (1.5 mm hemispherical tips) sliding in reciprocating motion against ground and lapped flat specimens of zirconia (ZrO_2) at a load of 0.39 N, speed of 0.17 mm/s, and stroke length of 3 mm. (From Demizu, K., Wadabayashi, R., and Ishigaki, H., *Tribol. Trans.*, 33, 505, 1990. With permission.)

tips) sliding in reciprocating motion against ground and lapped flat specimens of alumina (Al_2O_3) and zirconia (ZrO_2) at a load of 0.39 N, speed of 0.17 mm/s, and stroke length of 3 mm. From data such as these, we conclude:

1. Relative humidity affects the friction of some material combinations more than others.
2. Increasing relative humidity can either increase or decrease friction, depending on the material combination involved.

In many of the investigations performed by the author and Yust at Oak Ridge National Laboratory, and in the published literature, the wear behavior of ceramics such as alumina and silicon nitride was worse at low humidity levels. Often, friction is less affected by humidity than wear rate. Some of the best-known investigations of such effects on ceramics were published by Fischer. For example, Fischer's paper with Tomizawa[61] illustrated how water vapor can facilitate the formation of friction-modifying surface films on silicon nitride slid against the same type of material. In pin-on-disk tests, dry friction was only slightly higher than friction under humid conditions; however, the wear rate dropped about two orders of magnitude as humidity was varied from nearly 0 to 100% (sliding in water).

A comprehensive set of investigations on the effects of surrounding environment on the friction and wear of ceramics has been reported by Sasaki.[62] The effects of relative humidity on the friction of four ceramics: alumina (Al_2O_3), silicon nitride (Si_3N_4), silicon carbide (SiC), and partially stabilized zirconia (PSZ) were investigated using a pin-on-disk apparatus and a special bubbler apparatus to produce air with the desired humidity level. Results are summarized in Table 7.15. In each instance, increasing relative humidity was associated with decreasing friction; however, the magnitude

TABLE 7.15

Effects of Relative Humidity on the Sliding Friction of Four Ceramics

Material	0% RH	20% RH	50% RH	75% RH	100% RH
Al_2O_3	0.72	0.59	0.52	0.52	0.33
PSZ	0.75	0.71	0.55	0.52	0.49
SiC	0.48	0.32	0.30	0.19	0.17
Si_3N_4	0.74	0.70	0.66	0.63	0.60

Note: 0.4 m/s velocity, 10 N load, self-mated materials at room temperature; reported are
average μ_k values after 1.26 km of sliding distance.

Source: Sasaki, S., *Bull. Mech. Eng. Lab.*, 58, 1992. With permission.

of that trend differed. In the case of silicon nitride, the trend of Sasaki's data agreed
qualitatively with that of Tomizawa and Fischer, but Sasaki's friction coefficient values were somewhat lower at comparable humidity levels. More recent studies on such
effects by Gee and Butterworth[63] demonstrated that the effect of relative humidity on
the friction of ceramics can be reversed by testing at higher speeds.

The ambient environment can significantly affect friction, depending on the procliv-
ity for surface films to form by adsorption, by reaction of the material with its surroundings, or by the catalytic action of the friction surface in forming friction-modifying
deposits. These kinds of considerations become very important when it comes to ensur-
ing the reliability of electronic equipment such as switches, relays, and electrical contacts in general.[64] Therefore, our ability to control friction is often linked to our ability
to control the composition, properties, and distribution of thin films in the sliding interface. The lubrication methods described in Chapter 6 exploit that approach.

7.7   STIFFNESS AND VIBRATION

Both stiffness and external vibrations can affect frictional behavior. Conversely,
frictional interactions can themselves stimulate vibrations. In some cases, controlled vibrations can in effect reduce friction, but in others, vibrations can result in
mechanical instabilities and wear. The detailed mathematics of friction–vibration
interactions (FVI) is beyond the scope of this book; however, the subject has been
one of growing interest in the tribology community. In this section, we shall consider
the subject in broad terms, providing several illustrations to demonstrate how vibration in a mechanical system can be induced by friction, or conversely can induce
changes in friction.

Friction-induced vibrations can produce audible noise—a significant concern
for subway riders and designers of "quiet" submarines. However, vibration sensing
has been used to monitor the reliability and performance of machinery. In that case,
it is not only friction but also wear and surface damage that cause the disturbance.
Reviews of the use of vibration monitoring can be found in articles by Martin[65] and
Smith[66] in the proceedings of the Mechanical Failures Prevention Group[67,68] and in
the journal *Vibration*.

The general principle of vibration condition monitoring is to compare the sensor outputs of a system that is known to be operating satisfactorily with those of the current system. Acceptable tolerance levels and deviations from the normal are set to trigger alarms. In some instances, specific types of problems can be inferred from the characteristics signatures of the vibrational spectra. Coupled with other diagnostic techniques, such as temperature sensing and lubricant analysis, clues can be assembled to eliminate some possibilities and identify the cause(s) for tribosystem problems.

Two important applications requiring the control of FVI are brakes and clutches. Another important application is machining where tool vibrations and chatter can ruin the surface finish of parts. On the positive side, FVI have been used to construct vibrating conveyor belts and part feeders for manufacturing operations. In principle, it may be possible to control friction in tribocomponents such as bearings and gears using vibrations, as some have suggested on the basis of theoretical arguments based on atomic-scale interactions,[69] but engineering a practical physical means to implement this concept is problematic.

Figure 7.22 illustrates schematically the complexities associated with FVI. The potential for such effects to occur is present in every rubbing contact, but the specific nature of the contact will determine whether the conditions are right for such interactions to be felt strongly. There are two basic types of FVI: (a) self-excited and (b) externally excited. Self-excited vibrations can arise from such phenomena as the variation in friction force produced during stick-slip and when asperity junctions engage and disengage during continuous sliding. Externally excited vibrations can

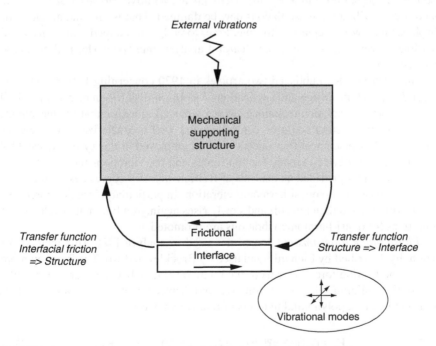

FIGURE 7.22　Complexities associated with friction–vibration interactions.

be produced by nearby equipment, or from the inherent vibrational characteristics of the machine (i.e., its design [stiffness and geometry], imposed operating parameters, and materials of construction). Friction-induced vibrations can be continuous, resulting in a steady squeal as a machine operates, or they may be intermittent if the machine operating parameters change with time.

The manifestations of FVI vary. For example, at slow sliding speeds, stick-slip may occur, and at high sliding speeds, dynamic instabilities can cause catastrophic failures. Machine chatter can also be induced by increasing the contact force or by the onset of nonuniform interfacial material transfer, scuffing, or galling. FVI can occur in all directions relative to the frictional interface. For example, those resulting from changes in shear tractions (due to transfer patches, momentary seizure, or heterogeneous surface deposits) can occur parallel to the surface. Those resulting from gross surface irregularities or external vibrations can act either normal or transverse to the contact surface, altering the normal load. One source of normal vibrations may be the surface undulations due to thermoelastic instabilities.

When a mechanical system of connected bodies of mass m and stiffness k is given a certain initial displacement and released, it will, under certain conditions, vibrate with a definite frequency f, called the *natural frequency*:

$$f = \frac{1}{2\pi}\sqrt{\frac{k}{m}} \tag{7.16}$$

If the system is continually forced to vibrate by a periodic forcing frequency, the forced and natural vibratory displacements are additive. However, if the system is not free to vibrate, the oscillations may be damped. That is, the magnitude of the displacements will decrease with time. Free, forced, and damped vibrations are all possible in frictional systems, and a careful analysis must be performed on a case-by-case basis.

Brockley and Ko published two articles in 1970 concerning FVI.[70,71] The first paper described the design and application of a pin-on-disk tribometer to study FVI. Special vibration isolation techniques were employed to assure that the mechanical behavior of the system could be controlled and well characterized. Acceleration, velocity, and displacement transducers were all employed in this system to establish the relationship between kinetic friction forces and the vibrations induced by them. The second paper provided an analysis of friction–velocity curves describing which conditions gave rise to quasi-harmonic vibration. In particular, if friction force rises to a maximum, and then falls off and slowly rises again, as a function of velocity, the conditions for quasi-harmonic vibrations are promoted.

Phase-plane diagrams are sometimes used to analyze FVI. This method was originally described by Lienard[72] and used by Brockley and Ko. The function representing the friction force in terms of the relative velocity between surfaces is determined. If the sliding velocity is v, time is t, and displacement is x, we can express the friction force as a polynomial in terms of relative velocity:

$$\mathbf{F}_t = C_n(v-y)^n + \cdots + C_{n-1}(v-y)^{n-1} + \cdots + C_0 \tag{7.17}$$

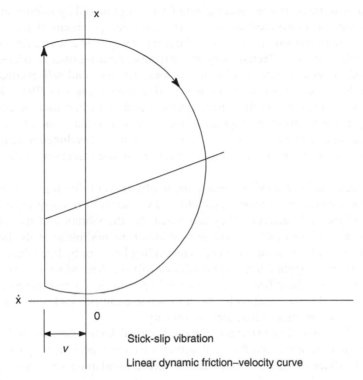

x

ẋ

0

v

Stick-slip vibration

Linear dynamic friction–velocity curve

FIGURE 7.23 A phase-plane diagram for stick-slip. (After Brockley, C.A. and Ko, P.L., *Trans. ASME*, October, 550, 1970.)

where $y = dx/dt$. The differential of y with respect to x can be written in terms of a friction function $F(v - y)$ and various constants r (the damping coefficient), m (the sliding object's mass), and k (the spring constant of the system) as follows:

$$\frac{dy}{dx} = \frac{F(v - y) - ry - kx}{my} \qquad (7.18)$$

Phase-plane diagrams show phase trajectories characteristic of the FVI in the system. Figure 7.23 shows the phase-plane diagram for stick-slip. At "stick" conditions, (dx/dt) is a constant and the displacement (x) increases. At slip, (dx/dt) increases rapidly to catch up. Then it decreases to v again and the cycle repeats.

Aronov et al.[73] constructed a special apparatus in which the stiffness, natural frequency, and damping could be controlled. After a careful set of experiments with AISI 01 tool steel, dry and lubricated with water, they found the following:

1. The breakdown of the lubricant film to produce a wear and friction transition occurred at a critical load that was independent of the system rigidity.
2. At a certain critical load, continuous, self-excited vibrations occurred. These were independent of the breakdown of the lubricating water film.
3. The mild wear rate increased with increases in either the normal load or system rigidity.

A series of three articles dealing with FVI was prepared by Aronov et al.[74-76] The first paper demonstrated, using cast iron and steel specimens, that the normal load at which the transition between a mild form of wear and a severe form of wear occurs is significantly affected by system stiffness. Four regimes of friction were distinguished: steady-state, nonlinear, transient friction, and self-excited vibrations. In the second paper, attention was paid to three classes of FVI: stick-slip, vibrations from random surface irregularities, and quasi-harmonic oscillations. Coupling between different degrees of freedom was found to be an important consideration in self-excited vibrations. The third paper developed a wear model based on system stiffness and slider oscillations in the direction of the normal force.

Bartenev and Lavrentev[77] discussed the implications of differences in the stiffness of the contacting asperities (exemplified by a series of microsprings) and the stiffness of the bulk material. They concluded that the vibration of the asperities is in general nonlinear and nonsymmetrical about the mid-plane of the interface. Therefore, normal forces are generated in the sliding interface by free vibrations that tend to oppose the applied force and result in a lifting effect, which in turn reduces friction. Increasing the stiffness of a system or improving its viscous damping capacity in the normal force direction has been shown to eliminate stick-slip behavior by making the friction force independent of velocity.

Tolstoi[78] showed how externally imposed normal force vibrations reduced the sliding friction between two flat steel surfaces having "medium roughness." The mass of the slider was about 1.5 kg, and the forced oscillations were provided by a

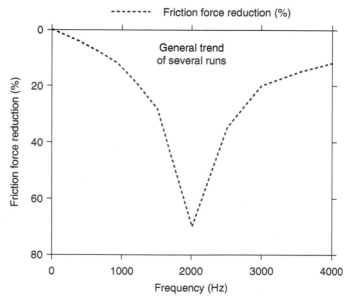

FIGURE 7.24   Tolstoi showed that a reduction in friction occurred over a narrow range of frequencies of normal force vibrations. (Adapted from Tolstoi, D.M., *Wear*, 10, 199, 1967.)

piezoelectric oscillator operating at 1–2 W power consumption. The data are sum-marized in Figure 7.24. The resonant frequency resulting in the most dramatic drop in friction could be estimated within about 10–15% using Equation 7.16.

Soom and Kim[79] studied the interaction between the normal and friction forces during unlubricated sliding of steel on steel. They defined the frictional behavior in terms of four parameters: F_F is the average friction force, f_f is the friction force fluctuation, F_N is the normal force, and f_n is the normal force fluctuation. The average forces were assumed to be constant, and the fluctuations were time dependent. The fluctuating portions were represented as functions of the slider mass and acceleration. Data were obtained for sliding at 0.12 m/s under a normal force of 6.1 N. A transfer function, $[f_f/f_n]$, was defined as the oscillatory friction coefficient μ_D to establish a relationship between the fluctuations in friction and normal forces, respectively. The value of μ_D varied as function of the frequency of frictional fluctuations in the range of 0.31–0.35 (cf. the average $\mu = 0.24$), but μ_D was even larger at the lower frequency friction fluctuations (say, <600 Hz in their experiments).

Work by Hess and Soom[80,81] considered the effects of nonlinear vibrations in cases of smooth Hertzian contacts and rough surfaces. Consideration was given to the effects of oscillations on the reduction of contact area, and it was found that dynamic loads produced a larger effect on surface separation than static loads of the same average magnitude.

In the past decade, the need to improve stability during read/write operations in computer drives has prompted growing interest in FVI modeling at micro- and nanoscales (e.g., Ref. 82). However, FVI becomes important in the design of machines of many types and sizes, in the analysis of bearing failures, and in the selection of friction test methods.

7.8 COMBINED EFFECTS OF SEVERAL VARIABLES

The effects of specific variables on frictional behavior have been discussed sepa-rately. However, the division of frictional behavior into the effects of single variables is, in truth, simply a convenience and may be misleading to some extent. During fric-tional contact, more than one frictional processes can occur simultaneously, and, as mentioned earlier, it is difficult to change just one variable at a time while conduct-ing a friction experiment. For example, trying to separate the effects of increasing sliding velocity from the effects of frictional heating may not be possible because the frictional temperature rise is proportional to velocity. Likewise, increasing the normal force on a contact may penetrate surface films that might otherwise mitigate friction, which would also affect the surface roughness due to increased wear. This interrelationship between factors makes laboratory studies of the effects of only one variable at a time difficult to interpret. Therefore, investigating the conjoint effects of multiple variables on either static or kinetic frictional behavior remains a chal-lenging and important area for friction research from both a practical and a funda-mental point of view. Recognition of the fact that simultaneous effects are going on in sliding interfaces underlies a better understanding of transient phenomena such as running-in, which happens to be the subject of Chapter 8.

REFERENCES

1. L. F. Coffin (1959). Some metallurgical aspects of friction and wear, in *Friction and Wear*, Elsevier, New York, NY, p. 42.
2. P. J. Blau (1992). Effects of surface preparation on the friction and wear behaviour of silicon nitride/silicon carbide sliding pairs, *J. Mater. Sci.*, 27, pp. 4732–4740.
3. *Friction and Surface Finish* (1940). Proceeding of the Special Summer Conference, June 5–7, Massachusetts Institute of Technology, Cambridge, MA.
4. A. M. Swigert, Ref. 3, remarks, p. 184.
5. M. B. Karamis and B. Selcuk (1993). Analysis of friction behaviour of bolted joints, *Wear*, 166, pp. 73–83.
6. P. J. Blau (2001) Simulation of cylinder bore surface finish parameters to improve laboratory-scale friction tests in new and used oil, in *Engine Systems: Lubricants, Components, Exhaust and Boosting System, Design and Simulation*, American Society of Mechanical Engineers, New York, ASME ICE Vol. 37-3, pp. 57–63.
7. C. A. Gowger, Ref. 3, discussion, 239 et seq.
8. N. S. Eiss Jr., and K. A. Smyth (1981). The wear of polymers sliding on polymeric films deposited on rough surfaces, *Trans. ASME, J. Lubric. Tech.*, 103, pp. 266–273.
9. Y.-R. Jeng (1990). Experimental study of the effects of surface roughness on friction, *Tribol. Trans.*, 33, pp. 402–416.
10. J. C. Roberts (1985). Surface morphology studies in polymer–graphite epoxy sliding, *ASLE Trans.*, 28(4), pp. 503–510.
11. I. V. Kragelskii and V. S. Kombalov (1969). Calculation of the value of roughness after running-in, *Wear*, 14, pp. 137–140.
12. I. V. Kragelskii, M. N. Dobichin, and V. S. Kombalov (1982). Running-in and equilibrium roughness, in *Friction and Wear Calculation Methods*, Pergamon Press, Oxford, pp. 297–316 (Chapter 9).
13. P. J. Blau (1989). *Friction and Wear Transitions of Materials*, Noyes, Park Ridge, NJ, pp. 237–238.
14. S. Liu and Q. Wang (2001). A three-dimensional thermomechanical model of contact between non-conforming rough surfaces, *J. Tribol.*, 123(1), pp. 17–26.
15. W. W. Chen, W. Chen, J. Cao, C. Xia, R. Talwar, R. Lederich, and Q. Wang, (2007). Elasto-plastic rough surface contact analysis for the effects of topographical characteristics, load, hardening parameters and material properties, *Proceedings of International Symposium on Computational Mechanics*, July 2007, Beijing, China.
16. F. P. Bowden and D. Tabor (1986). *The Friction and Lubrication of Solids*, Oxford Science Press, UK, pp. 5–32 (Chapter 1).
17. I. V. Kragelskii (1965). *Friction and Wear*, Butterworths, London, pp. 30–59 (Chapter 2).
18. V. A. Bely, A. I. Sviridenok, M. I. Petrokovets, and V. G. Savkin (1982). *Friction and Wear in Polymer-Based Materials*, Pergamon Press, Oxford, pp. 83–85.
19. B. Bhushan and M. T. Dugger (1990). Real contact area measurements on magnetic rigid disks, *Wear*, 137, pp. 41–50.
20. Y. L. Wu, B. Dacre, and R. F. Babus' Haq (1993). Measurements of real surface area by the conical method and fractal approach, *Wear*, 160, pp. 265–268.
21. F. B. Stupak and J. A. Donovan (1990). Fractal characteristics of worn surfaces, *Mater. Res. Soc.*, extended abstract (EA-25), p. 5.
22. F. P. Bowden and D. Tabor (1986). *The Friction and Lubrication of Solids*, Oxford Science Press, England, p. 31.
23. F. P. Bowden and D. Tabor (1986). *The Friction and Lubrication of Solids*, Clarendon Press, Oxford, pp. 94–101.
24. R. Holm (1958). *Electric Contacts Handbook*, Springer-Verlag, Berlin, p. 205.
25. P. W. Bridgman (1931). *The Physics of High Pressure*, Macmillan, New York, NY.

26. P. J. Blau (1985). Relationships between Knoop and scratch micro-indentation hardness and implications for abrasive wear, in *Microstructural Science*, Vol. 12, eds. D. O. Northwood, W. E. White, and G. F. VanderVoort, ASM International, Materials Park, OH, pp. 293–313.
27. P. J. Blau (1992). Scale effects in friction: An experimental study, in *Fundamentals of Friction*, eds. H. M. Pollock and I. L. Singer, Kluwer, Dordrecht, The Netherlands, pp. 523–534.
28. F. P. Bowden and D. Tabor (1956). *Friction and Lubrication*, Methuen Press, London, p. 37.
29. S. Kato, E. Marui, and K. Tachi (1985). Frictional properties of a surface covered with a soft metal film, *Memoirs of the Faculty of Engineering, Nagoya University*, 37(1), pp. 1–37.
30. M. B. Peterson and F. F. Ling (1966). Friction and lubrication at extreme pressures, in *Friction and Lubrication in Metal Processing*, eds. F. F. Ling, R. L. Whitely, P. M. Ku, and M. B. Peterson, American Society of Mechanical Engineers, New York, NY, pp. 39–67.
31. K. G. Budinski (1991). Friction in machine design, in ASTM STP 1105, *Tribological Modeling for Mechanical Designers*, eds. K. C. Ludema and R. G. Bayer, ASTM, Philadelphia, PA, pp. 89–126.
32. D. H. Buckley (1981). *Surface Effects in Adhesion, Friction, Wear, and Lubrication*, Elsevier, Amsterdam, pp. 576–577.
33. H. Czichos (1978). *Tribology: A Systems Approach*, Elsevier, Amsterdam.
34. B. Bochet (1861). Nouvelles recherches experimentales sur le frottement de glissement, *Ann. Mines*, 19(38), pp. 27–120.
35. I. V. Kragelskii and V. V. Alisin (1981). *Tribology Handbook*, Vol. 1, Mir, Moscow, pp. 220–221.
36. G. M. Bartenev and V. V. Lavrentev (1981). *Friction and Wear of Polymers*, Elsevier, Amsterdam, p. 53.
37. W. A. Glaeser (1980). Bushings, in *Wear Control Handbook*, eds. M. B. Peterson and W. O. Winer, American Society of Mechanical Engineers, New York, NY, pp. 590–592.
38. S.-C. Lim and M. F. Ashby (1987). Wear mechanism maps, *Acta Metallurg.*, 35, pp. 1–24.
39. M. F. Ashby, J. Abulawi, and H. S. Kong (1991). Temperature maps for frictional heating, *Tribol. Trans.*, 34, pp. 577–587.
40. H. S. Kong and M. F. Ashby (1991). Friction heating maps and their applications, *Mater. Res. Soc. Bull.*, 16(10), pp. 41–48.
41. H. Kong and M. F. Ashby (1992). Wear mechanisms in brittle solids, *Acta Metallurg.*, 40(11), pp. 2907–2920.
42. H. S. Kong and M. F. Ashby (1991). *Case Studies in the Application of Temperature Maps for Dry Sliding*, Cambridge University Report, CUED/C-MATS/TR 186.
43. J. S. Halliday and W. Hirst (1956). The fretting corrosion of mild steel, *Proc. R. Soc., Lond.*, A236, pp. 411–425.
44. R. B. Waterhouse (1972). *Fretting Corrosion*, Pergamon Press, Oxford, pp. 93–97.
45. D. Godfrey and J. M. Bailey (1954). *Lubr. Eng.*, 10, p. 155.
46. K. Endo, H. Goto, and T. Fukunaga (1974). Behaviour of frictional force in fretting fatigue, *Bull. Jpn. Soc. Mech. Eng.*, 17(108), pp. 647–654.
47. W. P. Mason and S. D. White (1952). *Bell Syst. Tech. J.*, 31, p. 482.
48. H. Goto and D. H. Buckley (1985). The influence of water vapour in air on the friction behaviour of pure metals during fretting, *Int. Tribol.*, 18(4), pp. 237–244.
49. S. Abarou, D. Play, and F. E. Kennedy (1986). *Wear Transition of Self-Lubricating Composites Used in Dry Oscillating Applications*, ASLE preprint 86-TC-4C-2, p. 12.

50. B. J. Briscoe and T. A. Stolarski (1982). The effect of the complex motion in the pin-on-disk machine on the friction and wear mechanism of organic polymers, in *Other Tribological Problems*, Vol. IV, eds. M. Hebda, C. Kajdas, and G. M. Hamilton, Elsevier, Amsterdam, pp. 80–99.

51. M. Turell, A. Wang, and A. Bellare (2003). Quantification of the effect of cross-path motion on the wear rate of ultra-high molecular weight polyethylene, *Wear*, 255(7), pp. 1034–1039.

52. C. S. Yust (1994). Tribological behavior of whisker-reinforced ceramic composite materials, in *Friction and Wear of Ceramics*, ed. S. Jahanmir, Marcel Dekker, New York, NY, pp. 199–223.

53. F. P. Bowden and D. Tabor (1964). *The Friction and Lubrication of Solids*, Vol. II, Oxford University Press, Oxford.

54. G. M. Bartenev and V. V. Lavrentev (1981). *Friction and Wear of Polymers*, Elsevier, Amsterdam, pp. 118–119.

55. R. L. Johnson and E. E. Bisson (1955). Bearings and lubricants for aircraft turbine engines, *SAE J.*, 63(6), pp. 60–64.

56. M. Woydt and K.-H. Habig (1989). High temperature tribology of ceramics, *Tribol. Int.*, 22(2), pp. 75–88.

57. M. B. Peterson, J. J. Florek, and S. F. Murray (1960). Consideration of lubricants for temperatures above 1000°F, *ASLE Trans.*, 2(2), pp. 225–234.

58. R. C. Wagner and H. E. Sliney (1986). Effects of silver and Group II fluoride solid lubricant additions to plasma-sprayed chromium carbide coatings for foil gas bearings to 650°C, *Lubr. Eng.*, 42(10), pp. 594–600.

59. G. D. Gamulya, T. A. Kopteva, I. L. Lebedeva, and L. N. Sentyurikhina (1993). Effect of low temperatures on the wear mechanism of solid lubricant coatings in vacuum, *Wear*, 160, pp. 351–359.

60. K. Demizu, R. Wadabayashi, and H. Ishigaki (1990). Dry friction of oxide ceramics against metals: The effect of humidity, *Tribol. Trans.*, 33(4), pp. 505–510.

61. T. E. Fischer and H. Tomizawa (1985). Interaction of tribochemistry and microfracture in the friction and wear of silicon nitride, *Proceeding of Wear of Materials '85*, ASME, New York, pp. 22–32.

62. S. Sasaki (1992). Effects of environment on the friction and wear of ceramics, *Bull. Mech. Eng. Lab.*, 58, pp. 109-129.

63. M. Gee and D. Butterworth (1993). The combined effect of speed and humidity on the wear and friction of silicon nitride, *Wear*, 163–164, pp. 234–245.

64. M. Antler (1985). Survey of Contact Fretting in Electrical Connectors, *IEEE Trans. Hybrids, Components, Packaging Manuf. Technol.*, 8(1), pp. 87–104.

65. H. R. Martin (1992). Vibration analysis, in *ASM Handbook, Volume 18: Friction, Lubrication and Wear Technology*, ASM International, Materials Park, OH, pp. 293–298.

66. J. D. Smith (1982). Vibration monitoring of bearings at low speeds, *Tribol. Int.*, 3, pp. 139–144.

67. J. G. Early, R. T. Shives, and J. H. Smith (1983). *Time-Dependent Failure Mechanisms and Assessment Methodologies*, Cambridge University Press, Cambridge.

68. R. T. Shives and L. J. Mertaugh (1988). *Detection, Diagnosis and Prognosis of Roating Machinery to Improve Reliability, Maintainability, and Readiness Through the Application of New and Innovative Techniques*, Cambridge University Press, Cambridge.

69. Y. Braiman, F. Family, H. G. E. Hentschel, C. Mak, and J. Krim (1999). Tuning friction with noise and disorder, *Phys. Rev. E*, 59, pp. R4737–R4740.

70. C. A. Brockley and P. L. Ko (1970). The measurement of friction and friction-induced vibration, *Trans. ASME*, 92, pp. 543–549.

71. C. A. Brockley and P. L. Ko (1970). Quasi-harmonic friction-induced vibration, *Trans. ASME*, 92, pp. 550–556.
72. A. Lienard (1928). Etude des oscillations entrenues, *Rev. Gen. Elect.*, 23, 901 pp.
73. V. Aronov, A. F. D'Souza, S. Kalpakjian, and I. Shareef (1983). Experimental investigation of the effect of system rigidity on wear and friction-induced vibrations, *J. Lubr. Technol.*, 105, pp. 206–211.
74. V. Aronov, A. F. D'Souza, S. Kalpakjian, and I. Shareef (1984). Interactions among friction, wear, and system stiffness—Part 1: Effect of normal load on system stiffness, *J. Tribol.*, 106, pp. 54–58.
75. V. Aronov, A. F. D'Souza, S. Kalpakjian, and I. Shareef (1984). Interactions among friction, wear, and system stiffness—Part 2: Vibrations induced by dry friction, *J. Tribol.*, 106, pp. 59–64.
76. V. Aronov, A. F. D'Souza, S. Kalpakjian, and I. Shareef (1984). Interactions among friction, wear and system stiffness—Part 3: Wear model, *J. Tribol.*, 106, pp. 65–69.
77. G. M. Bartenev and V. V. Lavrentev (1981). *Friction and Wear of Polymers*, Elsevier, Amsterdam, pp. 53–63.
78. D. M. Tolstoi (1967). Significance of the normal degree of freedom and natural normal vibrations in contact friction, *Wear*, 10, pp. 199–213.
79. A. Soom and C. Kim (1983). Interactions between dynamic normal and frictional forces during unlubricated sliding, *J. Lubr. Technol.*, 105, p. 221.
80. D. P. Hess and A. Soom (1991). Normal vibrations and friction under harmonic loads: Part 1—Hertzian contacts, *J. Tribol.*, 113, pp. 80–86.
81. D. P. Hess and A. Soom (1991). Normal vibrations and friction under harmonic loads: Part 2—Rough planar contacts, *J. Tribol.*, 113, pp. 87–92.
82. J. Kiely, T.-T. Hsia (2006). Three-dimensional motion of sliders contacting media, *J. Tribol.*, 128(3), pp. 525–533.

8 Running-In and Other Friction Transitions

Breaking-in, running-in, and wearing-in are examples of tribological transitions. They are characterized by changes in friction and wear with time, numbers of sliding cycles, or sliding distance. Transitions may occur under lubricated or unlubricated conditions. The attributes of friction transitions are (a) changes in the nominal magnitude of the friction force, (b) the time that a tribosystem requires to reach steady state or some other distinct condition such as seizure or coating wear-through, and (c) the characteristics of short-term fluctuations in the friction force.

When well-lubricated tribosystems are running at steady state, the Λ-ratio is high, there is minimal solid contact, and the friction coefficient remains low and steady. But there are numerous engineering situations when steady-state conditions are not achieved, and the friction force (and possibly the normal force as well) behaves erratically. Transient power demands on motors and transmissions can cause momentary changes in frictional torques, both at steady state and during start-up and shutdown, when thick lubricating films may not be present. In brakes and clutches, frictional response is a function of the contact pressure, the surface speed, and the cumulative number of contacts. Tires rolling over sandy, oily, icy, or hot surfaces will experience changes in their traction characteristics. A newly replaced bushing or seal may require several minutes of operation to acquire the steady-state frictional characteristics of its predecessor. Many similar examples of friction transitions exist.

Transitions in wear can accompany or precede changes in friction. The subject of wear transitions was discussed extensively in a previous book,[1] and here we shall focus mainly on friction transitions, but these can accompany changes in wear as well.

8.1 UNDERSTANDING AND INTERPRETING FRICTION TRANSITIONS

Friction transitions take a variety of forms. Some are induced by intentional changes in the operating conditions of a machine (e.g., by changing the contact force, speed or direction of motion, raising the temperature, substituting a different lubricant, repetitive braking, etc.). Other transitions occur as the materials and lubricants age under continued use (or disuse).

The more extensive and severe the conditions of solid contact, the faster the clearances, surface roughnesses, surface chemistry, and the extent and severity of subsurface damage can change. One might refer to this process as *tribodynamic aging*. Although the term tribodynamic aging is introduced here, its concept is not new. For example, Kerridge and Lancaster[2] analyzed the stages of metallic wear in the mid-1950s and proposed that a layer of metallic material forms and is transferred to the opposing surface to produce transitions in friction and wear. This transferred layer, also called a third-body layer or a tribolayer, can cushion the load or fracture and serve as the source of wear debris. Kerridge and Lancaster's experiments using a brass pin sliding on the circumference of a Stellite™ ring, with and without lubrication by cetane, illustrated that the onset of transfer could be delayed by using lubricants. Once transfer to the counterface began, however, the wear rate acquired a new value. Ten years later, Welsh[3] showed that the wear rate of steel surfaces varied from low to high values and back again as contact load was varied over a certain range of values. That was interpreted as the formation of iron oxide layers with different stoichiometries (Fe_2O_3 or Fe_3O_4) depending on the increasing amount of frictional work that had to be dissipated.

Table 8.1 shows a proposed coding system for friction and wear transitions that depends on their origins and characteristics.[1] It is based on whether the change is a one-time event or one that recurs at periodic or nonperiodic intervals. It also describes events as externally induced, like an intentional change in motor speed or noninduced (natural) as the materials wear or the lubricant additive package ages. Table 8.2 provides a few examples of using the system to describe tribological transitions.

TABLE 8.1
Classification of Tribological Transitions

Frequency	Induced	Noninduced (Natural)
Single occurrence	ISO	NSO
Multiple occurrence, random	ICR	NCR
Multiple occurrence, periodic	ICP	NCP

Source: Based on Blau, P.J., in *Friction and Wear Transitions of Materials*, Noyes, Park Ridge, NJ, 1989.

TABLE 8.2
Examples of Tribological Transitions

Description	Code
A prescribed running-in procedure for a metal bearing	ISO
Seizure occurring when conducting a step-loading test	ISO
Reduction in friction when sliding speed is increased	ISO
Change in friction due to the wear-through of a protective film	NSO
Sudden production of wear debris and surface roughening due to the culmination of subsurface fatigue processes	NSO
Intentional selection of a materials couple that exhibits transfer during sliding	ICR
Removal of debris at random intervals from a sliding contact	ICR
Changes in lubrication regime when turning a machine on or off	ICR
Stick-slip of a door hinge	NCP
Change in friction due to reversing the direction of sliding	NCP
Supply of lubricant to a contact drip by drip	ICP
Intentional removal of debris at specific intervals	ICP
Buildup and sudden loss of transferred deposits	NCR
Wear of a self-lubricating material when exposure of a new pocket of solid lubricant results in a sudden friction drop	NCR

The system described in Table 8.1 is not without ambiguities. For example, consider a tribosystem that exhibits stick-slip behavior that occurs at irregular intervals. In that case, using the classification defined in Table 8.1, one might designate it as 'NCR/NCP' indicating both the periodic nature of the recurrent phenomenon as well as the fact that its onset cannot readily be predicted.

Friction transitions can also be categorized in terms of continuity. For example, the onset of seizure may result in a rapid, relatively discontinuous, stepwise change in friction, but running-in may cause friction to increase or decrease monotonically and relatively continuously until steady state is finally reached. Discontinuous friction transitions, such as scuffing or seizure, are difficult to predict and control, but some NSO transitions are very reproducible for a given sliding arrangement and materials pair. For example, reciprocating wear tests of carbon–graphite materials[4] produced very repeatable transitions in friction that allowed the investigators to stop tests before, during, and after transitions had occurred. In that case, the transitions were induced by surface fatigue that involved a well-defined crack initiation and growth period followed by spalling and transfer to the counterface.

The friction versus time records for unlubricated sliding surfaces may display numerous peaks and valleys (noise). In practice, whether or not a system is claimed to exhibit stable sliding behavior depends on the investigator's definition of stability. It also depends on the experimental method.[5] As indicated in Chapter 3, the manner of sensing force, the kind of friction data acquisition system, and the data capture settings can affect the interpretation of friction records. In the current context, the terms *continuous* and *discontinuous* refer to the general trends in friction behavior as some variable is changed (e.g., time, load, speed). It is not intended to refer to short-term fluctuations (frictional noise).

Friction transitions can be modeled either empirically by simply fitting a curve to the data or by first principles (contact mechanics, materials properties, interatomic potentials, etc.). In the next section, equations are presented to describe frictional behavior during running-in. Some of these are intended as a convenience to enable characteristic behavior to be quantified. They allow one to assign rate constants and magnitudes to the friction changes. Other models explicitly account for material properties and applied variables. Discontinuous friction transitions can be associated with "critical values," such as the critical normal force to produce a jump in friction due to seizure or the critical sliding time to induce coating wear-through. In contrast, continuous transitions imply a gradual change in the balance of the dominant interfacial processes. It is harder to assign a duration for such transitions because the changes are more gradual, and one set of contributory sliding processes gradually evolves into another.

Figure 8.1 shows how the presence of thin coatings and exposure to the environment can alter the shape and regularity of friction coefficient versus time behavior.[6] All three curves were generated on a pin-on-disk machine with a sapphire ball (30 mm radius) sliding at 67 mm/s with a 20 gf load. The lowest curve is for a silicon substrate coated with a 20 nm thick carbon film. The next higher curve is for a noncoated Si wafer. The highest curve indicates what happened when the same coated specimen was retested after exposure to 90% relative humidity at 60°C for 7 days. The level of the friction force, the general shape of the curves, and the magnitude of frictional variations were all affected. Another example (Figure 8.2), from an early

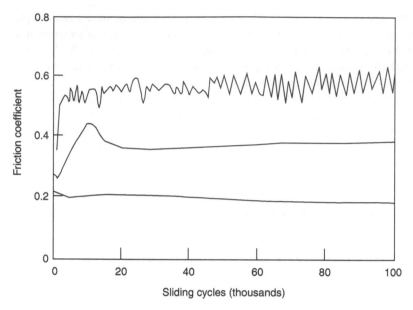

FIGURE 8.1 Effects of thin films and exposure to humidity on the characteristics of frictional behavior.

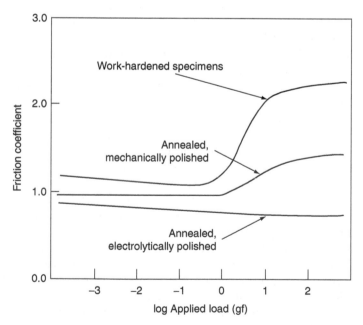

FIGURE 8.2 Adapted from Effects of load on the friction coefficient for surfaces subjected to different mechanical treatments. (Adapted from Bowden, F.P. and Tabor, D., *The Friction and Lubrication of Solids, Volume II: Contact of Metals*, Oxford University Press, Oxford, 1964.)

work on the friction of gold,[7] shows how the manner of polishing and work-hardening the surfaces can affect the magnitude of friction transitions and even determine whether they will occur at all.

Friction transitions occur over a wide range of contact sizes. Large rolling bearing components exhibit friction changes during running-in, yet there is reason to suspect that individual asperity contacts experience localized changes in microscale friction over much shorter times. Atomistic simulations of diamond-on-diamond friction have depicted periodic frictional (frictomic) changes from $\mu = 0.2$ to $\mu = 0.4$ due to crystallographic lattice periodicity.[8] This type of "natural transition" can be categorized as NCP.

Interpreting and understanding friction transitions requires a multidisciplinary approach and a comprehensive understanding of the possible causes. Friction transitions should be analyzed in the context of the specific tribosystem in which they occur. Consequently, the same type of friction–time behavior may arise for different reasons in different tribosystems. For example, a stepwise change in friction force and an increase in noise level could result from coating wear-through in one case and from the sudden introduction of abrasive contaminants into a lubricant in another. The spread of surface damage from one part of a surface to another or from one bearing component in a machine to another is facilitated by wear debris entrained in circulating lubricants. This can trigger friction or wear transitions; in a previous book, this process was called *tribocommunication*.[1]

ISO-type transitions are probably the most-studied transitions and, as noted in the previous chapter, normal load (or contact pressure) and sliding velocity are the most-studied variables in the tribology literature. Since they can be treated as independent variables, it is possible to create transition diagrams or friction or wear maps. During the 1970s and 1980s, Salomon and de Gee[9] developed what has become known as *IRG transitions diagrams*. ("IRG" refers to the International Research Group on Wear of Engineering Materials, sponsored by the Organisation of Economic Co-operation and Development in Europe.) Based initially on experiments with steel-on-steel contacts, these diagrams delineate regions for certain types of friction and wear behavior under boundary-lubricated conditions at various temperatures. An example of an IRG transitions diagram is shown in Figure 8.3. The normal load on a contact (e.g., a 10 mm diameter ball sliding on a continuously rotating 76 mm diameter ring) is plotted as the ordinate and the sliding velocity as the abscissa. To identify the boundaries for safe operation, runs are conducted at different loads, velocities, and in some cases, oil temperatures. The characteristics of the three regions shown in Figure 8.3 are described in Table 8.3. IRG transition diagrams have successfully been used to screen lubricants and materials combinations for a variety of applications.

Recognizing that time-dependent transitions can occur, especially when operating at conditions close to region II and III boundaries, Blau and Yust[10] added the time variable to IRG transitions diagrams. Figure 8.4 illustrates how this was done. The force versus duration plane in the figure is analogous to the graphical depiction of the fatigue life in terms of applied load and cycles to failure. The higher the load, the shorter is the time until failure. Transition boundaries then become surfaces rather than lines. Unfortunately, preparing empirical diagrams of this sort requires

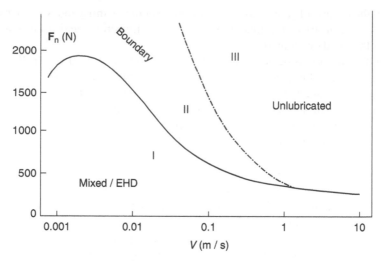

FIGURE 8.3 Example of an IRG transitions diagram showing the boundaries between different regimes of sliding, including elasto-hydrodynamic (EHD). (Adapted from Salomon, G. and de Gee, A.W.J., *The Running-in Process in Tribology*, Butterworths, London, 1981.)

TABLE 8.3
Characteristics of Regions in IRG Transitions Diagrams

Region	Lubrication Regime	μ_k	Frictional Behavior
I	(Partial) Elastohydrodynamic	0.04–0.10	Smooth, low, and steady
II	Boundary lubrication	0.2–0.4	Momentary rise and fall during running-in, but relatively steady afterward
III	Unlubricated behavior	0.2–0.5	Pronounced running-in and erratic variations during running

many long-term tests under a systematic matrix of loads and speeds. The procedure involves choosing extremes of conditions to determine whether transitions are present and if so, gradually targeting smaller volumes within the parameter space where the transition boundary surfaces lie. The enhanced version of IRG diagrams combines features of both ISO (i.e., load and velocity) and NSO (i.e., duration of contact) transitions.

Friction-related transitions have been graphically depicted in other ways. Ashby et al.,[11] whose work was discussed in Chapter 4 in regard to frictional heating, used dimensionless, normalized load and normalized velocity parameters to map transitions. Depending on how they are grouped, a number of variables can be portrayed in transitions diagrams. However, as stated previously, it is important to recognize that changing one test variable, such as contact pressure, may induce changes in others, such as surface roughness. Therefore, treating certain variables as if they were independent of one another may be misleading or fallacious.

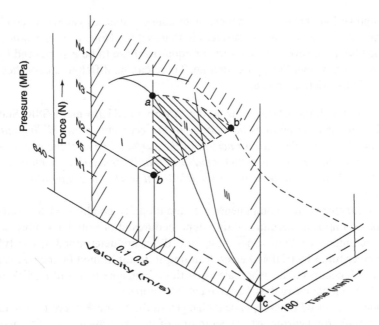

FIGURE 8.4 Example of a transition diagram in which the duration of sliding contact has been extended to give it three dimensions.

8.2 FRICTION TRANSITIONS DURING RUNNING-IN

When two surfaces are first brought into contact under load and moved relative to one another, changes in the condition of one or both surfaces occur. Sometimes these changes are detectable as changes in friction. After such initial perturbations, the friction force may reach a condition that is referred to here as the *steady state*, in which the various influences on friction have reached a balance, at least temporarily. The changes that occur between start-up and the acquisition of the steady state are termed *running-in* (also called breaking-in or wearing-in). Wearing-in, as its name implies, applies to changes in form and roughness between the initial condition and the steady state (i.e., reaching geometrical conformity). Several definitions, first presented elsewhere,[1] are as follows:

> *Steady state*—in tribology, that condition of a given tribosystem in which the average kinetic friction coefficient, wear rate, and other specified parameters have reached and maintained a relatively constant level. Note: Other parameters that could be used to define steady state include temperature, concentration of debris particles in a lubricant, and surface roughness.
>
> *Break-in*—in tribology, those processes that occur before steady state when two solid surfaces are brought together under load and moved relative to one another. This process is usually accompanied by changes in friction force and rate of wear. Synonym: run-in.
>
> *Run in (verb form with no hyphen)*—in tribology, to impose a set of conditions on a tribosystem to reduce the time required to achieve a steady state,

improve long-term performance, or to cause a state of geometric conformity to exist at the contact surfaces in that system. Note: Run-in conditions
may be more severe, less severe, or equal to those to be experienced by a
given component during its lifetime. There can be different approaches to
run in for friction or wear.

In the last definition, "run in" is used as a verb. GOST (former USSR) Standard
16429-70 defines running-in as *the change in the geometry of the sliding surfaces
and in the physicomechanical properties of the surface layers of the material during the initial sliding period, which generally manifests itself, assuming constant
external conditions, in a decrease in the frictional work, the temperature, and the
wear rate.*[12]

Some machines and components are intentionally conditioned for long-term
operation by imposing certain running-in procedures after initial assembly or after
periodic maintenance. Sometimes these procedures are determined by careful testing and other times by trial and error. Running-in is not limited by the size scale of
interaction. It occurs on the nanoscale asperities of a magnetic recording disk as well
as on full-size locomotive diesel engine cylinder bores.[13]

Braun[14] measured time-dependent changes in the Stribeck curve for a statically
loaded, porous iron bearing. At the beginning of the experiment, the friction coefficient (μ) versus speed relationship did not show the characteristic Stribeck drop-and-
rise behavior (see Chapter 6); rather, μ simply decreased as speed increased. As time
went on, the curve assumed the characteristic "bathtub" shape and the minimum μ
continued to decrease until at 1650 h the minimum μ was considerably more than
an order of magnitude lower than that observed at the start. Braun asserted that if
the same porous bearing shaft is cyclically loaded, conformity between the bearing
and shaft might not be achieved, and the mixed lubrication regime, with its higher
friction and wear, would result.

8.2.1 ANALYSIS OF RUNNING-IN BEHAVIOR

Frictional analysis of running-in requires attention to three attributes of friction versus time behavior: (a) the general curve shape, (b) the duration of certain features of
the curve, and (c) the magnitude of the fluctuations in the friction force at different
times. These attributes are illustrated in Figure 8.5. The curve shape depicts the
overall trend in friction within a given system. The duration of features implies the
speed at which individual friction processes tend toward steady state. The variation
of friction reflects the stability of the micro- or nanoscale events that contribute to
the overall frictional behavior.

Friction curves contain a wealth of information about the state of the tribosystem
and its tribodynamic processes, yet most published work on friction reports little
more than a nominal (average) friction coefficient and sometimes a level of variation
(e.g., $\mu = 0.5 \pm 0.1$). Comparatively little work has been published to interpret the
details of friction traces despite their clues to friction's fundamental processes.

Friction curve analysis (FCA) is the name given to measuring and interpreting
the features of friction versus time behavior.[15] It requires a combination of detailed
friction force recording, observations of system behavior, and studies of the contact

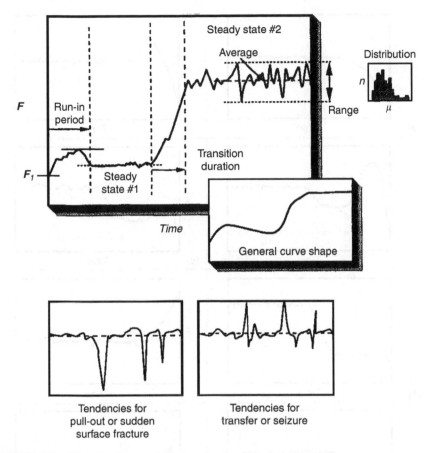

FIGURE 8.5   Three attributes of friction running-in curves.

surfaces so that each stage in surface evolution can be correlated with the friction at that time. Since there are several potential causes for features that develop on friction curves, supplementing FCA with microscopy and chemical analysis of contact surfaces is advisable. However, before embarking on an FCA, it is important to establish the degree of repeatability of the friction–time behavior in that particular tribosystem.

Eight common shapes of friction–time curves were identified by Blau based on a survey of the tribology literature in the early 1980s.[16] These forms are shown in Figure 8.6. Table 8.4 indicates some of the possible causes for each type of curve. More specific examples are given in the 1981 papers by the author (see Refs 16 and 17). In particular, the effects of the following four processes were discussed: metal transfer, film formation and removal, debris generation, and cyclic surface deterioration. The causes for each of the eight shapes shown in Figure 8.6 are not uniquely connected to a single set of cooperative mechanisms. In fact, similar curve shapes can be produced by different sets of interfacial processes.

The shape of a friction curve is affected not only by the materials involved but also by the applied load and other tribosystem characteristics. For example,

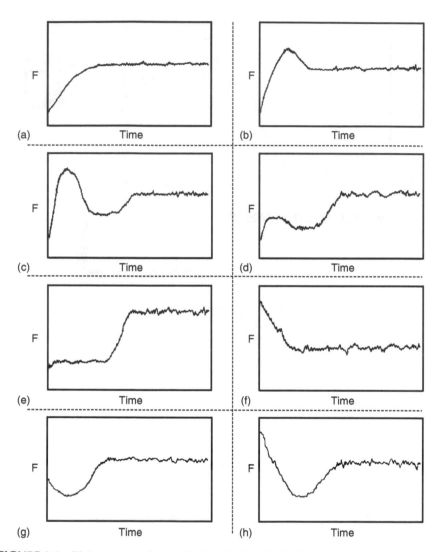

FIGURE 8.6 Eight common forms of initial frictional behavior.

Figure 8.7 shows how lowering the normal force triggered different interfacial processes and changed the shape of the friction curve from type (a) to type (e). These experiments involved unlubricated Cu sliding on 52100 steel in three ring-on-block tests run in argon gas. In type (a), copper oxide lubricated the surface until it was gradually worn off. In type (e), the effects of the oxide persisted because the load was low enough not to wear the film away too quickly.

    Different processes that contribute to the externally observed friction require different times to reach steady state. As various processes begin to operate during running-in, the superposition of their influences leads to complex friction changes until a balance is reached. Sometimes the processes continue to evolve, and the

TABLE 8.4

Possible Causes for Friction Break-In Curve Shapes

Type	Occurrence	Possible Cause(s)
a	Contaminated surfaces	A thin film of lubricious contaminant is worn off the sliding surface(s)
b	Boundary-lubricated metals (typical of IRG region II)	Surfaces wear in; initial wear rate high until the sharpest asperities are worn off and surface becomes smoother
c	Unlubricated oxidized metals, often observed in ferrous or ferrous–nonferrous pairings	Wear-in, as in type (b), but with the subsequent development of a debris layer or excessive transfer of material
d	Same as type (c)	Similar to type (c), but the initial oxide film may be more tenacious and protective
e	Coated systems; also, systems in which wear is controlled by subsurface fatigue processes	Wear-through of a coating; or subsurface fatigue cracks grow until debris is first produced—the debris then creates third bodies, which induce a rapid transition in friction. Sometimes a few initial spikes in friction signal the onset of this transition
f	Clean, pure metals	Crystallographic reorientation of regions in near-surface layers reduces their shear strength and lowers their friction. Alternatively, the initial roughness of the surface is worn off, leaving smoother surfaces
g	Graphite on graphite; metal on graphite	Creation of a thin film during running-in; debris or transfer produces a subsequent rise in friction
h	Hard coatings on ceramics	Roughness changes, then a fine-grained debris layer forms

steady-state period is either not achieved or only short-lived. Maintaining a continuous, extended steady-state friction is critical if the machine or component's functioning is to be well behaved over the service lifetime. The proper choice of running-in parameters is important, since the interfacial processes must be controlled for optimum tribosystem performance. Improper running-in can result in short component lifetimes or erratic behavior.

The time to run in can be estimated by identifying the steady-state portion of a μ versus time test record and extrapolating backward toward start-up until the first significant deviations from steady-state friction occur. Unfortunately, this method is not always practical, and it may require some judgment regarding when the running-in portion of the data blends into the steady-state portion. Sometimes the running-in portion of friction curves is distinct, as it is in type (b) (see Figure 8.6b). However, running-in may conclude gradually, and the time to reach a steady state may only be approximated. If there is no distinct steady-state behavior (e.g., the nominal magnitude of the friction force continues to change after an initial rise), it may not be possible to ascribe a specific value to the duration of the break-in period.

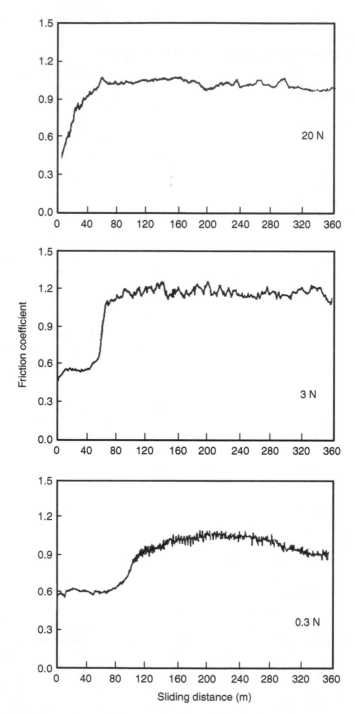

FIGURE 8.7   Effects of normal force on the shape of friction break-in curves.

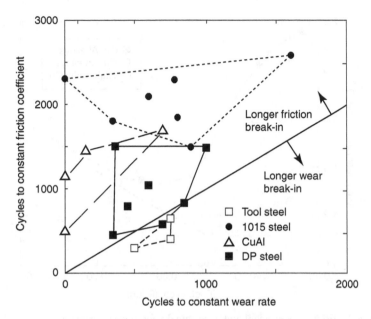

FIGURE 8.8 Ring-on-block sliding friction and wear data illustrating that the time to reach steady-state frictional behavior is not in general equal to that to achieve steady-state rate of wear for metallic material combinations.

Wear-in is not synonymous with running-in. The time to reach a steady-state rate of wear may not be the same as that required to reach a steady-state friction force behavior. This fact is illustrated from experiments that are summarized in Figure 8.8. Polished blocks of various materials (shown in the key to the figure) were slid against rotating rings of 52100 steel under similar contact loads and speeds. Displacement of the block specimen as it rubbed against the ring was used as a measure of wear. When the rate of downward displacement of the sliding block became constant with time, steady-state wear was assumed to have set in. The number of cycles to reach steady-state wear and those to reach steady-state friction were not in general identical. Each point represents one test. The ratio of the two durations varied depending on material combination. For Cu–Al alloys, the friction run-in was considerably longer than the wear-in, but tool steel blocks reached steady-state friction conditions before the wear rate became constant. With 1015 steel blocks, there is wide scatter in the data, suggesting little repeatability in behavior. In general, the observed behavior has important implications for comparing friction coefficients with wear rates, namely:

> The friction processes in a given sliding system may arise from different material and systems properties than the wear processes. These processes often do not reach steady state at the same time. High friction does not necessarily imply high wear because the frictional energy input to the system is partitioned differently from one tribosystem to another. This energy may be used, for example, to form oxides, grow cracks, plow through the surface, heat the surface, or shear debris layers. Thus, two systems may have the same coefficients of friction, but one may wear much more than the other. Friction and wear are related, but not directly.[18]

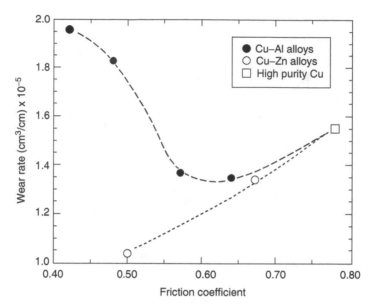

FIGURE 8.9 Friction coefficient and wear rate do not always correlate.

As further evidence of friction–wear relationships, Figure 8.9 shows pin-on-disk data in which polished blocks of several copper alloys were slid on rotating 52100 steel rings in an Ar gas environment at 5.0 cm/s sliding speed and 134 N load.[19] For Cu–Zn binary alloys, the wear rate increased with increasing friction coefficient, but this was not so for Cu–Al alloys. Therefore, friction coefficients cannot and should not be used to estimate the wear of materials *a priori* unless that relationship within a specific tribosystem has been established. The author has observed that there is a more consistent relationship between friction and wear under predominantly abrasive conditions than under sliding conditions that involve adhesion and other processes.

The duration of frictional running-in depends on more factors than simply which two materials are in contact. For example, pin-on-disk experiments with various alumina (ceramic) and aluminum (metal alloy) sliding couples indicated that both the shape of the friction curve and the duration of the running-in period depended on which material comprised the pin and which material comprised the disk.[20] Stroke-by-stroke sliding tests of 52100 steel, 2014-T4 aluminum, and poly(methyl methacrylate) polymer on abrasive papers showed that frictional break-in behavior could be affected by the ability of soft materials to fill the spaces between grits on the abrading surface.[21] Break-in behavior can also be strongly affected by the alignment between sliding surfaces. Misaligned or skewed bearing surfaces can lead to rapid, uneven wear-in and unstable frictional behavior.[22]

Two terms are particularly associated with running-in: *asperity truncation* and *elastic shakedown*. Although these primarily refer to wear, they also affect friction and are worthy of mention. Changes in surface roughness have historically been linked with running-in. In asperity truncation, one begins with a fresh, unworn surface in which the asperities are relatively sharp. Often, a Gaussian or normal

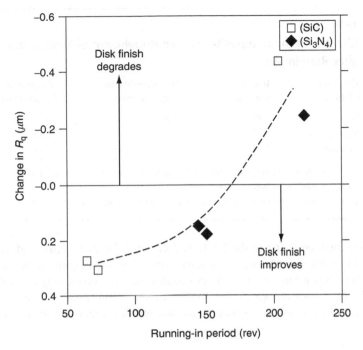

FIGURE 8.10 Depending on the initial surface finish and the conditions of imposed sliding, surface finish may either improve or degrade during running-in, as illustrated by pin-on-disk tests of silicon nitride and silicon carbide.

distribution of asperity peak heights is observed. The surface is placed against an opposing surface and sliding begins. The highly concentrated contact stress at the tips of the highest interacting asperities results in their rapid truncation, and that produces a state of surface conformity on the microscale. Thus, one would expect changes in surface roughness parameters such as peak-to-valley height, root-mean-square roughness, skewness, and the deviation of the slope of asperities from the average value (delQ). There is ample evidence that surface roughness changes during running-in; however, it is not always true that surfaces become smoother. If one starts with a polished surface sliding against a rougher surface, the smoother one may become rougher whereas the reverse is true for the counterface. Table 8.5 illustrates that the magnitude of the change in roughness of a steel gear tooth surface after running-in depends on the finishing method used.[23] Figure 8.10 shows data for pin-on-disk tests of three surface finishes on ceramics in which the surface roughness decreased after running-in for the ground surfaces, but it increased after running-in for the polished surfaces.[24]

Elastic shakedown is the term used to describe the process in which a (typically metal) surface that initially yields plastically during running-in eventually reaches an *elastic shakedown limit* at which the behavior of the near-surface layers is no longer plastic but has reached a work-hardened condition sufficient to support the contact pressure in an elastic manner. This type of analysis is typically applied to Hertzian contacts in bearings and gears. A detailed discussion on the subject of

TABLE 8.5
Changes in the Composite Surface Roughness[a] of Steel Gears after Running-In

Finishing Method	Initial R_q (μm rms)	Post Running-In R_q (μm rms)	Change in R_q (μm rms)
Hobbing	1.78	1.02	−0.76
Shaving	1.27	1.02	−0.25
Lapping	3.30	1.91	−1.39

[a] Composite surface roughness (R_q) is the square root of the sum of the squares of the roughnesses of both surfaces.

Source: Mobil EHL Guidebook, Mobil Oil Corporation, New York, NY, 1979.

shakedown may be found in the book by Johnson.[25] Work by Hills and Ashelby[26] has addressed the contribution of residual stresses to the magnitude of the shakedown limit. Depending on the contact conditions, shakedown may require only a few cycles of contact and would therefore occur very early in the running-in process, perhaps even before other factors, such as bulk temperature, reach their steady-state values.

8.2.2 MODELING OF RUNNING-IN

Although most published models for running-in center on changes in surface roughness, there are also other changes occurring both on and below the surfaces during running-in. We recall that the change in surface roughness during running-in affects the film thickness ratio (Λ) in lubricated systems and may move the operating point on the Stribeck curve.

The famous Russian tribologist Kragelsky (sometimes spelled Kragelskii or Kragelski) and his colleagues,[27] for example, described several approaches to modeling running-in. One approach considers the conditions necessary for reaching the optimum, "equilibrium" surface roughness Δ_{opt} at which friction coefficient will be lowest:

$$\Delta_{opt} = \frac{16\tau_o^{1.25}\Theta^{0.75}}{p_c^{0.5}\alpha_{hys}^{1.25}}$$

(8.1)

where τ_o is the shear strength of the asperities, p_c the initial contact pressure, α_{hys} the hysteresis loss factor (depicting a change in strength properties as a result of sliding), which is determined by uniaxial tension and compression experiments and Θ is expressed in terms of the elastic modulus E and Poisson's ratio v,

$$\Theta = \frac{1-v^2}{E}$$

(8.2)

Another discussion of this approach was contained in an article in *Wear* journal.[28]

A semiempirical running-in and transitions model has been published by Blau.[29] In its simplest form, the model can be represented as a product of two factors:

$$\mu(t) = L(t)S(t) \tag{8.3}$$

where $\mu(t)$ is the time-dependent coefficient of friction, $L(t)$ the time-dependent lubrication factor, and $S(t)$ the time-dependent contribution of the solid materials. Each factor in the model is further broken down into a form that permits the magnitudes and rates of change in frictional contributions to be incorporated.

The lubrication factor modifies the contribution of the friction of the solids. It varies from 0 to 1. If the lubricant is absent or fails to prevent solid contact, $L(t) = 1.0$. If the lubrication chosen for the system works well from the start and continues to do so, $L(t)$ may take on a constant value, typically between 0.001 and 0.1. If the state of lubrication changes during running-in, the time dependence of L can be written more generally,

$$L(t) = \frac{1}{1+l} \tag{8.4}$$

with l being in essence the lubricant effectiveness,

$$l = l_0 e^{-At} \tag{8.5}$$

where l_0 is the initial lubricant effectiveness, A a rate constant, and t the running time. If $L(t)$ is constant, then $A = 0$, and if the lubricant effectiveness is 0, then the $L(t) = 1.0$ and the behavior depends wholly on the solid materials. If A is positive, then the effectiveness of the lubricant degrades with time in a manner shown in Figure 8.11 (assuming $l_0 = 5.7$).

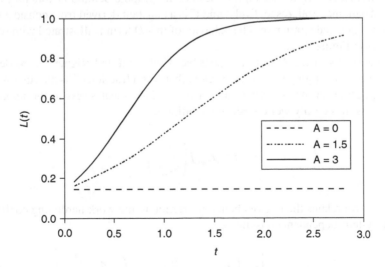

FIGURE 8.11 Contributions of the ratio of shear strengths and ratios of Schmid factors to frictional changes during running-in, as suggested by Equation 8.7.

In this simple model, the lubrication factor is treated as being independent of the behavior of the solid bodies. Clearly, when the surface roughness changes from wear during sliding, it affects the film thickness-to-composite roughness (Λ) ratio, which, in turn, affects the regime of lubrication, and then the changes in the solids could ultimately affect $L(t)$.

To describe the role of the solid bodies in modeling the time dependence of friction, several additional terms are needed. $S(t)$ can be represented as follows:

$$S(t) = D + T + V \tag{8.6}$$

where D is the initial deformation and texturing term, T represents the effect of longer-term transitions, and V is the variability contribution (i.e., the frictional chatter or noise contribution superimposed on the others). We can consider each of these separately.

Most metals and alloys, and some polymers, alter their subsurface crystal orientations as a result of friction-induced shear forces. When that orientation is altered, the shear strength of the near-surface layers may change accordingly. This change in shear strength affects friction. If orientation causes the layers to acquire an easy-shear orientation, the contribution to the friction force decreases. Kuhlmann-Wilsdorf[30] considered that the friction was governed by the properties of the softer of the two contacting bodies and suggested the following relationship:

$$D = C \left(\frac{\tau_s}{\tau_b} \right) \left(\frac{m_b}{m_s} \right) \tag{8.7}$$

where τ_s and τ_b are the shear strengths of the work-hardened surface and the bulk material, respectively; m_s and m_b are the crystallographic Schmid factors for the surface and bulk material, respectively; and C is a constant derived from considerations of contact area and strain rate. The behavior of this D term is illustrated with several examples in Figure 8.12.

Other forms of the D term are possible. Heilmann and Rigney[31] considered a monotonically changing friction coefficient that was a fraction of its maximum value μ_{max} depending on the changing ratio of the surface shear stress to the maximum shear stress in the fully work-hardened condition:

$$\mu = \mu_{max} f \left(\frac{\tau_s}{\tau_{max}} \right) \tag{8.8}$$

Another form relates the rate of change in friction to the work-hardening coefficient n, by a system-dependent constant m:

$$D = \mu_{max} \left\{ 1 - \left(\frac{\tau_{max} - \tau_0}{\tau_{max}} \right) \left[1 - \exp \left(\frac{-mn}{t} \right) \right] \right\} \tag{8.9}$$

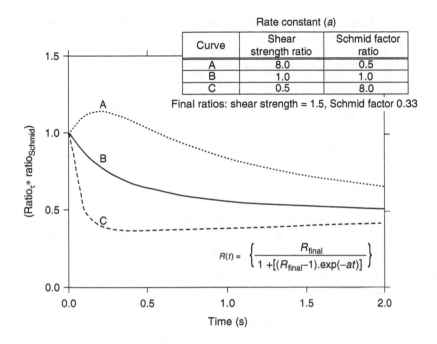

Rate constant (a)		
Curve	Shear strength ratio	Schmid factor ratio
A	8.0	0.5
B	1.0	1.0
C	0.5	8.0

Final ratios: shear strength = 1.5, Schmid factor 0.33

$$R(t) = \left\{ \frac{R_{final}}{1 + [(R_{final}-1).\exp(-at)]} \right\}$$

FIGURE 8.12 Contributions of the ratio of shear strengths and ratios of Schmid factors to frictional changes during running-in, as suggested by Equation 8.7.

The transition term T is intended to account for longer-term, noninduced changes in the friction force, that is, those that require an incubation period. One example is the development of subsurface fatigue damage, which eventually results in the production of wear particles, which in turn cause a rapid transition to three-body contact. A mathematical expression that permits one to model friction curves of type (e) is

$$T = \left(\frac{\mathbf{F}_p}{2} \right) \left\{ 1 + \left[\frac{t - t_i}{(|t - t_i|^b + 1)^{1/b}} \right] \right\} \tag{8.10}$$

where $\mathbf{F}_p$ the total frictional contribution due to the transition process, t_i the incubation time to the midpoint of the transition, and b a rate constant to describe how fast the transition occurs.

The form of Equation 8.10 is shown in Figure 8.13. One use of this is to portray the wear-through of coatings. Individual coatings may have specific values of t_i, and the rates of deterioration may be described by the constant b. It can be shown that by using various combinations of L, D, and T terms, all of the curve shapes shown in Figure 8.6 can be reproduced.

The variability in the friction trace as a function of time is represented by the V term in Equation 8.6. In a previous paper on scale effects in friction,[4] the statistical distribution of values of the steady-state friction coefficient was discussed. The more narrow the distribution, the steadier the friction and the smoother the sliding.

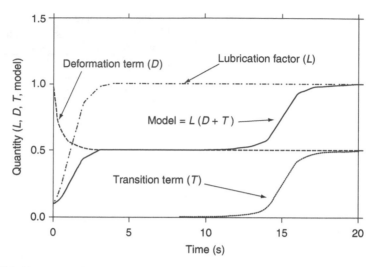

FIGURE 8.13   Illustration of how Equation 8.10 can be used to model longer-term transitions in frictional behavior.

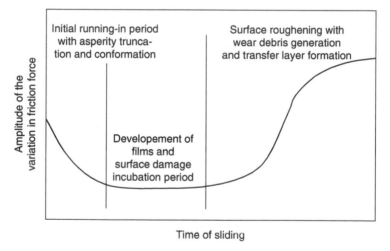

FIGURE 8.14   Changes in the amplitude of the friction force variation during running-in.

The variability of friction can change during running-in due to changes in surface condition (e.g., roughness, films, adhesive transfer, wear debris). Figure 8.14 shows schematically that the amplitude of the friction coefficient variation may be high initially, decreases when microconformity is achieved, and increases again as roughness increases due to wear-out or a transition to severe wear. Therefore, frictional response during running-in can be associated with wear.

Fractals have been used to study the frictional fluctuations and vibrations during running-in. Zhu et al.[32] conducted such experiments, using oil-lubricated pin-on-disk tests of AISI type 1045 steel pairs. They found that the fractal characteristics of both the friction force and vibrations during sliding could be correlated with the arithmetic average surface roughness of the steel, enabling one to monitor the progress of running-in.

8.2.3 MONITORING AND DEVELOPING RUNNING-IN PROCEDURES

If the friction force or torque in operating machinery cannot be measured directly, there are indirect measures available to monitor the running-in process. These include motor current signal analysis, vibrations, debris concentration in the lubricant, and bearing temperature. These and other techniques are described in a collection of papers in the ASM handbook volume on tribology.[33] Specific methods for monitoring the running-in of internal combustion engines were described in Neale's *Tribology Handbook*.[34] They include measuring oil temperature, coolant temperature, blow-by rate, crankcase pressure, exhaust smoke, specific fuel consumption, oil consumption, and compression pressure.

So-called *running-in accelerators* can be added to engines to facilitate the process, but with the possible exception of special piston ring treatments, today's automobile manufacturers do not use most of them anymore. Improvements in machining tolerances and lubricants have reduced the need for running-in accelerators. Neale listed the following historical approaches to introducing running-in accelerators: abrasive added to the intake air, additives to fuel, special oil formulations, special coatings for piston rings, and taper-faced rings in which the higher initial stress accelerated cylinder bore wear-in.

With the rapid expansion of information on the Internet, advice on break-in procedures for engines has also proliferated. Many of these were developed by trial and error but claim to be highly effective. For example, on a Web site called "Power News Magazine,"[35] there was a commentary on "break-in secrets." The claim was made that more aggressive break-in practices using a nonsynthetic oil ensured better engine power and less long-term wear for large and small internal combustion engine piston rings and liners. The rationale was that wear-in is critical for the first running period, so run the engine hard and change the oil after a very short period of initial use. Specifically,

> ... while about 80% of the ring sealing takes place in the first hour of running the engine, the last 20% of the process takes a longer time. Street riding isn't a controlled environment, so most of the mileage may not be in "ring loading mode." Synthetic oil is so slippery that it actually "arrests" the break in process before the rings can seal completely. I've had a few customers who switched to synthetic oil too soon, and the rings never sealed properly no matter how hard they rode. Taking a new engine apart to re-ring it is the last thing anyone wants to do, so I recommend a lot of mileage before switching to synthetic. It's really a "better safe than sorry" situation.

The foregoing approach seems reasonable, but different engine designs may require different approaches to running-in, and the guidance provided by either the manufacturer or a third party represents some combination of art and engineering.

With the running-in of gears, factors such as the use of very hard surfaces or wear-resistant coatings or treatments (nitriding and carburizing) and of antiwear additives in oils can reduce the rapidity of the running-in process. In other cases, special oils with EP additives are used to avoid scuffing during running-in.

The state committee for standards of the former USSR (i.e., GOST) produced a standard titled "Improving the Wear Resistance of Parts: Test Evaluation of the Run-In Capacity of Materials."[36] This method conducts a series of stepwise increasing contact pressure tests on conforming, curved surfaces. The test was conducted in three stages. First, the load on a fresh specimen was increased in steps to determine P_{MH}, the seizure pressure. Second, the run-in specimen was subjected to step loading to determine the maximum load capacity of the run-in surface, P_{MN}. The time to reach seizure when the limiting load was reached was called t_π. Third, the load was decreased until the friction coefficient dropped. The maximum running-in capacity was determined by the relationship between the critical seizure pressure for the unworn and the run-in specimens and by the time after reaching the maximum load that seizure occurred. The less is the time to seizure, the faster the material runs in.

Kragelsky et al.[37] devoted an entire chapter of their book on friction and wear calculation methods to running-in, and they proposed a method to estimate the proper running-in procedure for a journal bearing, as an example. At a given running-in load N, the running-in time t_{ri} is given as a function of the initial wear rate v_i, the mean value of the roughness of the roughest body R_z, and the factor of stepwise load increase Ψ (typically, 1.1–1.3). Thus,

$$t_{ri} = \frac{R_z}{v_i}\left(1 - \sqrt[3]{2 - \Psi}\right) \tag{8.11}$$

Next, one calculates the number of running-in steps n at a load N_i for each step. Given that the maximum running-in load N_{max} is 50% of the mean service load for the given component, during step i of the running-in process,

$$N_i = N_1 \Psi^{i-1} \tag{8.12}$$

$$n = \frac{\ln(N_{max}/N_1)\Psi}{\ln \Psi} \tag{8.13}$$

The total running-in time (t_Σ) is therefore:

$$t_\Sigma = t_{ri} n \tag{8.14}$$

Despite the importance of running-in, there are no ASTM, ANSI, or other Western standards to assess and develop running-in procedures for materials. Thus, industry usually develops its own running-in procedures for components such as brakes or gears.

8.2.4 FRICTION PROCESS DIAGRAMS

Friction process diagrams (FPDs) were introduced by the author[38] to depict the causes for friction transitions in materials. The approach combines experimental data and observations of contact surfaces. It is not necessary to address the fundamental mechanisms of friction to use FPDs. Rather, the construction of these diagrams is based on the recognition that friction response is in effect a mixture of contributions from concurrent processes, and that each process may in turn consist of a combination of more fundamental mechanisms. Examples of friction processes are debris layer formation, transfer layer formation, tribooxidation, and thin-film-dominated sliding. Construction of an FPD for a given tribosystem is based on knowing the friction coefficients characteristic of each of the friction processes that operate in that tribosystem.

Friction coefficients for the limiting cases of process domination can be determined either by specially designed experiments or from the extrapolation of friction data coupled with quantitative measurements of fractional surface areas covered by various deposits. In that analysis, it is assumed that the true contact area acted on by each friction process is proportional to the fraction of the apparent contact area that it occupies. For example, if the surface of a bearing is entirely covered by a debris layer, the friction coefficient would have a characteristic value μ_{debris}. Likewise, if the same system was wearing by severe metallic wear and 100% of the contact area was experiencing this type of contact, the friction will have another value μ_{metallic}. The observed friction coefficient for a mixture of metallic and debris-covered contacts would scale linearly. Thus,

$$\mu = f_{\text{debris}}\mu_{\text{debris}} + f_{\text{metallic}}\mu_{\text{metallic}} \tag{8.15}$$

where $(f_{\text{debris}} + f_{\text{metallic}}) = 1.0$. Thus, it is possible to construct a binary FPD in which the ordinal is the friction coefficient and the abscissa is the fraction of dominance of a second process. The value of the characteristic friction coefficient for one process is plotted on the left ordinate and the characteristic value for the second on the right ordinate. A straight line connects the two. One could then predict the friction coefficient by observing the contact surface periodically and measuring the fraction of the area occupied by either of the two competing processes. This technique should be suitable for estimating the friction of failed components from the appearance of their worn surfaces, provided that the limiting friction for each dominant process is known in that system, and if it is apparent that the load was evenly distributed across the apparent area of contact at the time of failure.

The concept embodied in FPDs was used in earlier work by Lee and Flom[39] to model the behavior of tungsten and molybdenum wires lubricated by colloidal graphite during ribbon drawing. Letting x be the fraction of the wire surface not covered by graphite, the friction coefficient during drawing could be expressed as

$$\mu = 0.5x + 0.1(1 - x) \tag{8.16}$$

This equation agreed within about 5% with the data.

In more complex systems, there may be more than two friction processes operating. For example, when a bronze block was slid against a rotating ring of the 52100 steel, the complex frictional behavior shown in Figure 8.15 was observed. This can be interpreted in terms of initial sliding on a copper oxide-covered layer. This layer

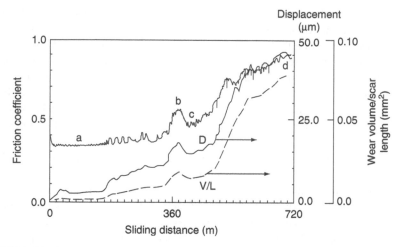

FIGURE 8.15 Frictional changes recorded during the initial sliding of an aluminum–bronze block on a steel ring.

wears through, leading to a rise in friction, but gradually a debris layer forms, reducing the contribution of metallic contact to friction. Eventually, the debris layer itself becomes too thick to act as a friction reducer and fragments of it are lost, exposing the metallic areas underneath. At last, the friction rises to a value typical of severe metallic wear.

The situation shown in Figure 8.15 can be represented by the ternary FPD shown in Figure 8.16. Here, each apex of the triangle represents the friction coefficient associated with a certain process. The series of parallel lines represent the fractions of each process that operate. There are three binary FPDs lying along each edge of the diagram. Within the large triangle, all possible combinations of the three competing processes can be represented. The friction coefficients are linearly scaled between apices, and similar values on all three sides are connected to produce *isotribes* (lines of equal friction) across the diagram. One can plot the instantaneous operating points for the situation depicted in Figure 8.15 on Figure 8.16. Thus, it is possible to trace the changing contributions of friction processes with time. Similarly to the binary case, the weighted friction coefficient is given as

$$\mu = f_1\mu_1 + f_2\mu_2 + f_3\mu_3 \text{ and } f_1 + f_2 + f_3 = 1.0 \qquad (8.17)$$

Ternary FPDs have interesting implications for the interpretation of frictional behavior of complex interfaces. For example, we see that the same friction coefficient can result from different combinations of the same three processes. Therefore, a given value of the friction coefficient does not imply a unique contribution of processes, but only defines a subset of possible combinations. When friction force or friction coefficient varies during sliding, the locus of operating points in the system constitutes a band across the FPD. The relative fractions of competing processes may vary widely even within a relatively small variation in observed friction coefficient.

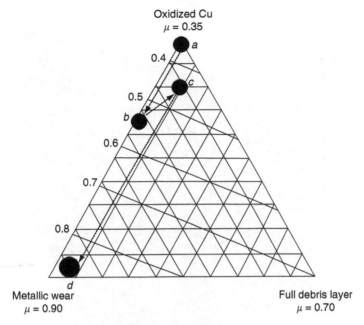

FIGURE 8.16 Representation of the frictional history of the couple from Figure 8.15 on a ternary FPD in which each apex represents the limiting case where the friction coefficient in the tribosystem is controlled by a single process only. The operating "point" is represented as a circle because instantaneous fluctuations limit its location to a region in friction process space.

FPDs represent the possible combinations of operating processes and the path taken by the operating point during the life of a sliding contact, but they do not depict the rates of tribodynamic changes in the system. To accomplish that, the approaches taken in the friction transitions model, described in Section 8.2.2, can be applied.

FPDs can be constructed to represent various situations in the same system. For example, it should be possible to establish ternary FPD for various temperatures, as shown in Figure 8.17. The characteristic apex values may not change to the same degree as temperature rises or falls. Therefore, isotribes may rotate in position from one plane on the diagram to another, depending on the effect of temperature on each of the apex values. The operating point of a tribosystem may move not only within a given plane, but also between planes as the system ages. As more is learned about the combination of fundamental mechanisms that result in characteristic friction coefficients for the apex processes, FPDs may someday be calculated entirely from first principles, but even without such first-principle approaches, FPDs can be a useful tool for understanding the characteristics of friction transitions.

In summary, the sliding friction of solids is a multidimensional phenomenon, and frictional behavior must be conceptualized as being time dependent. When we envision a sliding surface, we should expect to see an ever-changing pattern of instantaneous contact patches. The frictional contributions of some of the contacts may be governed by one dominant friction process, yet at the same time, other

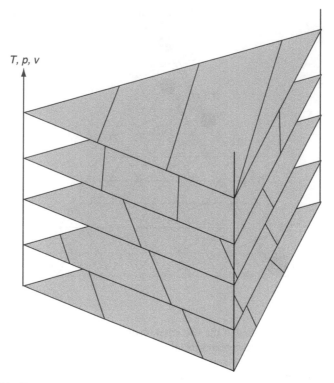

T, p, v

FIGURE 8.17 Ternary FPDs can be stacked up to represent the influence of other variables, such as temperature, on the characteristic apex values of the diagram.

processes may be at work on other contacts. Rotating our point of observation to look at the interface in cross section, we might observe velocity accommodation processes occurring at different places and with different magnitudes of shear strain at various distances from the mid-plane of the sliding contact. Finally, the variable of time is added. By visualizing such a rich, tribodynamic system, it becomes possible to gain a more mature insight into running-in and other types of friction transitions.

8.2.5 FLUCTUATIONS IN FRICTION FORCE

Rarely, if ever, does the friction force in an operating tribosystem remain completely steady. One might be tempted to assume that the less effective the lubrication scheme, the greater will be the variation in friction coefficient. For example, the sliding friction coefficient of an unlubricated steel pin on a steel disk in air might be 0.65 ± 0.15, representing a 60% difference between the minimum and maximum value. With boundary lubrication, it is possible to reduce μ below about 0.12. In well-lubricated steel bearings with a small amount of slip, $\mu \leq 0.01$. Yet even if μ varies during operation only from 0.002 to 0.003, this is still a 50% change. Fluctuations in friction are an indication that even in well-designed bearings, and even in well-controlled laboratory tests, it may not be possible to eliminate all factors that affect frictional stability.

In Chapter 4, it was asserted that since the friction force of tribosystems varies, any friction model that attempts to determine a single value of the friction force or friction coefficient for that tribosystem is false *ab initio*. In 1965, Lotfi Zadeh, a professor at the University of California at Berkeley and the modern founder of the field of fuzzy logic, wrote the following: "As the complexity of a system increases, our ability to make precise yet significant statements lose meaning and meaningful statements lose precision." In other words, "As complexity rises, precise statements lose meaning, and meaningful statements lose precision."[40] Fuzzy logic at first met with considerable criticism from traditional mathematicians, particularly in the United States in the 1960s and 1970s, since the historical prejudice against its concepts extends back to the truth/untruth arguments of Aristotelian logic developed 2600 years ago. But recently, it has gained more acceptance and has enabled significant commercial product successes, particularly in Japan. Its tenets suggest that some phenomena, among which the author includes frictional phenomena, can best be understood in terms of tendencies in behavior and not in precise, well-defined classes or single numerical values. Thus, if one truly wishes to understand the nature of frictional fluctuations, it is necessary to accept the fact that it may not be possible to ascribe a single, sharply defined value to the friction coefficient in sliding systems, but rather to accept the fact that some amount of frictional fluctuation is inevitable, and to treat the tribosystem and its modeling from that perspective.

REFERENCES

1. P. J. Blau (1989). *Friction and Wear Transitions of Materials*, Noyes, Park Ridge, NJ.
2. M. Kerridge and J. K. Lancaster (1956). The stages in a process of severe metallic wear, *Proc. R. Soc. Lond., A*, 236, pp. 250–264.
3. N. C. Welsh (1965). The dry wear of steels. I. The general pattern of behavior, *Phil. Trans. R. Soc. Lond., A*, 257, p. 31.
4. P. J. Blau and R. L. Martin (1994). Friction and wear of carbon-graphite materials against steel and ceramic counterfaces, *Tribol. Int.*, 27(6), pp. 413–422.
5. P. J. Blau (1991). Scale effects in steady-state friction, *Tribol. Trans.*, 34(3), pp. 335–342.
6. Y. Kokaku and M. Kitoh (1989). Influence of exposure to an atmosphere of high relative humidity on the tribological properties of diamond-like carbon films, *J. Vac. Sci. Technol., A*, 7(3), pp. 2311–2314.
7. F. P. Bowden and D. Tabor (1964). *The Friction and Lubrication of Solids, Volume II*, Oxford University Press, Oxford.
8. J. A. Harrison, C. T. White, R. J. Colton, and D. W. Brenner (1993). Atomistic simulations of friction at sliding diamond interfaces, *Mater. Res. Soc. Bull.*, 18(5), pp. 50–53.
9. G. Salomon and A. W. J. de Gee (1981). The running-in of concentrated steel contacts: A system oriented approach, in *The Running-In Process in Tribology*, eds. D. Dowson, C. M. Taylor, M. Godet, and D. Berthe, Butterworths, London, pp. 6–15.
10. P. J. Blau and C. S. Yust (1992). Sliding wear testing and data analysis strategies for advanced engineering ceramics, in *Wear Testing of Advanced Materials*, eds. R. Divakar and P. J. Blau, ASTM Special Technical Pub. 1167, ASTM, Philadelphia, pp. 161–170.
11. M. F. Ashby, J. Abulawi, and H. S. Hong (1991). Temperature maps for frictional heat in dry sliding, *Tribol. Trans.*, 34(4), pp. 577–587.
12. V. Kraghelsky, M. N. Dobychun, and V. S. Kombalov (1982). *Friction and Wear Calculation Methods*, Pergamon Press, Oxford, p. 297.

13. J. Maki and K. Aho (1981). Development of a running-in procedure for a locomotive diesel engine, in *The Running-In Process in Tribology*, eds. D. Dowson, C. M. Taylor, M. Godet, and D. Berthe, Butterworths, London, pp. 147–152.

14. A. L. Braun (1982). Porous bearings, *Tribol. Int.*, 15(5), p. 225.

15. P. J. Blau (1983). Monitoring the sliding contact conditions in laboratory wear tests of metals using time-dependent variations in friction coefficients, in *Time-Dependent Failure Mechanisms and Assessment Methodologies*, eds. J.G. Early, T. R. Shives, and J. H. Smith, Cambridge University Press, Cambridge, pp. 145–154.

16. P. J. Blau (1981). Interpretations of the friction and wear break-in behavior of metals in sliding contact, *Wear*, 71, pp. 29–43.

17. P. J. Blau (1981). Mechanisms for transitional friction and wear behavior of sliding metals, *Wear*, 72, pp. 55–66.

18. P. J. Blau (1998). Four great challenges confronting our understanding and modeling of sliding friction, in *Tribology for Energy Conservation*, eds. D. Dowson et al., Elsevier, UK, pp. 117–128.

19. P. J. Blau (1979). Interrelationships Among Wear, Friction, and Microstructure in the Unlubricated Sliding of Copper and Several Single-Phase Binary Copper Alloys, Ph.D. dissertation, The Ohio State University, Columbus, Ohio, 341 pp.

20. P. J. Blau and C.E. DeVore (1989). Interpretations of the sliding friction break-in curves of alumina–aluminum couples, *Wear*, 129, pp. 81–92.

21. P. J. Blau, E. P. Whitenton, and A. Shapiro (1988). Initial frictional behavior during the wear of steel, aluminum, and poly methyl methacrylate on abrasive papers, *Wear*, 124, pp. 1–20.

22. P. J. Blau and E. P. Whitenton (1984). The effect of flat-on-ring sample alignment on sliding friction break-in curves for aluminum bronze on 52100 steel, *Wear*, 94, pp. 201–210.

23. *Mobil EHL Guidebook* (1979). Mobil Oil Corporation, New York.

24. P. J. Blau (1992). Effects of surface preparation on the friction and wear behavior of silicon nitride/silicon carbide sliding pairs, *J. Mater. Sci.*, 27, pp. 4732–4740.

25. K. L. Johnson (1985). *Contact Mechanics*, Cambridge University Press, Cambridge (Chapter 7).

26. D. A. Hills and D. W. Ashelby (1980). A note on shakedown, *Wear*, 65, pp. 125–129.

27. I. V. Kragelsky, M. N. Dobychin, and V. S. Kombalov (1982). *Friction and Wear Calculation Methods*, Pergamon Press, London, pp. 193–198.

28. I. V. Kraghelsky and V. S. Kombalov (1969). Calculation of value of stable roughness after running-in, *Wear*, 14, pp. 137–140.

29. P. J. Blau (1987). A model for run-in and other transitions in sliding friction, *J. Tribol.*, 109(3), pp. 537–544.

30. D. Kuhlmann-Wilsdorf (1981). Dislocation concepts in friction and wear, in *Fundamentals of Friction and Wear*, ed. D. A. Rigney, ASM International, Materials Park, OH, pp. 119–186.

31. P. Heilmann and D. A. Rigney (1981). An energy-based model of friction and its application to coated systems, *Wear*, 72, pp. 195–217.

32. H. Zhu, G. Shirong, X. Gao, and W. Tang (2007). The changes of fractal dimensions of frictional signals in the running-in wear process, *Wear*, 263, pp. 1502–1507.

33. K. C. Ludema, H. R. Martin, F. E. Lockwood, M. Dalley, D. M. Eissenberg, and H. D. Haynes (1992). Wear monitoring and diagnosis: A collection of five papers, in *ASM Handbook, Volume 18: Friction, Lubrication, and Wear Technology*, ASM International, Materials Park, OH, pp. 290–332.

34. M. J. Neale (1973). Running-in procedures, in *Tribology Handbook*, Wiley, New York, NY (Section B30).

35. "Motoman", *Power News Magazine, How to Break-In Your Engine for More Power and Less Wear*, URL: http://www.mototuneusa.com/break_in_secrets.htm (accessed October 21, 2007). [Note: Internet Web sites are subject to rapid change, and the current reference may no longer be available. The reader is encouraged to use a search engine to find similar examples.]

36. GOST Standard 23.215-84 (1984). *Improving the Wear Resistance of Parts: Test Evaluation of the Run-in Capacity of Materials*, State Committee on Standards, USSR.

37. I. V. Kraghelsky, M. N. Dobychun, and V. S. Kombalov (1982). Running-in and equilibrium roughness, in *Friction and Wear Calculation Methods*, Pergamon Press, Oxford, pp. 297–316 (Chapter 9).

38. P. J. Blau (1994). Friction process diagrams for analyzing interfacially complex sliding contacts, *Tribol. Trans.*, 37(4), pp. 751–756.

39. M. Lee and D. G. Flom (1991). Lubrication mechanisms in tungsten and molybdenum wire drawing. Part 1. Simulated drawing experiments, *Lubr. Eng.*, 47(2), pp. 127–132.

40. D. McNeill and P. Frieberger (1993). *Fuzzy Logic*, Simon & Schuster, New York, NY, p. 43.

9 Applications of Friction Technology

Friction technology comprises the application of scientific concepts, engineering practice, and experience to the improvement of devices whose proper function depends in some manner on frictional contact. This multidisciplined field is too extensive to be covered in a single book, much less in a single chapter; however, the following sections illustrate by example how friction technology has improved energy efficiency, safety, and the quality of life.

9.1 APPLICATIONS IN TRANSPORTATION SYSTEMS

The history of tribology is strongly intertwined with that of transportation. Frictional interfaces abound in automobiles, trucks, trains, ships, and aircraft. Some interfaces, such as those in engines and drive trains, take part in propulsion and power transmission, but others, such as those in brakes, dissipate energy to enable vehicle control and deceleration. Other interfaces concern the cosmetic aspects of products, such as the feel of seat covers and upholstery. Windshield wipers, seat belts, adjustable sun visors, hinges, and door locks all involve friction. In this section, we will explore a few of the myriad frictional aspects of transportation system, using for example, brakes, tires, and internal combustion engine parts. Each of these has its own terminology and approaches to friction measurement and control.

9.1.1 FRICTION IN BRAKES

In heat engines, like steam engines and the internal combustion engine, thermal energy is converted into kinetic energy to produce motion. In mechanical brakes, however, kinetic energy is principally converted back into thermal energy that, in turn, is dissipated to the environment. Friction is an agent of the latter transformation. As described in Chapter 4, the temperature rise during frictional contact is proportional to the product of the tangential force and the relative velocity. In automotive, aircraft, or truck brakes, the total energy absorbed per stop varies directly with the initial speed at which the brake is applied and with the contact pressure on the braking surfaces (see, e.g., Figure 9.1 based on Ref. 1). Neglecting aerodynamic drag, tire rolling resistance, drive train friction, and the like, Buckman[2] once estimated that there is sufficient heat dissipated when braking a tractor-trailer truck weighing 80,000 lb (36,320 kg) from 60 miles/h (96.6 km/h) to zero to provide one-fifth of the heating needs for an average Michigan home in the wintertime. Heavily loaded commercial or military aircraft must sometimes brake suddenly during what is called a rejected take-off (RTO). Even assisted by engine thrust reversal and air braking, the brakes become red hot.

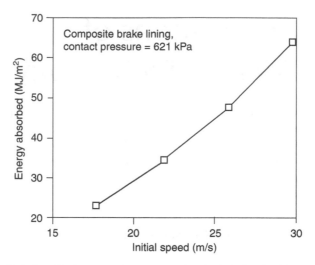

FIGURE 9.1 Relationship between the energy absorbed during braking and the initial speed for an aircraft brake lining material. (Adapted from Hooton, N.A., *Bendix Tech. J.*, Spring, 55, 1969.)

An elementary mechanics approach can be used to approximate stopping distance. If the rate of deceleration of a vehicle, initially at velocity v_o, is equal to a, the stopping distance x is[3]

$$x = v_o \left[t_r - \left(\frac{v_o}{2a} \right) \right]$$

(9.1)

where t_r is the reaction time of the driver. Note that a is negative for deceleration, producing a positive value for x.

Moore[4] described five braking modes whose characteristics are summarized in Table 9.1. In Table 9.1, the term *load transfer* refers to the redistribution of the normal force onto the tires depending on the inertia of the vehicle during braking. For a vehicle with a load per wheel of $\mathbf{W}$, a wheel base of L, and a center of gravity a distance h above the ground, undergoing a deceleration d, Moore gave the following first approximation of the friction force $\mathbf{F}$ at the tire/road interface during rear-wheel braking:

$$\mathbf{F} = \mu \mathbf{W} \left[1 - 4 \left(\frac{dh}{gL} \right) \right]$$

(9.2)

where $\mathbf{g}$ is the acceleration due to gravity. The second term in the brackets represents the load increment due to the inertial force. Its sign is negative for rear-wheel braking and positive for front-wheel braking. Unfortunately, as Moore points out, this expression limits the designer to using the same braking friction for the front and rear wheels—which is not the case when particularly severe braking conditions are

TABLE 9.1

Common Modes of Braking

Mode	Characteristics
Front-wheel braking	Causes severe load transfer, but is more stable than rear-wheel braking
Rear-wheel braking	Tends to be more unstable than front or diagonal braking and may cause the vehicle to spin
Four-wheel braking	Diminished tendency to spin, more even response in general
Diagonal braking (wheels at opposite corners of the vehicle)	Has several advantages in promoting the stability of braking in reducing load transfer between tires, but reduced stopping since two wheels are still free-rolling
Antiskid pulsed braking	Equalizes the braking forces to counteract the tendency for lock-up and to increase stability

Source: Adapted from Moore, D.F. (1975). *Principles and Applications of Tribology*, Pergamon Press, Oxford, p. 290.

in effect. Proportioning valves have been incorporated in hydraulic braking systems to help equalize the front and rear pressures during severe braking conditions, and more accurate expressions for the tire/road friction force have been derived for the use of brake system designers.

The major parts of a drum-type brake, used in older-model automobiles and some trucks, are shown in Figure 9.2. The shoes are pushed apart at the top by a cam or other actuator that provides the force **P** and results in a pressure distribution on the left and right shoes as they engage the lining. The braking torque **T** can be calculated as follows:[5]

$$\mathbf{T} = \mu w r^2 \left(\frac{\cos\theta_1 - \cos\theta_2}{\sin\theta_m} \right)(p_m - p'_m) \qquad (9.3)$$

where μ is the friction coefficient, w the face width of the shoe, r the internal radius of the drum, θ_1 the center angle from the shoe pivot to the heel of the lining (degrees), θ_2 the center angle from the shoe pivot to the toe of the lining (degrees), θ_m the center angle from the shoe pivot to the point of maximum pressure, p_m the maximum pressure on the right shoe, and p'_m the maximum pressure on the left shoe. This analysis is based on the assumption that the pressure p at any point having an angle θ with respect to the pivot is related to the maximum pressure p_m as follows:

$$p = p_m \left(\frac{\sin\theta}{\sin\theta_m} \right) \qquad (9.4)$$

and $\theta_m = 90°$ if $\theta_2 > 90°$ and $\theta_m = \theta_2$ if $\theta_2 < 90°$.

The friction coefficient is affected by temperature, and one should expect the torque generated in mechanical brakes to vary during operation. This phenomenon leads to responses that are called *brake roughness* and *brake fade*, depending on

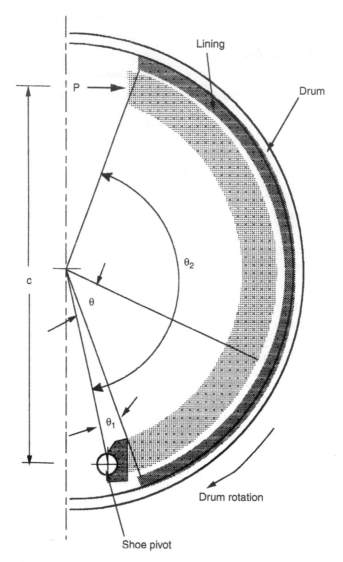

FIGURE 9.2 Diagram of a brake drum mechanism (only one-half of the unit is shown).

the brake pressure, how often or long the brakes are applied, and the initial vehicle speed. Modern vehicles are equipped with electronically controlled braking systems that can adjust the braking pressure to compensate for these effects.

9.1.1.1 Brake Materials

According to Hooton[1], friction brake materials must satisfy the following requirements:

1. They must operate at a moderately high and uniform coefficient of friction throughout the period of braking.

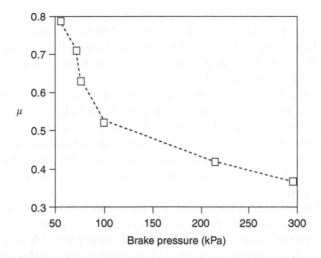

FIGURE 9.3 Effect of the contact pressure on the friction coefficient for a composite brake lining material. (From Hooton, N.A., *Bendix Tech. J.*, Spring, 55, 1969. With permission.)

2. Frictional interactions must be such that friction-induced oscillations that produce squeal or chatter are minimized.
3. Materials must be resistant to wear to ensure long life.
4. Materials must possess sufficient thermal diffusivity to prevent the interface from reaching a critical temperature for loss of performance.
5. Materials must be able to withstand the thermal and mechanical loads imposed during braking operations.

Figure 9.3, from Hooton, shows the effect of contact pressure on the friction coefficient for an experimental metal–ceramic, composite brake lining material. The drop in friction with increased pressure is consistent with the effects of high normal loads on the friction coefficients of materials.

There are five typical constituents of brake lining materials (also called "friction materials," a term that is also used in connection with clutch materials). These constituents can be categorized as binders, performance modifiers, abrasives, lubricants, and fillers.[6] Lining formulators argue that it has become necessary to use more additives in linings to make up for the removal of asbestos (more about that in another section). Typical automotive brake materials may contain more than 18 constituents, and original equipment front- and rear-brake linings for some automobiles may be formulated differently to compensate for differences in front and rear dynamics. In fact, front wheels tend to reach higher temperatures than rear wheels during severe braking.

Anderson[7] reviewed the friction and wear of automotive disc and drum brakes. Disc brakes offer better cooling but have a higher susceptibility to liquid or particulate contamination due to their exposed surfaces. Disc or drum materials are typically made from cast iron in various forms (3–3.5% C). The composition, microstructure, and heat treatment of the cast iron in commercial brake products is adjusted to avoid thermal distortion, cracking, or thermal conductivity. Since thermal issues, and other issues such as energy efficiency and corrosion resistance, are important in brakes, recent years have witnessed research into the use of nontraditional brake drum and rotor materials. These have included carbon-based materials, ceramic composites, and even aerospace metals like titanium.[8]

The potential for using titanium (Ti) alloys and metal matrix composites as brake rotor materials was investigated by Blau et al.[9] and a U.S.-based company has sold thermal-sprayed, Ti-based brake rotors for racing.[10] Titanium is less dense than cast iron and is much more resistant to corrosion from deicing salts that are used to melt snow and ice on roadways, but it has a much lower thermal conductivity than cast iron. The latter can affect heat buildup and the need for thermal management in hardware designs.

The author has used a friction heating parameter (Θ) as a means to measure the tendency of certain brake disc materials to convert frictional work into contact surface temperature rise during block-on-disc laboratory tests.[8] It is defined as the measured disc surface temperature rise (ΔT) per unit of frictional work at constant sliding velocity (v) and contact pressure. Thus,

$$\Theta = \frac{\Delta T}{Fvt} = \frac{\Delta T}{\mu Pvt} \tag{9.5}$$

where F is the average friction force during a sliding contact period of t seconds, μ the average friction coefficient during that time, and P the normal force. The units are (°C/N m) or alternatively (°C/J). The higher the Θ, the hotter the contact surface becomes for a given amount of frictional work input.

Values for Θ were determined for a series of 30 s drags on the subscale brake materials testing system (SSBT) apparatus described in Chapter 3, using an infrared detector aimed at the wear track. Discs were 127 mm in diameter and the 13 × 13 mm square-faced sliders were cut from commercial lining materials. Selected data are shown in Table 9.2. As might be expected, the values of Θ tend to increase as the disc thermal conductivity decreases. Adding carbide particles to a Ti-based composite raised the thermal conductivity and lowered the friction heating parameter, suggesting how disc compositions can be tailored to reduce heating. A similar relationship was seen between thermal conductivity and the temperature of aluminum-based metal matrix composite brake discs during simulated Alpine descents.[11]

Commercial brake lining materials are designed for different duty cycles and therefore vary in composition. Table 9.3, from Ref. 6, summarizes the range of compositions currently found in automotive brake linings. As many as 40 additives can be found in a single brake lining composition, but currently the number is between 12 and 20. The vague term "friction dust" is used in the industry to describe a proprietary additive formulation that is claimed to enhance brake performance. Depending on the source, this magical brew may contain lubricants, abrasives, fillers, or temperature stability enhancers.

Friction enhancers provide additional "grip." These are typically ceramics such as mullite, kyanite, alumina, and silica. Adding too much of such materials tends to accelerate abrasive wear. Metals such as lead, tin, bismuth, and molybdenum

TABLE 9.2
Friction Heating Parameters Obtained from Subscale Block-on-Disk Tests

Disk Material	Lining Material (Product Name)	Disk Thermal Conductivity, k_{th} (W/m K))	Θ (°C/J)
Cast iron	Jurid 539	42–57	1.18×10^{-3}
Cast iron	Armada AR4	42–57	1.04×10^{-3}
Cast iron	Performance Friction Carbon Metallic	42–57	1.23×10^{-3}
Ti-6Al-4V alloy	Jurid 539	6.83	8.33×10^{-3}
Ti-6Al-4V/10 wt% TiC composite	Jurid 539	8.21	6.33×10^{-3}
Thermal-sprayed Ti-6Al-4V (Red Devil Brakes)	High-metal content proprietary lining (Red Devil Brakes)	6.3	10.87×10^{-3}
Ceramic composite (SiC), Starfire Systems	Performance Friction Carbon Metallic	17.44	3.59×10^{-3}

Source: Blau, P.J. in *Research on Non-Traditional Materials for Friction Surfaces in Heavy Vehicle Disc Brakes*, Oak Ridge National Laboratory, Technical Report, ORNL/TM-2004/265, 2004, 36; Blau, P. J., Jolly, B. C., Qu, J., Peter, W. H., and Blue, C. A., *Wear*, 263, 1201, 2007.

TABLE 9.3
Typical Constituents Found in Automotive Brake Linings

Material	Functions[a]	Typical Content (%)	Range of Contents (%)
Phenolic resin	B	20–25	10–45
Barium sulfate	F	20–25	0–40
Mineral or organic fibers	PE	—	5–30
Cashew particles	PE	15–20	3–30
Graphite	PE, L	5–7	0–15
Metal sulfides	L	0–5	0–8
Ceramic particles	A	2–3	0–10
"Friction dust"	PE (?), L (?), A (?)	?	0–20

[a] Function codes: B, binder; F, filler; PE, performance enhancer; L, lubricant; A, abrasive.

Source: Blau, P.J. in *Compositions, Functions, and Testing of Friction Brake Materials and Their Additives*, Oak Ridge National Laboratory, Technical Memo, ORNL/TM-2001/64, 2001, 29.

(an oxygen getter), are used to form friction-modifying films on the brake surfaces. Antioxidants, such as graphite, may alter the bulk properties of the composite as well as lubricating the surface.

Studies of the effects of various additives to resin-based brake lining materials may be found in the work of Tanaka et al.[12] and more recently in the paper by Kim et al.[13] who used Taguchi design-of-experiments methods to optimize the formulation of lining materials. Figure 9.4 shows the effects of silica and alumina additions on the friction coefficient of pressed and sintered iron-based friction material containing 6% graphite, 14% barium sulfate, and tested against steel at a pressure of

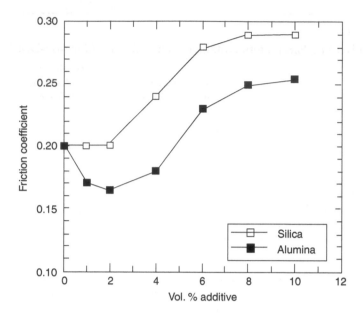

FIGURE 9.4 Effects of silica and alumina additions on the friction coefficient of pressed and sintered friction material. (From Kim, S.J., Kim, K.S., and Jang, H., *J. Mater. Proc. Technol.*, 136 202, 2003. With permission.)

15 kg/mm^2 and speed of 2.6 m/s.[14] Most of the benefits seem to occur between 2 and 6% additions of these materials, and note that alumina additions can either reduce or increase the friction depending on the amount used.

It has long been recognized[15] that brake performance depends on the nature and composition of the material that forms on brake surfaces during repeated use and that such material is not in general similar in composition to that of the disc, the drum, or the lining. Such deposits are sometimes referred to as *glazes, tribolayers, tribomaterials, transfer films*, or simply *friction layers*. Behavior of tribolayers is usually highly temperature-dependent because the properties of the film are determined by reactions with ambient oxygen. Thus, friction–temperature relationships can be improved by formulating materials to produce the desired tribolayer characteristics. Research on the evolution and behavior of tribolayers continues, and detailed characterizations of the structure of such layers continue to be published.[16–18] In fact, one manufacturer[19] has developed a truck drum brake lining using a technology that they call "cohesive friction," and claim that the superior performance in wear and friction is due to its unique transfer layer.

The last three decades of the twentieth century witnessed increased interest in development of new lining formulations. This was prompted in part by health concerns about asbestos. The Occupational Safety and Health Administration regulations of 1972[20] issued guidelines for airborne asbestos, and in 1987 the Environmental Protection Agency included it in its list of hazardous materials. Thus, brake manufacturers turned away from this material as a lining additive. Semimetallic materials and nonasbestos organics (NAOs) have been discussed by Jacko et al.,[21] who reviewed the compositions of various brake materials. Discussions may also be found in the book *Friction and Antifriction Materials*.[22]

In light of its historically empirical and secretive nature, the development of lining formulations and processing required a substantial financial investment, and therefore the detailed compositions and processing routes for commercial brake materials are rarely mentioned in the open literature. Hints of lining compositions can be found in the patent literature, but those who prepare such documents attempt to claim a broad range of possible compositions while at the same time protecting their knowledge of specific compositions that work best. The protection of intellectual property in the friction materials industry is a growing problem in international markets.

9.1.1.2 Brake Terminology and Jargon

A host of terms are used to describe the performance of brakes and brake materials. For example, the behavior of new brakes may be different than that for effectively worn-in brakes. This behavior is termed *green effectiveness*. After running for some time, a layer of char forms on the entire brake lining contact area (except for semimetallic linings). How well such surfaces perform is termed the *burnished effectiveness*. Brake *fade* refers to the loss of brake effectiveness due to any or all of the following: thermally induced fading, resin migration during brake cooling, formation of blisters, and external contamination. The ability of a friction material to regain braking effectiveness is called *fade recovery*. *Delayed fade* may occur during fade recovery. This is an insidious phenomenon because it can occur unexpectedly. *Blister fade* is caused by pockets of volatile materials remaining from fabrication causing momentary fade as they emerge on the worn surface. *Flash fade* occurs at high braking speeds due to excessive heat buildup. Except for the possibility of fade at near-stop and a small amount of thermal fading that causes progressive loss of effectiveness with decreasing speed, semimetallic linings tend to be fade-free. One important characteristic of automotive brakes is *wet friction*. Some additives, such as glass and *para*-aramid, are used to improve the wet friction of brakes. *Roughness* is a term used to describe the chatter or irregular vibrations that brakes may possess.

One major warranty issue for automobile and brake manufactures is brake noise. According to a recent survey, brake noise complaints represent about 20–25% of customer brake-related complaints, but the problem is far from new. In fact, Anderson[23] wrote that brake noise was documented as early as 3500 years ago in conjunction with a Sumarian quarry cart that used bronze chains wrapped around a wooden axle as a brake. A host of colorful descriptive terms have been used to describe brake noise, terms like groan, moan, growl, chirp, whistle, bark, gravel gurgle, and grunt. Sound and vibration engineers differentiate these terms still further (e.g., low-frequency squeal, high-frequency squeal etc). A full discussion of brake noise, its characterization, and its origins is beyond the scope of this book but can be found in the automotive engineering literature.

A certain kind of brake noise commonly occurs in the last few moments of braking just before the vehicle comes to a full stop. It results from friction–vibration interactions that are amplified and transmitted in the audible frequency range by the mechanical structure of the surrounding brake components and the wheel suspension system. Interestingly, there is no definitive evidence that brake noise affects either the safety or effectiveness of the brakes. Therefore, brake noise is more of a nuisance issue for the automobile companies than it is an issue of stopping performance. At least one automaker has attempted to sidestep the issue by stating in the owner's manual that a certain amount of noise is "normal" and should be ignored.

9.1.1.3   Aircraft Brakes

Aircraft brakes differ in configuration from automotive brakes. They usually consist of a stack of pairs of stators (carrying the friction material) and rotors. Both sets of rings are segmented. A torque tube projects through the stack. At the outer end is the pressure plate and housing.

Taxi stops are the frequent stops experienced by the brakes as a plane moves from the departure gate out to the runway. Many taxi stops are expected in the lifetime of a set of brakes. However, rejected take-offs (RTOs), also called aborted take-offs (ABOs), generally result in 100 times higher wear rates, and the Federal Aviation Administration requires changing the brakes after only one such event. Tatarzycki and Webb[24] have reviewed the friction and wear of aircraft brakes. Ho et al.[25–27] published a series of articles on the formulation and friction–temperature relationships of Cu-based aircraft brake materials.

High-performance aircraft brakes for military planes and large commercial aircraft are increasingly being made from carbon-based composites to withstand high temperatures and pressures. For example, Messier-Bugatti manufactures oxidation-resistant carbon brakes for the Boeing 777 passenger aircraft. They claim that each 10-disc brake must be capable of absorbing up to 144 MJ of energy to stop a fully loaded airplane at near take-off speed.[28]

Carbon-based brakes are made from coke aggregate and carbon binders. Since more than one form of carbon may be present, these are sometimes referred to as "carbon–carbon" materials. These are also used on high-speed trains, race cars, and upscale automobiles because they retain high strength and resist degradation at temperatures above $2200°C$.[29] Some carbon-based brakes contain a composite material of carbon with silicon carbide, a hard ceramic material. Advantages claimed for such compositions over traditional cast iron for automotive brakes include lower weight, high-temperature resistance (less fade), reduced wear, less brake dust, and reduced brake noise. Nevertheless, cost-versus-performance trade-offs for these materials—especially in regard to automotive applications where the price of premium brakes can be significant relative to the total vehicle cost—have been a subject of debate. There is also evidence that both carbon–carbon and C/SiC aircraft brakes can suffer from "morning sickness," a condition of low friction after standing overnight in damp conditions. Several braking applications are needed to restore the dry friction performance.

9.1.2   Friction in Tires

The friction of tires against the road surface is a second critical factor for the safety and control of motor vehicles. Most automobile and truck tires are pneumatic (inflated with air), but solid tires or tires filled with water are also used in off-road or heavy construction. Moore described tire performance in his book on tribology[30] and in a more recent review article.[31] He listed six functions of pneumatic tires:

1. Allows free and frictionless motion by rolling
2. Distributes vehicle weight over an area of the ground surface
3. Cushions the vehicle
4. Transmits engine torque to the road surface
5. Permits braking, steering, and driving loads to be generated
6. Ensures lateral and directional stability is maintained

To achieve these somewhat divergent aims (cf., 1 and 4), tire engineering embodies strategies that include the design of the tire structure, the materials of tire construction (up to 20 materials in some tires), the dimensions of the tire, the optimum inflation pressure, and the patterns of the tread. These factors combine to affect the shape of the tire/road "footprint" during acceleration, deceleration, constant speed rolling, and cornering conditions. Control of the footprint size and shape is an important consideration in tire design, especially as it affects vehicle response under wet driving conditions.

As noted in Chapter 5, rubber and elastomeric materials have rather unique friction and wear mechanisms involving adhesion, waves, roll formation, and hysteresis. Holmes et al.[32] listed 47 factors that can affect road-holding phenomena, and tread design is one of the most important. In aircraft tires, plain, straight ribs and unrestricted grooves are commensurate with good wet-grip requirements. High inflation pressures and severe braking requirements reduce the life of these tires dramatically. Wide automobile racing tires with little or no patterning are designed for improved dry cornering, traction, and braking, but possess poor wet-holding characteristics. The temperature of the racetrack significantly affects their traction and the ability to achieve high speeds in turns, especially if oil slicks are present. Changes in automotive tire tread patterns have been driven by a desire to improve the wet grip of the tires and to avoid "hydroplaning" that cause a driver to lose control. Radial tire construction permits more complex tread patterns than cross-ply tires. The texture of the tire surface can also be modified to alter water-absorbing and expulsion characteristics. Studded tires improve traction on snow and ice. Interestingly, the use of studs to improve traction is an idea whose evidence dates back more than 2000 years to the early Romans who pounded nails into the soles of their footwear to reduce wear and improve traction.

Modern tire treads contain channels and grooves, transverse slots and feeder channels, and sipes. *Sipes* are miniature cuts leading into the larger feeder channels. Channels and grooves are typically 3 mm wide and 10 mm deep. Transverse slots and feeder channels have smaller cross-sectional dimensions and are not continuous, but rather end abruptly within the tread rubber. They help to displace water from the tire footprint and permit gross movement of the tread. They permit localized tread "squirming" but do not contribute to drainage of water directly. Designers try to maintain a balance between the amount of squirming permissible and the water control characteristics afforded by wider channel and groove spacings.

In wet rolling, there are typically three zones of contact. The squeeze-film zone at the leading section of the contact length contains a wedge of pressurized water.

Farther back, in the "draping zone," the tread drapes over the larger asperities in the road surface to provide some traction. True contact is only achieved in the final traction zone. As speed increases, the size of the traction zone decreases until the critical speed for dynamic hydroplaning is reached. This occurs when the elastohydrodynamic action at the rear of the contact just overlaps the squeeze-film region at the leading contact zone.

Increasing the sliding velocity tends first to increase the friction coefficient of rubber tires to a maximum and then decrease it above that value. Bulgin et al.,[33] who advanced a theory of rubber friction based on energy loss in the surface area of the rubber through rapid making and braking of adhesive bonds, developed the following expression for the temperature of maximum tire friction T_F (in °C):

$$T_F = T_g + 10 \log v + 85 \tag{9.6}$$

where T_g is the second-order glass transition temperature for the rubber and v is sliding velocity in cm/s. Thus, the maximum friction temperature rises about 10°C for every order of magnitude increase in velocity. When two polymers of different T_g are blended, the friction–temperature curve either flattens out or tends to develop two peaks in friction at different temperatures.

Pillai[34] has discussed the complex system of forces that acts on the tire–road interface and has defined a number of terms in that regard. He noted that the lateral force on the tire F can be expressed in terms of vehicle weight $\mathbf{W}$, the vehicle velocity v, and the radius of curvature of the curve R (see also the Chapter 2 example for vehicle cornering):

$$F = \frac{\mathbf{W}v^2}{\mathbf{g}R} \tag{9.7}$$

where $\mathbf{g}$ is the acceleration due to gravity. This relationship, combined with another relationship developed by Schallamach and Turner[35] for the wear of tire treads due to cornering forces, permitted the development of a wear rate equation for the tires:

$$W = K\left(\frac{\mathbf{W}v^2}{R\mathbf{g}}\right) \tag{9.8}$$

where W is the wear rate and K is a constant. Moore stated that the mode of tire wear changes from contact fatigue wear to abrasive wear or wear by roll formation above a certain critical value of the friction coefficient (approximately, $\mu_{\text{crit}} = 1.2$–1.3).

Holmes et al.[32] pointed out that the grip of tires is affected not only by the tire design and materials used but also by the texture of the road surfaces and the presence of third bodies (gravel, grit, etc.) on the road. Coarse detritus can increase skid resistance on wet roads, although seasonal conditions, weathering, and pavement microtopography affect tire response as well. (Note: Interestingly, the roughness profiles of paved roads bear a remarkable resemblance to the stylus profiles of bearing components but about three orders of magnitude larger.[36])

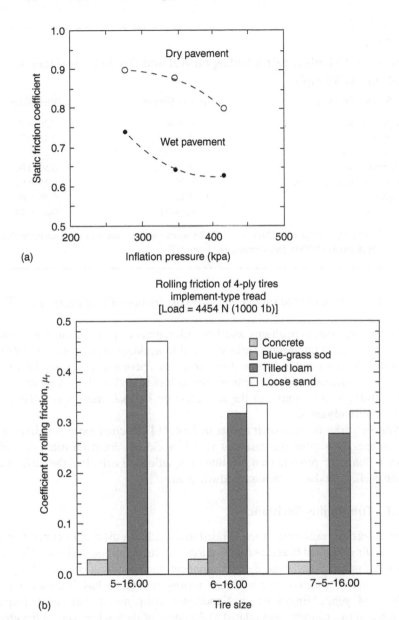

FIGURE 9.5 Effects of (a) inflation pressure and (b) road surface type on the frictional behavior of tires. (From *Marks' Standard Handbook for Mechanical Engineers,* 3–28, 1987. With permission.)

The design of general-purpose tires always involves a compromise of optimizing the features of the tire for different kinds of roads and weather. Figure 9.5a illustrates how tire inflation pressure affects the static and kinetic friction coefficients on dry and wet pavement.[37] A collection of papers concerning the action of tires on road surfaces can be found in ASTM Special Technical Publication 1164.[38] Figure 9.5b

TABLE 9.4
Friction Coefficients for Tires Sliding on Wet and Dry Road Surfaces at Speeds below 30 mph

Travel Surface Condition	μ (Dry Sliding)	μ (Wet Sliding)
Loose-packed dirt	0.40–0.60	0.30–0.50
Loose gravel	0.40–0.70	0.45–0.75
Truck escape ramps	2.5–3.5	—
New Portland cement	0.80–1.2	0.50–0.80
Traffic polished Portland cement	0.55–0.75	0.45–0.65
New asphalt	0.80–1.2	0.50–0.80
Traveled asphalt	0.60–0.80	0.45–0.70

Source: Parkka, D.J. in *Equation Directory of the Reconstructionist*, Institute for Police Technology and Management (IPTM), Jacksonville, FL, 1996, 212.

shows how the type of road surface can affect the friction of four-ply tires of different sizes.

Traffic accident consultants and law enforcement agencies maintain data on the friction of tires on various surfaces to aid in accident reconstruction. Table 9.4, for example, lists some representative ranges in friction coefficients for tires on various road surfaces with the vehicle wheels locked and sliding. A more comprehensive listing can be found in the handbook by Parkka, from which the data in Table 9.4 were obtained.[39]

Although friction data such as that in Table 9.4 are commonly acquired for flat road surfaces, it is often the case that vehicle accidents occur on tilted or inclined surfaces. Bonnett[40] provided a derivation of equations to calculate the difference in friction coefficient due to various roadway grades.

9.1.2.1   Tire Rolling Resistance

The term rolling resistance, when applied to tires, has a different connotation than that for rolling element bearings (see Section 2.5). Its definition varies within the tire literature. Luchini[41] reviewed the measurement and modeling of tire rolling resistance and noted that concern with the rolling resistance has been an issue for hundreds of years. Engineers in Napoleon's army noted that the horsepower required to move cannons was related to the depth of their wheel ruts in the ground. Over the years, engineers working through the Society of Automotive Engineers have developed and refined a variety of different methods to measure the rolling resistance of tires. Some, such as SAE J-2450, involve transient operation, and others, such as SAE J-2469, involve steady-state operation at single speeds with various loads and inflation pressures. Luchini pointed out difficulties in assessing the repeatability and reproducibility of rolling resistance measurements, noting that variability in test results is not uncommon from one batch of tires to another. Apparently certain brands and models of tires have more variability in rolling resistance than others, and the extent to which this is observed is dependent on the test method. Like friction reduction in engines and drivetrains, the rolling resistance of a tire can affect the fuel economy of a ground vehicle. In fact, it has been estimated that 4–7% of a car's fuel consumption is the result of tire rolling resistance.

Mechatronics technology is enabling the development of "smart tires" that contain built-in sensors to interface with real-time handling control systems in vehicles. Two recent European Union (EU) projects faced that challenge. One named APOLLO was focused on that topic.[42] The multipartner effort included the demonstration of several kinds of sensors mounted on the inner surface of a tire that were coupled by an infrared diode detector to a receiver on the inside of the wheel rim. A second EU project, called FRICTION, was initiated in 2006 to create an on-vehicle system for measuring friction and road slipperiness. Under leadership of a team at VTT, Finland, the sensing system is intended to be interfaced with other vehicle monitoring and control systems to improve highway safety under a range of operating conditions.

9.1.3 FRICTION IN INTERNAL COMBUSTION ENGINES

Friction in internal combustion engines is generally considered a "parasitic energy loss," and as such, efforts at friction reduction are linked to improved efficiency and fuel economy.[43] Figures 9.6a and 9.6b indicate that the contributions of various subsystems to the total engine friction depend on how fast the engine is running.[44,45] The pistons and connecting rods contribute the largest proportion. Therefore, more effort has gone into determining friction losses between piston rings and cylinder bores than in other parts of the engine. In Figures 9.6 a and 9.6b, note how the crankshaft friction contributions increase with engine speed while valve-train contributions are highest at low speeds and when the engine is idling.

The regimes of friction and the magnitudes of the friction coefficients experienced by different parts of the internal combustion engine can vary widely. Rosenberg[46] pointed out that the friction coefficient could range from $\mu = 0.2$, in boundary-lubricated situations (e.g., the valve train), to $\mu = 0.001$, in hydrodynamic situations (e.g., engine bearings and some sections of the piston stroke).

The term *friction mean effective pressure* (FMEP) has been used to quantify the friction level in engines. The power absorbed by mechanical friction in an engine is the difference between the power developed in the cylinders and the output or braking power. Since most of the friction power (FP) in an internal combustion engine comes from the piston rings and cylinders, we can write

$$FP = \mu_{pc}Nv \qquad (9.9)$$

where N is the ring–cylinder normal force, μ_{pc} the effective friction coefficient between the ring and cylinder, and v the mean piston speed. The FMEP is defined as follows:

$$FMEP = \frac{C(FP)}{Av} = \mu_{pc}\left(\frac{CN}{A}\right) \qquad (9.10)$$

where A is the piston ring contact area and $C = 2$ for two-stroke engines and 4 for four-stroke engines.

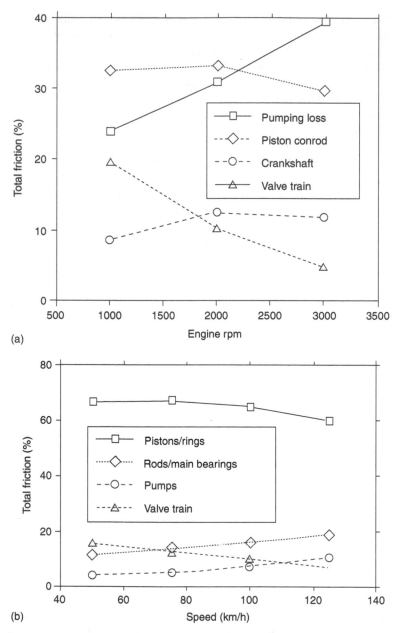

(a)

(b)

FIGURE 9.6 Effects of (a) engine rpm and (b) vehicle speed on the contributions of various engine subsystems to the total engine friction. [(a) Adapted from U.S. Congress, Office of Technology Assessment, *Improving Automotive Fuel Economy: New Standards, New Approaches*, OTA-E-504, U.S. Government Printing Office, Washington, DC, 1991. (b) Adapted from Cleveland, A.E. and Bishop, I.N., *Several Possible Paths to Improved Part-Load Economy of Spark-Ignition Engines*, Society of Automotive Engineers, Paper 150A, 1960.]

Isolating and measuring the frictional contributions of individual parts of an internal combustion engine are not trivial. Some investigators use instrumented "motored engines" in which the main shaft is rotated by an external power source; however, others use "fired" engines.[47] Now, sophisticated commercial software exists to compute frictional contributions of engine systems. Such models can be customized by combining submodels for subsystems such as the piston ring pack or the valve train.[48]

In the early 1990s, Hinein[49] conducted a study of the accuracy of engine friction methodology with attention to the effects of load, oil temperature, oil grade, engine speed, crankshaft balance, ring pack, and ring coatings. He described four methods that have been used to measure average frictional torque of internal combustion engines. The motoring test uses a bench-mounted engine driven by an electric dynamometer. This method does not provide accurate friction measurements because the cylinder pressures in a motored engine are much different than those in a fired engine, and under motored conditions, the frictional contributions due to lower back pressures on the compression rings, lower side thrust on the piston skirt, and lower loads on the main bearings are all smaller. In Willan's method, the rate of fuel consumption rate is plotted against engine torque at constant speeds and the resulting curve is extrapolated back until it intercepts the abscissa (at torque <0). The difference between that intercept and the origin is assumed to be the frictional torque. In Morse's method, designed for multicylinder engines (n cylinders), one of the cylinders is misfired, resulting in $n - 1$ firing and one motored cylinder. The engine load is reduced until the speed returns to its original value, and the difference in the torque between n and $n - 1$ cylinders firing is considered the frictional torque of the misfired cylinder. This procedure is repeated for the other cylinders to obtain the total. In the *indicated work method*, the actual pressure in the cylinder is measured to determine the power, as described earlier.

All four of the foregoing methods use mean conditions rather than measuring the instantaneous frictional torque M_{IFT}. Hinein developed what he termed the "$P-\omega$ method," which balances three torques that are assumed to be at equilibrium at constant speed:

$$M_{\text{IFT}} = M_{\text{P}} - M_{\text{L}} - M_{\text{I}} \tag{9.11}$$

where M_{P} is the instantaneous torque due to the cylinder gas force, M_{L} the instantaneous torque due to the load on the engine, and M_{I} the instantaneous torque due to the inertia of the moving parts of the engine. The method, once calibrated correctly, enables one to investigate the variation of frictional torque on an engine at various stages of the piston stroke.

The friction between the piston ring and the cylinder varies during the stroke because the relative sliding velocity ranges from zero at top-dead-center of the stroke to over 10–15 m/s at mid-stroke, depending on engine speed and stroke length. Within this range of velocities, several lubrication regimes—from boundary to mixed to hydrodynamic lubrication—can be experienced during the same stroke. The oil film thickness between the ring and cylinder may vary by more than a factor of 10 within a stroke, and likewise, the friction coefficient may vary from 0.001 to >0.10.

Some wear testing machines are designed to simulate piston ring–cylinder interactions only at lower speeds, since most wear occurs near the top-dead-center position where the sliding speed is low.

Piston rings are typically composed of cast iron or plated ferrous alloys, but the use of lubricative coatings and ceramics has been also tried (see, e.g., Ref. 50). As engine designs move toward lighter-weight components, lower emissions, and higher power density, speeds and operating temperatures will increase, placing higher demands on the materials and lubricants in frictional interfaces.

Ting[51] and Fuchsluger and Vandusen[52] provided reviews of lubricated and unlubricated piston rings, respectively. Nautiyal et al.[53] developed models for the friction and wear of boundary-lubricated piston rings. Let Z be the ratio of the shear strength of the lubricated contacts (S_i) to that of the metallic junctions (S_m). Then,

$$\mu = \frac{Z}{[\alpha(1 - Z^2)]^{1/2}} \tag{9.12}$$

The value of α comes from the relationship between the two shear strengths and flow pressure of the surface p:

$$\alpha = \frac{p^2}{S_m^2 - S_i^2} \tag{9.13}$$

Nautiyal et al. expressed the piston ring load per unit length of ring periphery **P** as a function of Hertzian contact width a, ring width h, and contact pressure P_i.

$$\mathbf{P} = P_i(h - a) \tag{9.14}$$

The Hertzian contact width a is given in terms of the mean radius of curvature of the ring profile R_m as

$$a = 1.493 \times 10^{-3}(\mathbf{P}R_m)^{1/2} \tag{9.15}$$

Finally, the maximum surface pressure P_{max} due to the ring load **P** is

$$P_{max} = \frac{2\mathbf{P}}{\pi a} \tag{9.16}$$

Nautiyal et al.[53] also conducted a series of low-speed reciprocating friction experiments to study the effects of temperature on the transition of friction coefficient from lower values to higher values as the boundary lubricant began to fail. They used a cast iron ring–cast iron liner couple in a modified Bowden–Leiben machine with a contact load of 54 N, a speed of 0.03 mm/s (simulative of the low speeds near top-dead-center position of the piston), and various concentrations of hexadecylamine additive to cetane. Table 9.5 compares their results with data for a fully formulated oil tested under the same conditions.

TABLE 9.5
Effects of Hexadecylamine Additives to Cetane on Friction Transition Temperature

Lubricant	μ Just Below Transition T	Transition Temperature (°C)
Crank case oil	0.105	160[a]
Cetane + 0.1% additive	0.11	120
Cetane + 0.2% additive	0.15	130
Cetate + 0.3% additive	0.11	130
Cetane + 0.5% additive	0.11	145

[a] Slope of the rise in friction after transition was lower than that for the cetane-based lubricants.
Source: Nautiyal, P.C., Singhal, S., and Sharma, J.P., *Tribol. Int.*, 16, 43, 1983.

Taylor[54] suggested a simpler method for computing frictional power loss in the piston–cylinder contact using the definition of viscosity. The total friction force is

$$\mathbf{F} = A\eta\left(\frac{u}{y}\right) \tag{9.17}$$

where η the viscosity, u the average piston velocity, A the area of contact (of all pistons in the engine), and y the thickness of the film (~1/16 the clearance of piston).

Ring–cylinder friction is controlled not only by the ring cross-sectional shape (contact shape and size), material choice, and lubricant but also by the pattern of machining grooves (lay) on the cylinder bore surfaces (typically crosshatched at an angle of 22–32°). Plateau honing is used to produce sufficient bearing area on the cylinder surface while still allowing crevices for oil retention. The author developed a simple method to simulate cylinder liner surface roughness and honing patterns on flat test specimens,[55] and it was used in the development of ASTM G 181, a laboratory test method for piston ring and cylinder friction.[56]

Not only friction in the piston rings but also friction in the piston skirt at the bottom end can affect fuel economy and engine efficiency. Figure 9.7, adapted from Rosenberg,[57] shows the effect of engine speed on the percent change in friction with piston skirt length. As might be surmised, the longer the skirt, the more significant is its contribution to friction. Additional reviews of the friction of internal engine parts can be found in papers by Davis[58] and Hoshi.[59]

There are at least 11 frictional interfaces on a piston:

- Cylinder interfaces with the top ring, scraper ring, and oil sealing ring (three places)
- Upper and lower faces of the ring groove (each of the three rings, equals six places)
- Piston pin bearing surface (one place)
- Piston skirt area (one place)

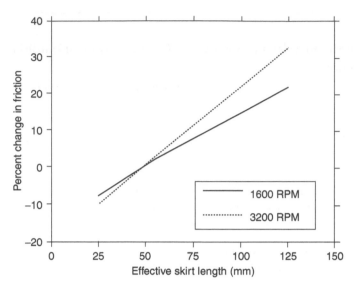

FIGURE 9.7   Effects of piston skirt length on the change in friction associated with different engine speeds. (From Rosenberg, R.C., *General Friction Considerations for Engine Design*, Society of Automotive Engineers, Warrendale, PA, Paper No. 821576, 1982. With permission.)

Conditions vary from well-lubricated to lubricant-starved, and from long stroke lengths along the cylinder bore to small-amplitude fretting in the ring grooves. In some cases, such as the ring and cylinder interface, the friction generated has a major effect on engine efficiency. In others, such as the lower ring groove friction, the effects are minor. Various strategies have been employed in the selection of materials for piston rings. The upper ring sees the most severe extremes in temperature. Some upper rings are alloy steel and others are cast iron with various coatings. Hard-plated or thermally sprayed rings for diesel engines have been developed.

The valve train is another area where friction reductions to improve engine efficiency have been studied.[60] More than one type of cam/tappet/roller follower configuration is used in the valve trains. The shape of the cam lobes has been shown to affect the friction conditions as well as the temperature distribution on the lobe surfaces.[61] For a given set of engine operating conditions, for example, the calculated flash temperature on a cam surface can vary from 10 to 70°C, the contact pressure can vary from about 0.2 to 0.6 GPa, and the lubricant film thickness can vary from nearly 0 to 0.7 μm within a single rotation.

Various coatings and surface treatments for cams and tappets have been tried. Softer coatings, such as phosphates and lubricative oxides, are specially tailored to assist in the running-in process. Hard coatings, such as TiN and TiC produced by chemical vapor deposition, are used to modify friction and enhance surface durability.

In summary, a great deal of effort has been expended to reduce parasitic friction losses in internal combustion engines. It has involved the development of additives for lubricants, new surface coatings, light-weight materials, and innovative component design. During the 1980s and 1990s, a great deal of attention was paid to the potential of ceramics, owing to their low density (weight savings), high hardness, and high-temperature resistance. At first ceramics were thought to be naturally low in friction, but research revealed that if ceramics were to be used on bearing surfaces, they would necessitate new lubricant formulations and the optimization of engine designs to make best use of their properties. Traditional engine lubricants were formulated to work with iron-based surfaces, and that limits the use of nonferrous materials such as ceramics and light-weight composites.

There have been limited successes in the application of ceramics to engines. For example, in some diesel engine cam roller followers, steel needle bearings with their many small moving parts and crevices were replaced by one-piece ceramic rollers. Ceramics for valves, valve guides, fuel injector plungers, cam roller followers,[62] and turbocharger rotors have been the subject of engine research and development programs.

As the twenty-first century unfolds, transportation technology is being increasingly driven by a desire to replace fossil fuels with alternative energy sources such as bio-fuels and to reduce the emissions from combustion-related power systems. These will bring a new set of challenges in friction and wear technology. As more and more electrically powered vehicles appear on streets and highways, long-standing friction problems associated with the internal combustion engine will recede; however, it is likely that a complete transition from internal combustion engines to electric vehicles will not occur for many years to come.

9.2   FRICTION IN BEARINGS AND GEARS

Bearings and gears are essential in transportation, military hardware, and industrial machinery. A bearing is a guide or a support for a rotating, oscillating, or sliding mechanical element. Considering this broad definition, bearings have a long and interesting history, dating back before the dawn of recorded history. Dowson[63] noted that handheld bearings of stone, wood, bone, and shell date back to the earliest known civilizations in Mesopotamia, 3500 BC, but evidence for crude pivoting devices for making tools, starting fires, and forming weapons dates back thousands of years before that. The long history of innovation in bearing design, lubrication, and materials could easily be the subject of a massive volume in itself.

There are dozens of types of bearings in use today. The OECD glossary of terms relating to tribology[64] contains definitions of 92 types of bearings. For the present purposes, we consider three types of bearings: rolling element bearings, sliding bearings, and noncontact bearings. All three types of bearings serve to support contact pressures and reduce friction between moving objects. Bearings range in size from tiny instrument bearings, such as watch jewels barely visible to the eye, to huge rollers for supporting the propeller shafts of aircraft carriers and for generators in electricity generating plants. In fact, there has even been work to design bearings that consist of several rings of atoms.[65] Pressurized gas bearings and magnetic bearings are examples of the noncontact type (although it could be academically argued that gas bearings are simply variants of sliding bearings in which the lubricant is a gas).

There are two types of rolling elements: balls and rollers (including needles). Ball bearings are probably more universally used, but roller bearings are common for the support of heavily loaded components. Two types of ball bearings, radial and angular contact, are common. In the former, the load is primarily applied to the balls perpendicular to the center shaft, and in the latter, a force component parallel to the rotating shaft is present providing a measure of thrust. Ball bearings consist of an outer ring, an inner ring, the balls, and usually some means to separate the balls (a cage). As Harris[66] pointed out, the principal source of friction in roller bearings is due to sliding in the contact zones, arising from the following:

- Sliding in rolling element raceway contacts resulting from the geometry of contact
- Sliding due to the deformation of contact surfaces
- Sliding between the cage pockets and the rolling elements
- Sliding contact between roller ends and guide flanges in roller bearings
- Sliding contact between seal lips and mating surfaces in sealed bearings

In addition, some bearings are designed with cages that slide on the outer or inner ring, and this is a source of friction as well. Other sources of friction losses in bearings include hysteresis losses due to the damping capacity of the raceway and "oil churning" losses due to excessive lubricant.[67]

Palmgren's experimental work[68] established certain relationships for the friction torque M_μ in rolling element bearings:

$$M_\mu = f_1 F_\beta d_\mathrm{m} + z\left(\frac{\mathbf{P}_\mathrm{o}}{C_\mathrm{o}}\right)^y \tag{9.18}$$

where the first term on the right-hand side is the friction torque arising from contact between the rolling elements and the raceway and the second term is the friction torque due to hydrodynamic fluid friction, usually relatively small. The symbols are defined as follows: F_β, the load on the bearing; d_m, the bearing pitch diameter; $\mathbf{P}_\mathrm{o}$, the static equivalent load; C_o, the basic static load rating; and z, y, and f_1, empirically derived constants, usually obtained from tables. More detailed analyses of bearing torques and refinements to their calculation may be found in the review by Harris,[66] the handbook chapter by Sibley,[69] and the textbook by Khonsari and Booser.[70]

Todd[71] described and modeled some of the detailed motions that balls in a ball bearing might exhibit so as to affect the frictional torque. The torque components involved in the special case of unlubricated ball bearings (as might be used in some aerospace or high-vacuum applications) are attributed to Coulombic friction:

- Conformity microslip due to differential velocities of the ball and raceway across the contact ellipse
- Spinning of the ball about a normal to the Hertzian contact area
- Subsurface hysteresis losses due to the cyclic stressing of material beneath the contact

TABLE 9.6
Friction Coefficients for Several Types of Rolling Element Bearings

Bearing Type	Friction Coefficient
Self-aligning ball bearings	0.0010
Thrust ball bearings	0.0013
Single-row, deep-groove ball bearing	0.0015
Needle bearing	0.0045
Tapered and spherical roller with flange-guided rollers	0.0018

Source: Palmgren, A. in *Ball and Roller Bearing Engineering*, SKF Industries, King of Prussia, PA, 1959.

In some aerospace applications, bearings can be oscillated through relatively large angles of arc, and this motion can, after several hundreds of cycles, lead to a progressive increase in the frictional torque. This phenomenon is called *blocking* and was noted in early experiments on bearings to be used for the Hubble space telescope.

Typical friction coefficients for various types of lubricated rolling element bearings, referenced to the bearing bore, are given in Table 9.6. With few exceptions, these values are more than a hundred times lower than those for those for dry sliding of most solid materials. It should be realized, however, that these values are the friction coefficients for tribosystems, not simply for one material against another. Bearings are tribosystems that were designed to optimize contact geometry, lubrication regimes, pressure distributions, material combinations, and load-bearing characteristics.

A novel approach to the design of bearings is called *Rolamite*. The Rolamite concept was developed in the mid-1960s by D. Wilkes of Sandia Laboratories, Albuquerque, New Mexico. Wilkes was working on a miniature device to sense small changes in the inertia of a small mass. At first, he tried a simple, S-shaped metal foil to support the surfaces, but that approach was not suitable since the assembly was too unstable. He then discovered that by inserting rollers into the S-shaped bends of the band, he could produce an interesting mechanical assembly with very low friction.[72] This system, which had very low friction in one direction and high stiffness transversely, became known as Rolamite (see Figure 9.8). One might suspect that the friction of the band along the contact with the roller would be high, but analysis of the device showed that all friction arises from rolling, not sliding, so the effective friction coefficients of the operating device remained <0.001. The effective friction coefficient could be varied from 0.0001 to 0.0008 by varying the tension on the beryllium–copper bands and their thickness. A typical size for the device used 12.7 mm diameter × 19.0 mm long rollers and foil band thicknesses of from 12 to 50 μm.

9.2.1 SLIDING BEARINGS

For some situations, rolling element bearings with their added weight and complexity may not be the best choice, and simpler sliding bearings can meet the need.

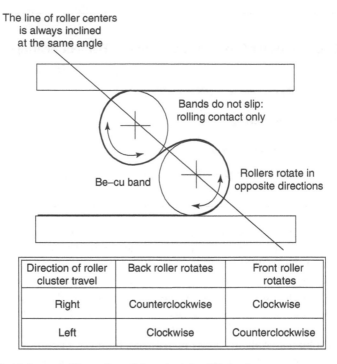

FIGURE 9.8   Schematic illustration of the principle of Rolamite.

Friction in sliding bearings can vary from high values at high temperatures, where liquid lubricants cannot be used, to low values approaching those of rolling element bearings. Plain or journal bearings, such as those used in engine big-end bearings, have a central shaft rotating in a sleeve. For hydrodynamic conditions in which a lubricant film of thickness h is developed, the friction force $\mathbf{F}$ is given by

$$\mathbf{F} = \frac{\eta A v}{h} \tag{9.19}$$

where η is the viscosity, A the bearing area, and v the velocity. Thus, the thinner the film, the higher the friction force (the operating point moves upward on the Stribeck curve beyond the minimum point).

The shaft in a journal bearing may not be perfectly centered in the bore. The difference in these positions is called the "eccentricity." An expression for the friction force in a noncentered bearing of bore radius R and length L is

$$\mathbf{F} = \frac{2\pi\eta v R L}{C(1+\varepsilon^2)^{1/2}} + \left(\frac{C\varepsilon\mathbf{W}}{2R}\right)\sin\phi \tag{9.20}$$

where $\mathbf{W}$ is the load, C the radial clearance, and the eccentricity ratio $\varepsilon = [C - h/C]$. The angle ϕ is called the "attitude angle" and is the angle between the vertical and the

TABLE 9.7
PV Limits for Several Bearing Materials Used under the Same Conditions

Bearing Material	State of Lubrication	μ	PV (MPa · m/s)
Cast iron and lead bronze	Dry	0.2–0.5	0.1
Porous bronze containing graphite	Dry	0.1–0.4	0.2
PTFE	Dry	0.05–0.3	0.05
Cast iron and lead bronze	Greased	0.05–0.15	1.0
Porous bronze, oil impregnated	Oil	0.01–0.1	1.0

Source: Pike, R. and Conway-Jones, J.M. in ASM *Handbook, Volume 18: Friction, Lubrication, and Wear Technology,* ASM International, Materials Park, OH, 1992, 515–521.

TABLE 9.8
Effects of Velocity on *PV* Limits for Four Polymeric Bearing Materials

Polymer	$v = 0.05$ m/s	$v = 0.5$ m/s	$v = 5.0$ m/s
Acetal	0.14	0.12	0.09
Nylon 6/6	0.11	0.09	<0.09
PTFE	0.04	0.06	0.09
PTFE with 15 wt% glass	0.33	0.39	0.5

Source: Eiss, N.S. in *CRC Handbook of Lubrication,* CRC Press, Boca Raton, FL, 1984, 197.

diameter of the center shaft, which passes through the point of minimum film thickness on the bore inner diameter. The first term on the right-hand side of Equation 9.20 accounts for viscous friction, and the second accounts for eccentricity.

It is common to categorize the severity of conditions for the use of sliding bearings by the *PV factor,* the product of the nominal bearing pressure and the sliding speed (commonly expressed in psi ft/min or MPa m/s). It is a measure of the energy input rate to the sliding interface. The maximum permissible value of *PV* for a material is called the *PV limit.* The *PV* limit for a given application or tribosystem depends on whether it is dry or lubricated, among other factors. Table 9.7 illustrates, for example, the differences in the permissible *PV* factors for several types of sliding bearings against a ground steel shaft.[73] Note that the maximum permissible *PV* factor for cast iron and lead bronze varies by a factor of 10 due to the state of lubrication.

PV factors are a common method of selecting polymeric sliding bearing materials. Table 9.8 shows the effects of velocity on limiting *PV* factors for several polymer bearing materials, as adapted from Eiss.[74] *PV* limits for a number of plastics are tabulated in the book by Bayer[75] and are also published by manufacturers of plastic bearing materials.

Since the *PV* factor is a measure of the energy input to a contact surface, one might expect the friction coefficient of sliding bearing systems to be related to its *PV.*

TABLE 9.9

Effects of *PV* Value on Friction Coefficient in Steel Thrust-Washer Tests

Material	μ, PV = 0.281 GPa m/s	μ, PV = 1.12 GPa m/s
Epoxy resin-bonded graphite/MoS$_2$/PTFE	0.21	0.05
Phenolic resin-bonded graphite/MoS$_2$/PTFE	0.19	0.14

Source: Pratt, G.C. in *Lubrication and Lubricants*, Elsevier, Amsterdam, 1967, 376–426.

An example of this relationship is illustrated in Table 9.9, which presents data for thrust-washer tests of various carbon-based composite bearing materials against EN 58 A steel.[76] The effects of *PV* on friction in that instance differed markedly with the type of resin used in the composite. Quadrupling the *PV* for the epoxy-based resin reduced friction by about 75%, but the same *PV* increase with the phenolic-based resin reduced friction of the couple by only about 26%.

9.2.2 GEARS

Gear teeth are bearing surfaces that, depending on their design, experience a combination of rolling and sliding (slip). Common types of gears that have rolling–sliding tooth behavior include the spur and helical gears. Worm gears, however, have almost all sliding and relatively little rolling. Gear tooth friction is of interest in optimizing the performance of gear trains and minimizing the wear of the gears by lowering surface tractions as the load is transferred between the teeth. Dudley[77] has provided an extensive discussion on issues involved in designing and lubricating gears. He indicated which lubrication regimes were associated with different types of gears, ranging from boundary lubrication in low-speed gears to full-film in high-speed gears. In a spur gear, when the profiled gear teeth first touch, sliding occurs, but that decreases until the contact point coincides with the pitch circle, where there is nearly pure rolling. Then the slip increases again until the tooth breaks contact. Thus, sliding friction is highest at both ends of the tooth engagement, and that leads to variations in the type of surface damage that occurs along the contact faces of gear teeth.

The classical form of the Hertz contact stress p_0 for spur gears is given by

$$p_0 = \left(\frac{WE^*}{\pi\rho} \right) \qquad (9.21)$$

where W is the load per unit face width, ρ is the relative radius of curvature of the teeth at the pitch point, and E^* is computed from the elastic moduli $E_{1,2}$ and Poisson's ratios $v_{1,2}$ of the two gear materials, respectively:

$$E^* = \left(\frac{1 - v_2^2}{E_1} \right) + \left(\frac{1 - v_2^2}{E_2} \right) \qquad (9.22)$$

The *disk machine* was developed to study gear tribology. It uses two circumferentially contacting circular disks (rollers) whose speed of rotation can be independently varied to produce various slip conditions. One such device was described by Johnson and Spence.[78] It uses two side-by-side rollers of unequal diameter, the larger of which passes through a lubricant bath. The ratio of the dissipated power to the transmitted power for a pair of gears is equal to the ratio of the friction torque (T_f) to the load torque (T_l), and this ratio is proportional to the friction coefficient:

$$\frac{T_f}{T_1} = K\mu \tag{9.23}$$

where K is constant for the gear geometry. In practice, however, the effective friction coefficient is not independent of tooth load, and friction forces vary with location on the tooth face. Chen and Tsai[79] computed the effective friction coefficient for involute gears during approach and recession. Assuming that the torque transmission efficiency of two gear sets is equal,

$$\mu_{eff}\mathbf{P}v = (1 - \eta)T_i\omega \tag{9.24}$$

where $\mathbf{P}$ is the normal contact force on the tooth, v the velocity of the contact point, ω the angular velocity, and η the torque transmission efficiency. When the teeth approach,

$$\mu_{eff} = \frac{r_b(1-\eta)}{(2-\eta)b} \tag{9.25}$$

but when the teeth recede,

$$\mu_{eff} = \frac{r_b(1-\eta)}{\eta b} \tag{9.26}$$

where r_b is the radius of the base circle of the gear, and b the distance between the point of contact and the teeth base along the line of action of the contact force.

Johnson and Spence compared mean friction values obtained from their disk machine with those obtained on actual gears and found that values agreed within 2% for cases where the grinding marks were transverse on the disks but were 23% higher than those for gears when the disks were ground circumferentially. Table 9.10 exemplifies their results. They also found that the friction coefficient decreased as Λ increased (i.e., the ratio of film thickness to composite surface roughness), but the rate of decrease differed with the direction of the grinding marks.

In 2007, Kleemola and Lehtovaara[80] published the description of an improved, twin-disk test that enabled them to measure friction versus percent slip under contact stresses, surface finishes, and Λ-ratios that are typical of industrial gearing.

The friction of bearings and gears has been addressed on the nanoscale level, in which entire bearings, bushings, and gears are composed of a few hundred to a few

TABLE 9.10
Comparison of Effective Friction Coefficients for a Disk Testing Machine and for Gears

Rolling Speed (m/s)	μ, Disk Machine (Transverse Ground)	μ, Disk Machine (Circumferential Ground)	μ, Gear Test
0.43	0.086	0.057	0.056
0.86	0.068	0.053	0.052
1.25	0.058	0.049	0.050

Source: Johnson, K.L. and Spence, D.I., *Tribol. Int.*, 24, 269, 1991.

TABLE 9.11
Types of Dynamic Seals

Seal Type	Characteristics and Uses
Lip seals	Used to seal low pressures (up to about 138 kPa [20 psi]); the lips are loaded against the shaft by the pressure differential or by light springs; elastomers or self-lubricating polymers are typical materials
Face seals	Higher pressure use, for example, in pumps; generally have a mating ring, a seal ring, and an O-ring; many different designs and a variety of materials
Piston ring seals	Almost always metallic; used in piston engines, reciprocating pumps, and compressors
Labyrinth seals	Provide tortuous paths for fluids or gases to help seal; generally ineffective at low shaft speeds, but good for low-pressure differential sealing
Hybrid seals	For example, packing integrated with a face seal; labyrinth seal followed by a floating ring seal

Source: Dray, J.F. in *ASM Handbook, Volume 18: Friction, Lubrication, and Wear Technology*, ASM International, Materials Park, OH, 1992, 546–552.

thousand atoms. Drexler[81] has evaluated the potential for such atomic-scale components as spur gears, atomic point bearings, helical gears, worm gears, bevel gears, and planetary gears. While certain aspects of the dynamics of nanoscale bearings and gears have been developed, no actual components have been created to date.

9.3 FRICTION IN SLIDING SEALS

Sliding seals are used for rotating shafts, bushings, pumps, and other mechanical components. The frictional contributions from such seals can far exceed the torque of rolling element bearings. As with bearings and gears, there is a large variety of seal types. In fact, there is an entire published glossary of seal terminology.[82] For static seals such as O-rings and gaskets, friction generally is not a primary concern, but for dynamic seals friction is important. Dynamic seals can experience either rotary or reciprocating motion. Table 9.11 lists several types of seals described by Dray.[83]

TABLE 9.12
Pressure and Temperature Limitations for Some Common Seal Materials

Material	Maximum Pressure (MPa)	Maximum Service Temperature (°C)	Typical Application
Natural rubber	62	95	O-rings, gaskets
Flax, asbestos, PTFE	7	95	Main shaft packing
Cu, Al, asbestos	69	>540	Engine/turbine manifold gaskets
Carbon–graphite, SiC	<34.5	<260	High-pressure pump face seals
Cr-plated cast Fe	<20.7	<260	Engine piston ring

Source: Dray, J.F. in *ASM Handbook, Volume 18: Friction, Lubrication, and Wear Technology*, ASM International, Materials Park, OH, 1992, 546–552.

Ideally, seals should have low friction and low wear while exhibiting good sealing. Thus, a lubricating film or surface layer should be formed on properly functioning dynamic seals. Low-friction polymers, such as poly-tetrafluoroethylene (PTFE), have been successfully used in lip and face seals, but hard ceramics such as silicon carbide and high-alumina ceramics have also been used in applications like water pump seals. Table 9.12 lists selected seal materials and their limitations.

The most popular seal material for general fluid power reciprocating seals in high-volume applications in machinery is rubber. That material has an interesting property of being able to achieve nearly 100% real contact area at relatively low contact pressures (~100 MPa).[84] Elastomers are also commonly used in rotary shaft seals, as discussed in a review by Johnson.[85]

Like sliding bearings, face seal materials are often rated in terms of *PV* limits. In the case efface seals that surround a shaft (e.g., a ship propeller shaft), however, a slightly different definition of *PV* can be written:[86]

$$PV = [\Delta P(b - k) + P_{sp}]V_m \qquad (9.27)$$

where ΔP is the differential pressure to be sealed, b the seal balance (the ratio of the hydraulic closing area to the hydraulic opening area), k the pressure gradient factor (for water $k \approx 0.5$, for oil slightly more, and for light hydrocarbons slightly less than 0.5), P_{sp} the mechanical spring pressure, and V_m the mean face velocity. This expression for *PV* can also be related to the friction power per unit area of seal (N_f):

$$N_f = \mu A(PV) \qquad (9.28)$$

where A is the apparent contact area of the seal face. Table 9.13 lists several seal material combinations and their friction coefficients at constant *PV* (35.1 MPa m/s, or 1 ksi ft/min).

9.4 FRICTION IN MANUFACTURING PROCESSES

Manufacturing processes of many kinds involve frictional considerations. Several examples have been chosen to illustrate these: fine slitting, orthogonal metal cutting,

---

TABLE 9.13

Friction Coefficients for Several Face Seal Materials

Rotating Face	Stationary Force	μ
Resin-filled carbon–graphite	Cast iron	0.07
Resin-filled carbon–graphite	Tungsten carbide	0.07
Resin-filled carbon–graphite	Silicon carbide	0.02
Silicon carbide	Tungsten carbide	0.02
Silicon carbide	Silicon carbide	0.02
Tungsten carbide	Tungsten carbide	0.08

Source: Johnson, R.L. and Schoenherr, K., in *ASME Wear Control Handbook,* American Society of Mechanical Engineers, New York, NY, 1980, 727–753.

---

extrusion and rolling, and friction welding. These discussions are not intended to represent comprehensive reviews of each subject, but rather serve to compare friction analyses in different kinds of manufacturing processes and to illustrate how the basic concepts of friction have been applied to help improve manufacturing methods.

9.4.1 FRICTION CUTTING

The process of using a rapidly rotating disk to cut iron and steel was first reported around 1830 and used by the early Shakers in Pennsylvania. Childs[87] has provided a review of the process of friction cutting, and Hong and Ashby[88] have described how wear maps can be used with friction cutting. Childs supplemented his review with data from more recent experiments involving the slitting of difficult-to-cut alloys using thin, water-cooled disks of mild steel spinning with rim speeds of between 50 and 100 m/s. The high rim speeds are the essential ingredient for successful application of the technique. It might be expected that higher hardness steels would cut better, but high carbon disks tended to roughen significantly when used for friction cutting. On the contrary, rims of lower carbon steel disks tended to glaze and provide better cutting action. Improvements have also been achieved by raising the conductivity of the cooling water with sodium carbonate and with the application of a current between the disk and workpiece.

The friction force **F** on a disk of thickness t, cutting distance d (Figure 9.9), and frictional stress τ may be written as

$$\mathbf{F} = \tau \text{ (slot contact area)} = \tau (td) \qquad (9.29)$$

Childs derived an expression for calculating the frictional temperature T during cutting:

$$T = T_0 + 0.75 \left(\frac{\alpha \mathbf{F}}{tK} \right) \left(\frac{U\kappa}{d} \right)^{1/2} \qquad (9.30)$$

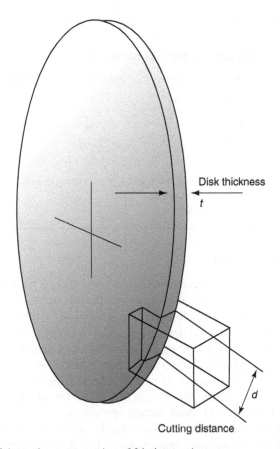

Disk thickness

t

d

Cutting distance

FIGURE 9.9   Schematic representation of friction cutting.

where T_0 is the ambient temperature, α the heat partition coefficient (i.e., relative amount flowing into the disk [close to 1.0]), K the thermal conductivity of the disk, U the disk rim speed, and κ the thermal diffusivity of the disk.

Transitions in the slitting rate k (i.e., the volume of material removed per unit time divided by the product of the applied load and rim speed) were observed due to either temperature effects on the materials or the formation of lubricious oxides on the grove/disk surfaces. The units of k are the same as those often used to report wear rates, and therefore the rates of slitting could then be compared to the wear rates of the materials when tested against the slitting blade material in traditional laboratory sliding wear tests. Childs provided a variety of slitting rate data for Ni–Cr alloys, titanium alloys, cast irons, and steels. He investigated the effects of load, velocity, pH, and disk thickness. Slitting rates were classified as either "mild," if they were less than about 1.0×10^{-17} mm²/N, or "severe," if they were greater than that. Table 9.14 gives several examples of slitting conditions for different materials and whether, within range of those conditions, mild or severe slitting rates were observed. Actual rates can be found in Ref. 87.

TABLE 9.14

Friction Cutting Conditions for Various Materials

Material	d (mm)	t (mm)	Velocity (m/s)	Load (N)	M or S[a]
Inconel 601	3.0	0.25	95	5–50	M/S
Nimonic 75	2.6	0.25	45–95	5–180	M
Titanium + 6Al + 4V	2–10	0.38	95	5–110	S
Mild steels	2–5	0.25	23–95	5–120	M
Cast iron, flake type	4–9	0.25–1.0	95	5–115	M/S

[a] M, mild cutting rate; S, severe cutting rate; M/S, both types of cutting observed depending on cutting conditions within the reported range of variables.

Source: Childs, T.H.C., *Tribol. Int.*, 16, 67, 1983.

9.4.2 MACHINING OF METALS

H. Ernst and M. E. Merchant became interested in the friction between cutting tools and workpieces in the 1930s when the former was the research director of Cincinnati Milling Machine Company and the latter was a cooperative fellow with that company while working on a doctoral degree at the University of Cincinnati. They published a paper in 1940,[89] describing experiments that were designed to elucidate the role of friction in producing the observed surface finishes on machined parts and how friction related to the formation of built-up edges on cutting tools. Early works, like that of Ernst and Merchant, recognized that friction is an important element in the quality and effectiveness of metal-cutting operations.

During orthogonal cutting, the friction along the rake face of the tool contributes to the power required to machine materials and the dynamic stability of the cutting process. Figure 9.10 shows an idealized situation for orthogonal metal cutting, adapted from the discussion by Schey.[90] Assuming that there are no contributions to the friction from the sliding contact of the worn "wear land" of the tool across the freshly cut material behind the tool tip, that the tool is perfectly sharp, and that there is no built-up edge of material clinging to the tip of the tool, one can vectorally resolve the various forces shown in Figure 9.10 to estimate the friction coefficient for the chip material against the rake face of the tool. Letting $\mathbf{P_c}$ be the cutting force tangential to the surface of the stock, $\mathbf{P_t}$ be the downward force on the tool, then knowing the rake angle α and the cutting forces (from a two-axes dynamometer) permits one to resolve the normal force $\mathbf{P}$ and the friction force acting along the rake face $\mathbf{F}$. The mean friction coefficient μ is related to the friction angle ψ as follows:

$$\mu = \frac{\mathbf{F}}{\mathbf{P}} = \frac{\mathbf{P_c}\sin\alpha + \mathbf{P_t}\cos\alpha}{\mathbf{P_c}\cos\alpha - \mathbf{P_t}\sin\alpha} = \tan^{-1}\psi \qquad (9.31)$$

Furthermore, the relationship between the tangential and normal forces are related to α and ψ as follows:

$$\tan(\psi - \alpha) = \frac{\mathbf{P_t}}{\mathbf{P_c}} \qquad (9.32)$$

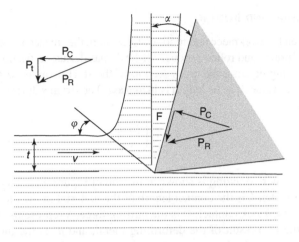

FIGURE 9.10 Frictional interfaces in orthogonal metal cutting.

The angle of the primary shear plane with respect to the horizontal is φ. As Schey pointed out, investigators have used criteria for maximum shear stress or minimum strain energy to determine that, for frictionless cutting:

$$\varphi = \frac{\pi}{2} - \frac{1}{2}(\psi - \alpha) = \frac{1}{2}(\pi - \psi + \alpha) \qquad (9.33)$$

In real metal-cutting operations, deformation may not be as uniform as Figure 9.10 depicts. The tool geometry changes as the tool wears, a built-up edge may form at the tool tip, and friction along the rake face may not be steady but rather may experience periodic seizures resulting in tool chatter. Furthermore, as the tool tip wears, a horizontal "wear land" may form behind the tip, and this added friction surface drives cutting forces higher. The wear features on the wear land can mar the finish of the cut part. For more comprehensive discussions of orthogonal cutting, the reader should consult the books by Schey[90] or Armarego and Brown.[91]

Cutting fluids are used for both lubrication and cooling. In low-speed cutting, lubrication can increase the shear angle, decrease the power consumption, and decrease the chip thickness while at the same time tightening the chip curl. The use of lubricants also helps delay the onset of built-up edge formation to higher cutting speeds or even prevent its formation. In 1977 Williams and Tabor[92] considered the role of lubricants in metal cutting, emphasizing the mechanisms of friction between the rake face of the tool and the emerging chip. In particular, they described the formation of fine, interconnecting capillaries that serve to feed the lubricant into the interface. The efficiency of that process largely determines whether friction at the rake face will be low or more like the friction of clean materials in a vacuum.

9.4.3 DRAWING AND ROLLING

Friction between the workpiece and tooling affects both the surface finish and the internal integrity of drawn and rolled products. Schey[93] has discussed friction during drawing. In the drawing process, the cross section of the stock is reduced from an initial value A_o to a final value A_1 as its length is increased. The strain ε during drawing is

$$\varepsilon = \ln\left(\frac{A_o}{A_1}\right)$$

(9.34)

The situation for wire drawing is shown at the top of Figure 9.11. The angle α is the cone angle of the die, L is the contact length of the material in the die, h is the mean diameter of the die/stock contact, $\mathbf{P}$ is the pulling force, S represents the forces generated in the die as a result of the deforming metal, and μ_d is the die/stock sliding friction coefficient. Assigning a mean flow stress to the wire material of σ_m,

$$\mathbf{P} = \sigma_m(1 + \mu_d \cot \alpha)\,\phi\varepsilon$$

(9.35)

and for round wire,

$$\phi = 0.88 + 0.12\left(\frac{h}{L}\right)$$

(9.36)

For tube drawing (bottom of Figure 9.11), there is an additional frictional contribution from the interface between the inside of the tube and the floating plug with a taper angle β. Equation 9.35 can be modified to account for this by replacing the term in parentheses with

$$\left(1 + \frac{\mu_d + \mu_p}{\tan \alpha - \tan \beta}\right)$$

(9.37)

Various lubricants are used to reduce the friction during wire and tube drawing. Compounded soap powder is a common one. It is sometimes applied to the wire by drawing the strand through a box filled with this powder, but it is also possible to apply lubricant to the die or to immerse the drawing die and wire in a bath of lubricant. The material being drawn and the speed of the operation influence the composition of the selected lubricants.

Unlike drawing, where friction is not generally desirable, friction is an essential element in the rolling of metal strip and plate stock. Friction draws the stock into the gap between the rollers. Initially, when the edge of the plate just contacts the rollers, problems can occur if the rolls are well lubricated or smooth and refuse to "bite." Then an additional horizontal force must be applied to initiate the engagement. When the stock first enters the contact zone, the roll surface is moving faster than the surface of the stock, producing a zone of backward slip, but beyond a certain point (the *neutral point*) the situation is reversed (see Figure 9.12). Thus, the frictional stresses along the arc of contact point toward the neutral point.

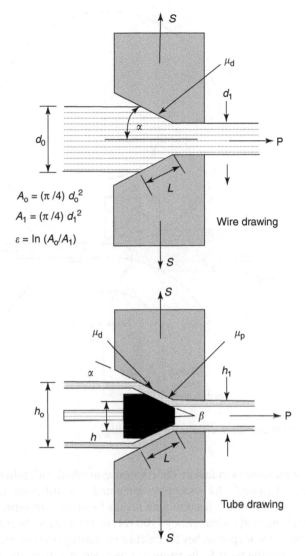

$$A_o = (\pi/4)\, d_o^{\,2}$$
$$A_1 = (\pi/4)\, d_1^{\,2}$$
$$\varepsilon = \ln (A_o/A_1)$$

Wire drawing

Tube drawing

FIGURE 9.11 Schematic representations of wire drawing and tube drawing.

Material enters the zone and contacts the rolls along an arc producing a so-called *friction hill*. The friction hill, a rising and falling distribution of friction forces within the confines of the contact, depends on the magnitude of the friction coefficients involved and on the geometry of the contact. Schey[94] described the development of friction hills and presents an extensive discussion on the theory and analysis of rolling. He showed that the shape and "steepness" of the friction hill change with different ratios of the contact length to the stock thickness (reduction ratios). In some situations, like hot rolling, the friction hill may develop a double hump, making analysis of the rolling process complex.

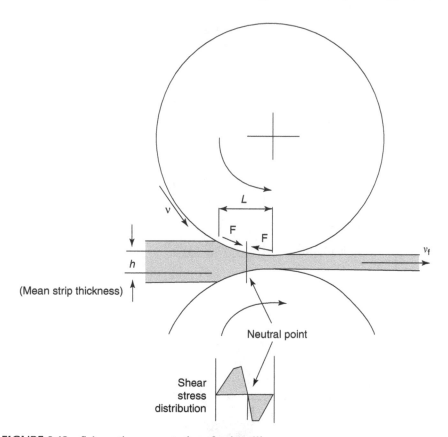

FIGURE 9.12 Schematic representation of strip rolling.

Attempts to measure or calculate the friction associated with rolling are fraught with difficulties. Lenard,[95] for example, compared five different approaches to calculating the average rolling friction with results from his own experimental studies of the cold rolling of aluminum alloy 1100 H4 strip and found them all lacking. In general, most theoretical approaches for calculating rolling friction coefficients with *forward slip* contain the term S_f. Its value can be calculated from the surface speed of the roll v and the exit velocity of the strip v_f, or from the distance (L_f) between impressions left on the strip by two lines spaced a distance L apart on the roll:

$$S_f = \frac{v_f - v}{v} = \frac{L_f - L}{L} \tag{9.38}$$

Letting h_1 be the entry thickness of the sheet, h_2 the exit thickness, and R' the radius of curvature of the flattened roll. Then the arc of contact α is

$$\alpha = \sqrt{\frac{h_1 - h_2}{R'}} \tag{9.39}$$

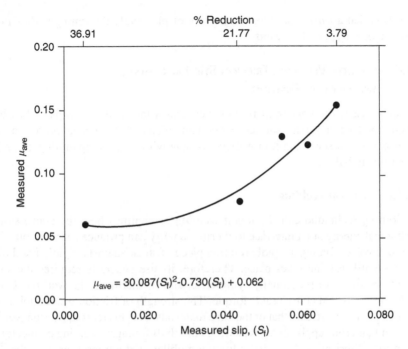

FIGURE 9.13 Relationship between the average measured friction coefficient for rolling aluminum strip and the measured slip. (Based on Lenard, J.G., *Tribol. Trans.*, 35, 423, 1992.)

and the angular distance along the arc to the neutral point is

$$\phi_n = \sqrt{S_f\left(\frac{h_2}{R'}\right)} \qquad (9.40)$$

Figure 9.13 shows Lenard's experimental data for the relationship of S_f and μ_{ave} in rolling aluminum strips initially 3.17 mm thick and 50 mm wide. As shown in the figure, a second-degree polynomial fit reasonably approximates Lenard's data. The friction coefficient of metallic couples can be reduced by applying high pressures. Therefore, rather than expect that a single value of the friction coefficient will be maintained within a rolling contact, the friction coefficient may vary along the arc of contact, owing to the pressure distribution.

Schey[96] provided a comprehensive review of the role of lubrication in hot and cold rolling. Factors such as the roll diameter, the surface speed of the rolls, the viscosity and additive composition of the lubricants, the force on the rolls, and the surface finish of the rolls can all affect the power requirements needed for rolling as well as the surface quality of the finished sheet and plate. Analyses of friction and lubrication in rolling often involve determination of the lubricant film thickness and the regime of lubrication present at various points in the contact arc. Avitzur[97] has also provided an analysis of friction in metal forming. In that treatment, he described what he called a *wave model* for the friction of plastically deforming bodies in the presence of entrapped lubricants.

9.4.4   FRICTION WELDING, FRICTION STIR PROCESSING, AND FRICTION DRILLING

In recent years, the heat from friction between a tool and a workpiece has been used to bond metals, create surface layers with enhanced mechanical properties, and produce holes in sheet and thin-walled castings without creating cutting chips like conventional drilling.

9.4.4.1   Friction Welding

The joining technique called *friction welding* exemplifies how the conversion of mechanical energy at an interface to thermal energy can produce a strong joint. The process involves forcing a rapidly rotating piece of stock against a rigidly fixed piece of stock until welding takes place. Therefore, its use generally requires the workpiece to be round or rotationally symmetric, and furthermore, the workpiece must be able to withstand considerable torque. The strength of friction-welded joints can be equal to or greater than that of the base materials. For this reason, friction welding has seen numerous applications including ball-shaft linkages, welding connectors to piston rods, fabricating drive shafts for automobiles, and joining gears to hubs.[98]

There are two variations of friction welding processes: *continuous drive* (CD) and *inertial welding* (IW).[99] In CD friction welding, the workpiece is rotated at constant speed while the contact pressure is increased. When a predetermined axial shortening or heating time is reached, the drive is engaged and rotation is braked. In IW, one workpiece is mounted on a flywheel and the other is held fixed. The flywheel is accelerated to a set speed, the drive is disengaged, and the workpieces are forced together, causing the stored energy in the flywheel system to dissipate into the weld zone. Variables in the operation of friction welding machines include rotational speed, heating pressure, forging pressure, heating time, braking time, forge delay time, and forging time. Typical friction-welded joint designs are axial rod-to-rod, tube-to-tube, rod-to-tube, rod-to-plate, tube-to-plate, tube-to-disk, and shaft-to-disk. Duffin and Bahrani[100] described the process for mild steel tubing with an outside diameter of 19 mm and an inside diameter of 12.7 mm. Rubbing speeds ranged from 0.42 to 3.36 m/s and axial pressures ranged between 2.4 and 19 kN. Production friction welding machines used for steels generate tangential speeds of from 300 to 650 ft/min (1.5–3.3 m/s).

Despite its being considered a solid-phase welding technique, the temperatures in the sliding interface during friction welding, especially the flash temperatures at the tips of the asperities at initial engagement, can easily exceed the melting point of one or both materials. The friction welding cycle can be thought of as consisting of a frictional heating stage and an upsetting (forging) stage. The first stage involves asperity interaction to produce localized softening and transfer of material. Torque on the couple increases momentarily and frictional heating increases until a continuous layer of plasticized material is formed from a coalescence of the transfer fragments. As the surface speed decreases, the hot layer thickens, and the torque eventually drops to zero. In IW, the thickness of the interfacial layer is related to the relative speed. The higher the engagement speed, the thicker the layer. As hot metal is forced from the joint, axial shortening increases until joint cooling brings it to a stop.

For IW, the energy stored in the flywheel at any instant can be estimated from the following expression:

$$E = 10.46W(ks)^2 \tag{9.41}$$

where E is the stored energy (J), W the weight of the flywheel system (kg), k the radius of gyration (mm), and s the speed in rpm. For example, when a solid 32.0 mm diameter pump shaft of AISI 1018 steel was inertially welded to a similar diameter austenitic stainless steel bar, the weld energy was 113.9 kJ.[101]

Two important criteria for the friction welding of metals are: (a) the alloy should be capable of hot forging and (b) the alloy should be a poor dry bearing material. Metal alloys that contain inclusions to make them free-machining should be welded with caution because the inclusions can form stringers during friction welding and weaken the joint. Self-mated metals friction weld better than dissimilar metals, but for special cases, like self-mated Ag and Co, this is not always true. Table 9.15 lists examples of metals combinations that do and do not friction weld easily.

As with any tribosystem, the cleanliness of the interface affects the friction coefficient. In friction welding, the strength of the resulting joint is also affected by contaminants trapped in the interface. Therefore, surfaces to be friction-welded should be prepared by grinding and cleaning thoroughly to remove scale and other sources of inclusions in the bond.

Kong and Ashby[102] derived the heat diffusion length l_b after time t from the interface during friction welding of self-mated metal cylinders of radius r_0 that have a thermal diffusivity of a.

$$l_b = \frac{r_0}{\pi^{1/2}} \tan^{-1}\left(\frac{4at}{r_0^2}\right) \tag{9.42}$$

TABLE 9.15
Friction Welding Combinations

Good Metallurgical Bonding	Not Suitable for Friction Welding
Most self-mated metals (except Ag and Co to themselves)	Cemented carbides to cemented carbides
Ni alloys to stainless steels	Brass to most kinds of steels
Cu to carbon steels	Pb to anything but itself
Al to Ti and its alloys	Ag to anything except Cu
Valve steels to Ni alloys and various steels	Mg alloys to most metals, except Al

Source: American Welding Society in *Welding Handbook*, American Welding Society, New York, NY, 1968, 239–261.

TABLE 9.16

Calculated Thermal Diffusion Distances for Friction Welding

Materials	Conditions	t (s)	l_b (mm)
Steel to itself	r_o = 10 mm, a = 9.11 mm^2/s, velocity = 1.6 m/s, pressure = 50 MPa	1.0	3.1
Al to itself	r_o = 20 mm, a = 88.5 mm^2/s, velocity = 3.1 m/s, pressure = 8 MPa	2.0	6.8
Cu to Ti	r_o = 20 mm, a(Cu) = 109.8 mm^2/s, $a(\beta\text{-Ti})$ = 6.27 mm^2/s, velocity = 3.1 m/s, pressure = 80 MPa	5.0	5.5 (in Cu) 2.1 (in Ti)
High-density, polyethylene to itself	r_o = 10 mm, a = 0.22 mm^2/s, velocity = 0.26 m/s, pressure = 0.63 MPa	10.0	1.0

Source: Kong, H.S. and Ashby, M.F. in *Wear Mechanism Modeling*, Argonne National Laboratory, Final Subcontract Report, ANL/OTM/CR-3, Argonne, IL, 1993.

Sample calculations from Kong and Ashby are given in Table 9.16. Separate diffusion lengths into each material are calculated if the materials are dissimilar.

A further discussion of the fundamentals of friction welding, including the effects of pressure and temperature on the changes in the friction coefficient that occur during the process, can be found in the review by Hazlett.[103] Hazlett pointed out that the friction coefficient between sliding metals may either increase or decrease as a function of sliding speed, depending on the applied contact pressure. Therefore, friction welding parameters like contact pressure, rubbing velocity, and axial travel must be tailored to provide just the right amount of energy into the interface to permit the development of strong metallurgical joints.

9.4.4.2   Friction Stir Welding, Friction Stir Processing, and Friction Drilling

Like friction welding, friction stir welding (FSW), friction stir processing (FSP), and friction drilling (FD) use the heat generated by a rapidly spinning tool (usually tool steel or tungsten carbide) to soften, deform, and rearrange workpiece material. Figure 9.14 compares the geometries of FSW, FSP, and FD. The tool shapes, fixtures, and direction of relative motion differs for these three processes, but there is no melting, so workpiece microstructures, while refined, textured, and differing in properties from those of the starting material, do not have the appearance of a casting.

A book edited by Mishra and Mahoney[104] provides overviews of FSW and FSP and exemplifies the kinds of materials that have been joined and surface modified successfully. FSW is of interest in aerospace and military applications and has been used to join plates of aluminum, steel, and titanium alloys at thicknesses exceeding 25 mm. However, FSP has been used to modify the surfaces of materials and to improve their durability. It is interesting to reflect that using friction itself can be used to process surfaces in such as way as to enhance their friction and wear behavior. FSP has been applied to bulk material[105] and to produce nanocomposite microstructures with attractive wear characteristics.[106]

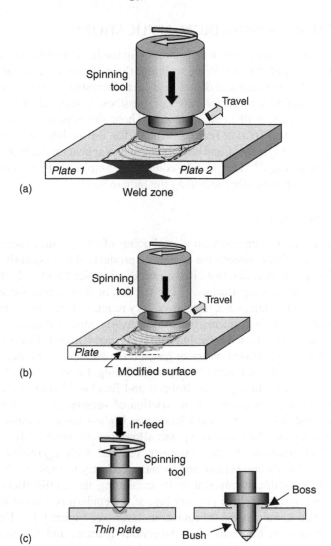

FIGURE 9.14 Comparison of the geometries for (a) friction stir welding, (b) friction stir processing, and (c) friction drilling.

FD is also known as thermal drilling. A spinning tool is brought into contact with a surface, and the friction of its tip begins to soften the material, eventually allowing the imposed down-force to displace material to form a hole with an exit end extrusion that can be tapped. It has been shown to work on certain high-strength steel sheet, plain carbon steel sheet, and thin-walled castings of aluminum.[107,108] FD has advantages over conventional drilling processes in that it produces few chips, if any, does not require cutting fluids that must be disposed of, and can increase the thickness of a sheet so that a sufficient number of additional threads can be cut to avoid the need for a back-up nut or a weld-nut.

9.5   FRICTION IN BIOMEDICAL APPLICATIONS

Friction plays an important role in the operation of the human body, whether it is the friction of the soles of the feet during walking, the grinding of teeth, or the movement of joints. The medical and dental fields are interested in friction since it can affect the installation and performance of prostheses, surgical procedures, and lubricants for appliances like contact lenses. Those developing root canal files or bone saws can reduce friction heating through the use of lubricating treatments. Conferences, topical symposia, and books continue to appear on the ever-broadening subject of biotribology.[109] This section contains examples of biotribology studies related to friction, principally concerning the human body.

9.5.1   FRICTION OF SKIN

The friction of skin is a concern from several points of view. Some concern cosmetic formulations but others concern the design of products that frequently come into contact with skin. As was described in the introduction (see Chapter 1), it can affect the feel of piano keys to a performer.[110] The friction of skin against surgeon's glove materials is also important, since manual surgery requires a measure of tactile feedback, the ability to grip instruments, and the ability to remove the gloves. Sometimes a starchy powder has been used to facilitate insertion and removal of surgeon's gloves, but the powder can cause skin rashes or get into the patient and produce irritation. An alternative to the loose powder is a hydrogel lining. Using a specially designed, finger-on-flat-ended beam apparatus, Roberts and Brackley[111] studied the effects of coatings and surface topography on the friction of surgeon glove materials. They noted that for hydrogel coatings, the friction was highest when the coating was wet (about 0.9) and lowest when the coating was allowed to dry (about 0.4).

The feel of cosmetics to the consumer, coupled with aggressive marketing practices of that highly competitive and lucrative industry to various ethnic groups, has prompted a significant investment in measuring the tactile friction of skin products. Skin consists of three primary layers: the epidermis, the dermis, and the hypodermis, also known as the subcutaneous layer (see Figure 9.15). Each layer is further subdivided on the basis of its structural features and functionality. For example, the epidermis itself is composed of five strata: stratum germinativum, stratum spinosum (prickly layer), stratum granulosum (granular layer), stratum lucidum, and stratum corneum (horny layer). The complexity of the skin and its many sublayers makes friction modeling problematical. The problem is further complicated by the presence of lubricative perspiration originating from two types of glands: the eccrine glands that secrete mainly water containing dissolved salts and the apocrine glands that secrete fatty substances responsible for sweat's odor.

Skin comprises a tribosystem with a stack of connected layers that vary in elasticity, shear strength, and continuity. If there is minimal lubrication on the surface then the epidermis may experience no motion relative to the counterbody, and the skin layers below would have to accommodate the imposed shears. With a large number of soft, shearing layers operating, the stack would behave like a viscous fluid; however, all the layers do not have the same compliance. As the lateral displacement increases, one-by-one the sublayers in the friction stack reach their

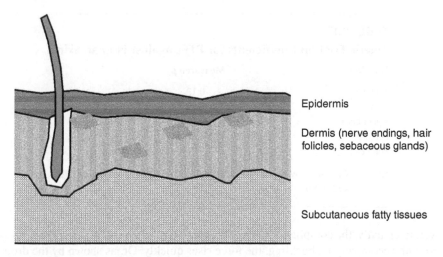

FIGURE 9.15   Layers of human skin.

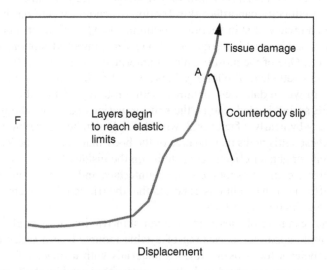

FIGURE 9.16   Change in friction of skin resulting from the transfer of control from one compliant layer to another (hypothetical).

elastic limits and transfer the load to the remaining layers. As deformation increases at the surface, there may be buckling of the skin ahead of the traction area, while the shear strength of subsurface layers eventually rises due to exhaustion of their elastic compliance. At that point, two possibilities could occur: either there will be slip at the counterbody–skin interface or there may be tissue trauma. Thus, one might hypothesize a friction force/lateral displacement behavior of the kind shown in Figure 9.16. The rise in friction is nonuniform due to the complexity of the various sublayers reaching their limits of strain and transferring shear to the next weakest

TABLE 9.17
Kinetic Friction Coefficients for PTFE against Human Skin

Location	Measured μ_k
Abdomen	0.12
Palm	0.21
Upper back	0.25
Forearm	0.26
Forehead	0.34

Source: Cua, A.B., Wilhelm, K.P., and Maibach, H.I., *Br. J. Dermatol.*, 123, 473, 1990.

layer. Eventually, the compliance limit of the stack is reached (point A in Figure 9.16) and with accompanying buckling, the force rises quickly. Or, as shown by the drop-off in the figure, slip occurs.

Measurements of the friction of human skin have not been standardized. It is recognized that the properties of skin vary from one place to another on the human body and that ethnicity can affect skin friction depending where on the body it is measured. In a review of skin friction measurements, Gitis[112] points out the variety of conditions used by past investigators. Probe sizes, imposed motion, and materials vary widely. One of the most popular counterfaces used for skin friction studies is polytetrafluoroethylene (Teflon™). Table 9.17 provides data from a study by Cua et al.,[113] in which data were obtained with a rotating 15 mm diameter circular disk of PTFE spring-loaded against the skin on various parts of a body. Values in Table 9.17 vary by nearly a factor of 3, with the abdomen being the lowest and forehead the highest. Gitis notes that the area of the body that seems to be least sensitive to either ethnic or gender effects seems to be on the inside of the forearm; however, data from Elsner et al.,[114] using the same counterface and test method as Cua et al. but under different skin conditions, reported that the friction coefficient for the forearm could be as high as $\mu_k = 0.48$.

As a final example of measuring human skin friction, Gee et al.[115] devised a relatively simple apparatus consisting of a table mounted on a series of flexures. Friction coefficients for stroking various materials with a finger (of a 23-year-old male volunteer) were compared with those using either a rubber ball or a steel ball secured to the end of a rod. Skin was found to produce the highest values, much different from those of the other slider materials. Selected results, read from figures in the cited reference, are given in Table 9.18. The authors noted that the direction of stroking affected the friction coefficients as well.

Owing to skin's complexity, its sensitivity to ambient conditions, and the debate over what to use as a suitable counterface, there is a lack of standardization and repeatability in the measurement of human skin friction. Relative friction measurements within a given study that used good laboratory practices seem to be more useful than attempting to compare results from one study to another. Nevertheless, work continues to develop improved apparatus for skin friction measurements on materials of various kinds.

TABLE 9.18
Sliding Friction Coefficients of Skin, Steel, and Rubber on Various Surfaces (Stroking in One Direction)

Surface	μ, Finger	μ, Rubber	μ, Steel
Paper	0.6	—	0.2
Polyethylene	1.6	—	0.4
Mild steel (0.3 μm Ra)	1.8	0.8	0.2
Polycarbonate	2.6	1.6	0.4

Source: Gee, M.G., Tomlins, P., Calver, A., Darling, R.H. and Rides, M., *Wear*, 259, 1437, 2005.

9.5.2 FRICTION IN CONTACT LENSES

Another important application in biotribology is in the friction and lubrication of contact lenses. Introduced around 1950, contact lenses have influenced the lives of many people, both as aids to sight and as cosmetic enhancements. According to Valint of the State University of New York, Buffalo,[116] eyelids blink at velocities ranging from 0 to 0.3 m/s, and the typical normal force on the eye varies from 0.2 to 0.8 N. Similarly, Conway and Richman[117] estimate blinking speeds at about 0.12 m/s and contact pressures ranging from about 35 to 40 g/cm^2 (3.4–3.9 mN/mm^2).

Normally, eyes are lubricated by tears that are composed of three layers.[118] The highly viscous outer layer is about 0.1 μm thick and composed of lipids. The second layer, called the tear layer, is about 10 μm thick and has a lower viscosity. The inner layer is made of viscous mucus. Although there is some debate on the precise mechanism of fluid response, recent proposals suggest that hydrodynamic conditions can prevail during some portion of a blink.

As with attempts to measure the friction of human skin, the choice of an appropriate counterface to simulate either the surface of the eye or the inside of the eyelid is problematic. Investigators have often relied on polymers as solid bodies on which to evaluate candidate lubricants for contact lenses. Nairn and Jiang[119] used a pad-on-rotating disk apparatus to study the lubrication characteristics of various compact lens solutions. They applied a contact pressure of 3.5 mN/mm^2 and several speeds. Commercial contact lenses were mounted on the pad and polycarbonate (PC) or polymethylmethacrylate (PMMA) were used for the disk materials. In addition to tests under dry sliding, five commercial brands of lens lubricating solutions were ranking in order of lubricity. Friction coefficients without lubrication were as high as 0.64. In the manner of Stribeck, plots of friction coefficient versus Sommerfeld Number [$S \propto$ (velocity $\times$ viscosity)/normal force] indicated the presence of boundary or mixed-film lubrication behavior. Data read from Figures 4 and 5 of Ref. 119, for two commercial contact lenses sliding on PMMA and lubricated by saline solutions, are provided in Table 9.19. Note that the friction coefficients decrease with S, as would be expected for a transition between boundary and mixed-film lubrication conditions.

TABLE 9.19

Effects of Sliding Speed (Sommerfeld Number) on the Kinetic Friction Coefficient of Two Types of Contact Lenses against PMMA (Saline Solution, Nominal Contact Pressure = 4.0 mN/mm²)

S	μ_k, Bausch and Lomb SeeQuence® 2 Lens	μ_k, Johnson and Johnson NewVue® Lens
10^{-7}	0.52	0.17
10^{-6}	0.48	0.09
10^{-5}	0.32	0.05

Source: Nairn, J.A. and Jiang, T.-B. in *Proceedings of ANTEC*, Boston, MA, May 7–11, 1995, 5.

9.5.3 FRICTION IN ARTIFICIAL JOINTS

The aging population, its desire to retain an active lifestyle into old age, and an explosion of new orthopedic surgical techniques and materials have prompted increased research on the tribology of prostheses. In joint implants, such as hips and knees, friction seems to be less important than wear because it is wear debris that can lead to osteolysis, implant loosening, and great cell formation. Nevertheless, friction has been included in many tribology studies of implant behavior since it affects the smoothness of the implant's operation.

In hip and knee prostheses, the static friction between the implant and the bone cement is needed for finite element modeling of implant loosening. Nuño et al.[120] conducted friction experiments in both dry and bovine serum environments to estimate this quantity and determined that the values of μ_s in the range 0.3–0.4 were reasonable for modeling purposes.

A parameter defined similarly to the friction coefficient, but called the friction factor, has been used to characterize prostheses materials. It was defined by Unsworth[121,122] as follows:

$$f = \frac{T}{rL} \tag{9.43}$$

where T is the torque, r the radius of the femoral head, and L the applied load. Using a hip function simulator, Scholes et al.[123] measured the friction factors and calculated lubricant film thicknesses (using the method of Hamrock and Dowson[124]) for metal/metal, metal/ceramic, and metal/polymer hip joints lubricated with bovine serum or aqueous solutions of carboxyl methyl cellulose. Calculated minimum film thicknesses ranged from 0.05 to 0.09 μm and for only one combination, alumina-on-alumina, did the Λ-ratio exceed 3. The latter had a friction factor range of 0.002–0.05, in contrast with a CoCrMo-on-CoCrMo combination with Λ < 1 and a friction factor range of 0.15–0.26. The CoCrMo-on-ultrahigh-molecular-weight polyethylene combination had intermediate values between the two extremes.

Additional research on implant tribology can be found in a book edited by Hutchings.[125]

9.5.4 FRICTION IN STENTS

Stents are medical appliances composed of an expandable wire mesh or perforated tube that are inserted into body passages to maintain the flow of blood, bile, urine, or air. Friction during the insertion or removal of stents must be controlled to avoid tissue damage and enable the device to be inserted smoothly into the proper location. One approach is to coat the outside of the device with a hydrophobic polymer,[126] and another more recent method is to introduce several milliliters of a special lubricant called RotaGlide™ into the guiding catheter.[127]

Laroche et al.[128] developed a finite element model to aid clinicians in patent-specific prediction of friction between a stent and the arterial wall during balloon angioplasty.

9.6 OTHER APPLICATIONS OF FRICTION SCIENCE

The remainder of this chapter samples a potpourri of applied friction technology ranging from footwear to cables, wires, and amusement part rides. The degree to which friction science has been used in the development of these applications varies widely. In many cases, much remains to be done to connect experimental observations with their fundamental causes.

9.6.1 FRICTION OF FLOORING

One of the simplest experiments in friction involves the measurement of the traction of shoes on flooring. No self-respecting student of friction should escape the opportunity to place weights in a shoe and drag it across different types of flooring to measure the friction coefficient. The bottom of a well-worn dress shoe is schematically illustrated in Figure 9.17. Two areas, one composed of leather and one of rubber, bear the contact load in that case. A wire was used to connect a similar shoe to a spring scale (i.e., fish scale), and the scale was used to pull the shoe along various types of flooring in a laboratory and office building under a total normal force of 18 N. The resulting data are given in Table 9.20. Note that the friction coefficient in different parts of the building varied by nearly a factor of 3. Clearly, the human body is an excellent friction-compensating mechanism, able to adjust its position and direction of force application in such a way as to maintain balance on those different surfaces.

The obvious problem with this method is that people do not walk by sliding along flat-footed. Sometimes the corner of the heel entirely bears the load in a rocking motion, and sometimes the friction is static (as when the ball of the foot pushes off against the floor with no slip). Therefore, simple experiments should not be used to obtain accurate assessments of the traction of flooring where safe building designs and the protection of the public are involved. More sophisticated, articulated strut devices, such as those discussed in Section 3.4.11, or full-size flooring platforms mounted on dynamometer stages are used for such studies.

Despite the interest in friction of flooring, the definitive study of friction coefficients for people slipping on banana peels—a comic event portrayed in numerous movies and a potential liability issue for food markets—is yet to be published.

FIGURE 9.17   These U.S.-sized 26 shoes are 0.44 m long and were custom-made for the 265 lb (120.3 kg) world championship boxer, Primo Carnera. Their leather soles show scuff marks where the metatarsal bones meet the phalanges (where push-off friction force is highest), and at the leading edge of the heels (arrows).

TABLE 9.20

Friction Coefficients of a Shoe Sliding on Various Floor Surfaces (Average Sliding Speed = 25 mm/s, Normal Force = 18 N, Drag Length = 150 mm)

Surface	Average μ
Rough red granite in a foyer area	0.77
Low-nap carpeting	0.68
Rubberized laboratory flooring	0.50
Raised button surface in a service hallway	0.46
Concrete floor in a utilities room	0.30
Linoleum flooring in a hall area	0.27

A recent flooring friction concern about New York City subway (underground) tiles was posted on the Internet by an organization called the Voices of Safety International (VOSI).[129] Headquartered in the New York metropolitan area, VOSI advocates its own standards for measuring the static friction of tile materials and has critiqued ASTM's efforts in that area. They published data comparing friction measurements from ASTM standard C1028[130] with their own procedure, VOSI V41.21, on recently installed tiles in the New York subway that were allegedly causing a number of slips and falls. It was proposed to replace the problematic new tiles with a version of different composition. Table 9.21 provides selected data from the VOSI study for the same footwear material, a synthetic called Neolite™. VOSI asserts that its method better reflects tile slipperiness, but irrespective of the test method used,

TABLE 9.21
Static Friction Coefficients for Neolite on Floor Tile under Dry and Wet
Conditions

Surface	Condition	μ_s, VOSI Test	μ_s, ASTM C1028
Current tile (Orion)	Dry	0.54	0.78
	Wet	0.57	0.70
Proposed replacement tile (Pegaso)	Dry	0.51	0.76
	Wet	0.41	0.67

Source: NYC Subway Tile Study, Voices of Safety International (VOSI), 2003, http://www.
        voicesofsafety.com.

Table 9.21 data show that the proposed replacement tile seems less desirable than the original. At this writing, despite a long-standing and sometimes passionate debate, there is no universal agreement on which is the best method to determine the traction of footwear on flooring.

9.6.2 Friction in Cables

Cables and ropes are subjected to constant or variable tensions as they function. The stress states within a cable are also affected by bending the bundled strands around pulleys or rollers. Depending on the design of a cable, the interlayer friction can affect not only its ability to be bent or twisted, but also its durability and vibrational self-damping characteristics.[131] External friction is important when cables or ropes must slide over a pulley or sheave. In aircraft carrier arresting gear, for example, the demands might require rapid accelerations from zero to 150–180 knots (77.2–95.6 m/s) and back to zero, with sufficient tension to stop an aircraft weighing upwards of 50,000 lb (22,700 kg) in as little as 320 ft (91 m).

One very practical, friction-related characteristic of cables or cable bundles is the tension required to pull a cable through a tube or jacket during installation or maintenance. Some companies specialize in cable lubricants designed for this purpose. They are available as liquids, gels, or sprays depending on the type of cable and conduit involved. Since cables often contain several materials and the conditions of use vary widely, it is difficult to ascribe any particular friction coefficient to a given type of cable; however, a first approximation of the force (tension) required to pull a cable through a conduct (T_{out}) can be estimated using the following expressions, the second of which is the same as that discussed in Chapter 2 for a wire or cable draped over a pulley:[132]

$$T_{out} = T_{in} + LW\mu \tag{9.44}$$

$$T_{out} = T_{in}e^{\mu\theta} \tag{9.45}$$

where T_{in} is the tension in, L the length of the straight run, W the weight of the cable per unit length, μ the friction coefficient, and θ the angle of bend. Since the static friction coefficient is usually greater than the kinetic friction coefficient for cable materials, it is recommended to keep the cable moving during insertion and not to stop, if possible. Friction control by lubrication is also important to avoid cable jamming when several cables are fed into the same duct.

An interesting application of cable lubrication is for the process known as "cable blowing" used to install optical fibers. There are two basic types of cable blowing. In one case, a small piston head is attached to the leading end of the fiber and high-pressure, high-flow-rate air is injected into the feed end to cause the piston to pull the fiber through the tube. Sometimes a mechanical feed roller helps to push the fiber through as well. A second variant of the technique involves high-pressure air to simply push the fiber through the duct. Ducts must be prepared for the high pressures involved (>100 psi), and specially formulated lubricants can increase the distance that a fiber can be inserted by this method.

Small-amplitude oscillations (fretting) between strands within a cable can lead to fracture. Some test methods to evaluate internal strands for friction and chafing use two wires twisted together or looped over a tensioned strand, but results can be complicated by wire stretching or kinking. Therefore, some investigators prefer to stretch wires over orthogonally oriented cylindrical surfaces to provide a support for the load. Such a device was described by Urchegui et al.[133]

9.6.3 FRICTION IN FASTENERS, JOINTS, AND BELTS

Mechanical fasteners such as nuts, bolts, and screws are manufactured and used by the millions. Friction in tightening and loosening such fasteners is critical if they are to function properly. The solutions to friction control can be as simple as rubbing the threads of a wood screw on a piece of soap to ease its insertion, or applying a thread-holding compound to prevent bolt-loosening in a structure subject to vibration. Sometimes special torque wrenches must be used to ensure that the proper degree of tightness is obtained without over-stressing the bolt and causing failure. Karamis and Selcuk[134] analyzed the friction in bolted joints and conducted experiments to calculate the friction coefficient for various bolt lubricants. They also noted that control of the surface finish on the overlapping portions of the joint was critical to avoid loosening. Bolted joint surfaces with higher arithmetic average roughnesses tended to loosen more easily. Table 9.22 lists friction data for triangular threads with various surface treatments, as reviewed by Karamis and Selcuk.

TABLE 9.22
Friction Coefficients for Threaded Steel Bolts

Thread Surface Treatment	μ (Unlubricated)
Unpolished	0.16
Phosphate-coated	0.18
Zinc-coated	0.14
Cadmium-coated	0.10

Source: Karamis, M.B. and Selcuk, B., *Wear*, 166, 73, 1993.

Belts and pulleys are used in machinery to transmit power between shafts. Friction between the belt and the shaft must be high enough to transmit the torque effectively, but the friction at the sides of V-belts can reduce the efficiency of the system since they require energy when they wedge into and pull out of the grooves on the pulleys.[135] V-belts and chain drives can be used to transmit more power than flat belts, but flat belts can be operated at speeds of up to 200 m/s. Some flat belts are constructed using a sandwich structure consisting of an inner traction or tension member, flanked by a bonding agent and polyamide fabric, and finally, outer layers of an elastomeric "friction cover." Nylon belts with leather or synthetic leather covers are sometimes used in oily environments. The friction of belts and webs is typically measured in wrap tests using tensioned drums, as explained in Chapter 3.

Ball splines are mechanical joints that combine both linear and rotary motion. They are used, for example, to join a motor to a shaft when the shaft must also move in and out axially. Shafts can be up to 5 m long, and some units can transmit torques up to 1240 kgf m while carrying loads of 17,500 kgf. Modern designs feature a linear bearing mounted on a three-splined shaft to provide good torque transmission as well as steady axial motion. The spline nut can traverse the shaft while it rotates about it. Therefore, rotary motion can be transmitted at any point along the shaft length. Spline nuts come in four types: nuts with a flange on the end, nuts with a flange at the center, rigid angular nuts with a flat side containing four bolt holes, and cylinder nuts that have a slot along one side. Pelser[136] has reviewed the design considerations involved in computing ball spline life and performance characteristics. For example, the force F needed to get the spline nut moving is given in terms of the equivalent radial load $\mathbf{P}_e$ and friction coefficient μ by

$$F = \mu \mathbf{P}_e + 1.0 \tag{9.46}$$

where

$$\mathbf{P}_e = \mathbf{P}_r + \frac{4000T_c}{3d_p \cos \alpha} \tag{9.47}$$

in which $\mathbf{P}_r$ is the calculated radial load (kgf), T_c the calculated torque load (kgf m), d_p the nominal shaft diameter (cm), and α the rolling element contact angle (degrees). The friction coefficient is obtained from a design chart, which plots μ as a function of the ratio of the equivalent load to the basic dynamic load rating C. Typical values of μ range from 0.015 at low $(\mathbf{P}_e/C)$ ratios (~0.02), to values less than 0.005 at $(\mathbf{P}_e/C)$ ratios above 0.1. In addition, values of μ for sealed bearings are slightly higher than for bearings without seals.

9.6.4   FRICTION IN PARTICLE ASSEMBLAGES

The friction of particle assemblages is an important area of technology. Applications in this area include the transport of powders in copying machines, the fabrication of pills, the flow of metal and ceramic powders into pressing dies, the use of powdered lubricants,[137] and soil mechanics computations, which are important in the construction of buildings and bridges.[138] Most practical situations involve not only the friction between the individual particles in a powder assemblage, but also the friction between the powder and the wall or walls of the container. The situation is complicated because various complex kinds of particle motions can occur within the assemblage and because the characteristics of the powder layer may not be uniform throughout its volume. Adams[139] has reviewed friction in granular particle assemblages. The assumed particle shapes, size distribution, and packing density of the powders have strong effects on friction. Higher-density powders tend to have the higher shear stresses, but many other factors have to be taken into consideration to understand how powder friction obtains. Studies of hard powders intentionally introduced into frictional interfaces have shown how the particle mass can undergo changes as it densifies under compression, shears, and breaks up into clumps.[140] The densification of layers of spherical carbon molecules (C_{60}, fullerenes) under slider conditions caused them to form a high-shear-strength, high-friction layer, rather than to lubricate like tiny bearing balls, as some theorists incorrectly predicted. In fact, the friction coefficient of a fullerne powder layer between stainless steel and aluminum was five times higher than for no lubricant at all under the same conditions.[141] Clearly, studies of powder friction have important implications in both fundamental and applied control of friction.

Richard et al.[142] have modeled the frictional energy dissipation in idealized tribosystems in which powder layers are present. They used what they refer to as a discrete element method consisting of three parts: (i) rigid, nondeforming, and degrading solid bodies whose boundaries are modeled as single sheets of spheres, (ii) an interposed granular third body of equal-sized spheres, and (iii) boundary conditions between the third body and the two solids. The relative amounts of power dissipation at the interface and within the simulated granular layer were calculated for differing degrees of wall adhesion and shown to correlate with macroscopic experimental results.

Fine particles in a frictional interface can spell disaster for magnetic recording media, which must operate at high speeds at very close separation distances (flying heights), and for electrical contacts, which must maintain good conductivity and signal transmission capabilities over relatively small contact areas. In fact, it was the latter problem that prompted the classic studies of Ragnar Holm in the 1940s and 1950s.[143] Antler[144,145] spent many years at Bell Laboratories studying the friction and wear of electrical contacts. More recently, Burton[146] reviewed the subject of friction and wear in electrical contacts for motor brushes. Friction is important in read–write heads for magnetic tape systems[147] and for magnetic rigid disks.[148] As long as some form of moving physical contact is involved in information storage, friction will continue to be a focus of research in the computers and electronics industries.

9.6.5   FRICTION IN MICROTRIBOLOGY AND NANOTRIBOLOGY

The terms *microtribology* and *nanotribology* grew increasingly popular during the last decade of the twentieth century. There are two connotations for these terms. One connotation concerns the study of tribological phenomena that operate on the dimensions of individual atoms and molecules. This interest was a long-standing one in tribology but was rekindled and enabled by the availability of new instruments such as atomic force microscopes and nanoindentation devices. It concerns the physics of friction and harkens back to the signal work of Tomlinson and others mentioned in the first chapter. The other connotation involves the design of mechanical assemblies whose dimensions range from one or two hundred micrometers across to a few nanometers. Innovations in manufacturing processes have enabled the fabrication of mechanical devices such as miniature actuators and gears that are too small to see with the unaided eye. These are commonly called micro-/nanoelectromechanical systems (MEMS/NEMS). It is interesting that the scale of entire tribocomponents has now approached the dimensions of the surface features envisioned by early friction theorists.

The number of conferences and books devoted to micro- and nanotribology has expanded greatly since 1992 when the Microtribology Research Committee of the Japanese Society of Tribologists held the first two-day International Workshop on Microtribology. More recently, organizations like the Materials Research Society in the United States have held a series of symposia on micro-/nanocontact and friction, and there are a growing number of books and proceedings on the subject.[149,150] There is even an international nanotribology forum.[151] In contrast with the mechanical engineers, chemists, and metallurgical engineers who designed and investigated macrotribological systems, such as bearings, brakes, and gears, the field of micro- and nanotribology has attracted a new segment of the technical community: surface chemists, solid-state physicists, and molecular architecture theorists. Although this influx has brought new perspectives and enthusiasm to friction science, sometimes there is a conceptual gap between the quasi-idealized surfaces studied by micro/nano community and the surfaces that are observed in practical tribosystems, especially when the scale of wear damage that occurs in macroscale engineering systems far exceeds that of micro/nano phenomena.

In addition to the computer and electronics industries, applications of microtribology include small precision bearings and "micromotors" whose tiny gears are only tens of micrometers in diameter. As mentioned earlier in this chapter, Drexler's book on the subject of nanosystems[65] describes a number of molecule-sized components such as sleeve bearings, gears, rollers, belts, and cams. In view of its enabling characteristics, friction technology will continue to play an important role in the advancement of a host of related technologies.

9.6.6 AMUSEMENT PARK RIDES

Amusement park rides such as roller coasters and ferris wheels are obvious applications of friction technology. Roller coasters began as ice slides that were popular near St. Petersburg, Russia, during the seventeenth century. Sliders were constructed of lumber and slid on a layer of ice that covered an inclined surface. It is not known exactly when and where wheels were introduced to create the first true roller coasters, but some of the earlier designs with wheels used what were called side-friction designs in which the wheeled cars ran in grooves with side plates to keep the cars from leaving the track. Some of the larger coasters required a brakeman to control the stability of the trains. Beginning around 1920, "up-stop" wheels situated under the tracks were introduced and the side-friction approach fell out of favor.[152]

In 2005, the largest roller coaster in the world at the time, called the Kingda Ka, opened in Jackson, New Jersey. It reached a maximum height of 142 m (465 ft), boasted a maximum 90° angle of descent, and used hydraulic motors to accelerate its riders to a speed of 57.2 m/s (128 miles/h) within the first 3.5 s of the initial horizontal run. Undoubtedly, such performance records will continue to be broken as the public seeks more and more extreme adventures.

A less jarring, but tribologically interesting recreational application of combined rolling and sliding friction may be found in the design of *alpine slides* for amusement parks. The patron of the ride sits in a specially designed carriage (sled) that slides down a long winding channel, which can exceed 1 km in length. The channel is made of a composite fiberglass-type material. To provide the rider with a measure of control, the underside of the sled is designed to operate in a "fail-safe manner." That is, if the rider's hand is not bearing down on a lever that raises the sled onto a pair of free-turning wheels, the sled wheels are retracted below the level of two parallel skids, and the carriage slides to a stop under the retarding action of friction. Even at full speed, the sled is both sliding on the rear corner pads (nylon) and rolling on the wheels. Figure 9.18 shows the arrangement of pads, skids, and wheels on an alpine slide carriage. Friction coefficients may be relatively high for alpine slide

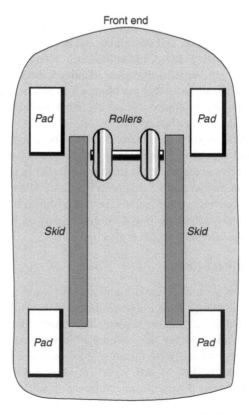

FIGURE 9.18 Diagram of the underside rolling–sliding contacts used to control the performance of alpine slide vehicles.

materials ($\mu \gg 0.5$), compared with the luge, bobsled, and toboggan whose runner/ice interface friction coefficients can be more than a factor of 10 lower. Water slides have also been gaining popularity. In this case, the friction between the fabric of the slider's bathing suit and the channel is greatly reduced by copious quantities of water, leading ultimately to thick-film, hydrodynamic lubrication.

9.7 CONCLUSION

The examples in this chapter have only hinted at the rich spectrum of applied friction problems confronting our society. Some of these are mundane while others stretch the limits of understanding. The success of some applications is based on applying established principles of friction and lubrication science, but for others they work mainly on the basis of experience and empiricism. Friction science and technology will continue to enable advances in the quality of life while challenging our understanding of physical phenomena, great and small.

REFERENCES

1. N. A. Hooton (1969). Metal-ceramic composites in high-energy friction applications, *Bendix Tech. J.*, Spring, pp. 55–61.
2. L. C. Buckman (1998). *Commercial Vehicle Braking Systems: Air Brakes, ABS, and Beyond*, Society of Automotive Engineers, STP 1405, Warrendale, PA.
3. D. C. Giancoli (1991). *Physics*, 3rd ed., Prentice Hall, Englewood Cliffs, NJ, pp. 26–28.
4. D. F. Moore (1975). *Principles and Applications of Tribology*, Pergamon Press, Oxford, p. 290.
5. A. S. Hall, A. R. Holowenko, and H. G. Laughlin (1961). *Machine Design*, Schaum's Outline Series, McGraw-Hill, New York, NY, p. 180.
6. P. J. Blau (2001). *Compositions, Functions, and Testing of Friction Brake Materials and Their Additives*, Oak Ridge National Laboratory, Technical Memo, ORNL/TM-2001/64, p. 29.
7. A. E. Anderson (1992). Friction and wear of automotive brakes, in *ASM Handbook, Volume 18: Friction, Lubrication, and Wear Technology*, 10th ed., ASM International, Materials Park, OH, pp. 569–577.
8. P. J. Blau (2004). *Research on Non-Traditional Materials for Friction Surfaces in Heavy Vehicle Disc Brakes*, Oak Ridge National Laboratory, Technical Report, ORNL/TM-2004/265, p. 36.
9. P. J. Blau, B. C. Jolly, J. Qu, W. H. Peter, and C. A. Blue (2007). Tribological investigation of titanium-based materials for brakes, *Wear*, 263(7–12), pp. 1201–1211.
10. G. Martino, Red Devil Brakes, Inc., Mount Pleasant, PA.
11. D. G. Grieve, D. C. Barton, D. A. Crolla, J. L. Chapman, and J. T. Buckingham (1996). Light-weight disc brake materials, in *Advances in Automotive Braking Technology*, ed. D. C. Barton, Professional Engineering Publishing, London, UK, pp. 87–106.
12. K. Tanaka, S. Ueda, and N. Noguchi (1973). Fundamental studies on the brake friction of resin-based friction materials, *Wear*, 23, pp. 249–365.
13. S. J. Kim, K. S. Kim, and H. Jang (2003). Optimization of manufacturing parameters for a brake lining using Taguchi method, *J. Mater. Proc. Technol.*, 136(Iss. 1–3), pp. 202–208.
14. V. Prochazka and E. Navara (1970). Additions to sintered Fe-based materials, in *Friction and Antifriction Materials*, eds. H. H. Hausner, K. H. Roll, and P. K. Johnson, Plenum Press, New York, NY, pp. 61–72.

15. S. K. Rhee, M. G. Jacko, and P. H. S. Tsang (1991). The role of friction film in friction, wear and noise of automotive brakes, *Wear*, 146(1), pp. 89–97.
16. M. Eriksson and S. Jacobson (2000). Tribological surfaces of organic brake pads, *Tribol. Int.*, 33(12), pp. 817–827.
17. P. J. Blau (2003). Microstructure and detachment mechanism of friction layers on the surface of brake shoes, *J. Mater. Eng. Perform.*, 12(1), pp. 56–60.
18. W. Österle, H. Kloß, I. Urban, and A. I. Dmitriev (2007). Towards a better understanding of brake friction materials, *Wear*, 263(7–12), pp. 1189–1201.
19. Carlisle Motion Products, Charlottesville, Virginia, Internet address: http://www.carlislebrake.com.
20. Occupational Safety and Health Administration (1972). Rules and regulations for exposure to asbestos dust, 37 *Fed. Reg.* 11318.
21. M. G. Jacko, P. H. S. Tsang, and S. K. Rhee (1984). Automotive friction materials evolution during the past decade, *Wear*, 100, pp. 503–515.
22. H. H. Hausner, K. H. Roll, and P. K. Johnson (1970). *Friction and Antifriction Materials*, Plenum Press, New York, NY.
23. A. Anderson (1995). Basics about friction excited oscillations, Link Engineering Report, FEV1, Link Engineering, Plymouth, MI, http://www.ifriction.com/reports.101.htm.
24. E. M. Tatarzycki and R. T. Webb (1992). Friction and wear of aircraft brakes, in *ASM Handbook, Volume 18: Friction, Lubrication, and Wear Technology*, 10th ed., ASM International, Materials Park, OH, pp. 582–587.
25. T.-L. Ho, M. B. Peterson, and F. F. Ling (1974). Effect of frictional heating on brake materials, *Wear*, 30, pp. 73–91.
26. T.-L. Ho and M. B. Peterson (1976). Development of Aircraft Brake Materials, ASLE preprint Bo. 76-LC-1B-3, presented at the ASLE/ASME Conference in Boston, October 5–7.
27. T.-L. Ho and M. B. Peterson (1977). Wear formulation for aircraft brake material sliding against steel, *Wear*, 43, pp. 199–210.
28. T. Haug and K. Rebstock (2003). New material technologies for brakes, in *Advanced Brake Technology*, eds. B. Breuer and U. Dausend, SAE International, Warrendale, PA, pp. 37–49.
29. Messier-Bugatti advertising brochure for SEPCARBIII® or http://www.messier-bugattiusa.com/IMG/pdf/777LR.pdf.
30. D. F. Moore (1975). *Principles and Applications of Tribology*, Pergamon Press, Oxford (Chapter 13.1).
31. D. F. Moore (1980). Friction and wear in rubbers and tyres, *Wear*, 61, pp. 273–282.
32. T. Holmes, G. Lees, and A. R. Williams (1972). A combined approach to the optimisation of type and pavement interaction, *Wear*, 20, pp. 241–276.
33. D. Bulgin, D. G. Hubbard, and M. H. Walters (1962). *4th International Rubber Conference*, London.
34. P. S. Pillai (1992). Friction and wear of tires, in *ASM Handbook, Volume 18: Friction, Lubrication, and Wear Technology*, 10th ed., ASM International, Materials Park, OH, pp. 578–581.
35. A. Schallamach and D. M. Turner (1960). The wear of slipping wheels, *Wear*, 3, pp. 1–25.
36. R. R. Hegmon (1993). A close look at road surfaces, *Public Roads*, 57(1), 7 pp. (online at http://www.tfhrc.gov/pubrds/summer93/p93su4.htm).
37. E. A. Avallone, T. Baumeister, A. Sadegh, ed., (1987). *Marks' Standard Handbook for Mechanical Engineers*, McGraw Hill, New York, pp. 3–28.
38. ASTM (1992). *The Vehicle, Tire, Pavement Interface*, American Society for Testing and Materials, Spec. Tech. Pub. 1164, Philadelphia, PA.
39. D. J. Parkka (1996). *Equation Directory of the Reconstructionist*, 2nd ed., Institute for Police Technology and Management (IPTM), Jacksonville, FL, p. 212.

40. G. M. Bonnett (1992). *Effect of Grade on Kinetic Friction in Accident Reconstruction*, http:///www.rec-tec.com/Grade.html.
41. J. R. Luchini (1999). Measuring and modeling tire rolling resistance, Presented at the workshop on Opportunities of Heavy Vehicle Energy Efficiency Gains through Running Resistance and Braking Systems R&D, Oak Ridge National Laboratory, August 18–19, Knoxville, TN.
42. T. Schrüllkamp, H. Goertz, and T. Hüsemann (2005). Development of an intelligent tire—Experiences for the APOLLO Project, *Intelligent Tire Technology Conference*, Frankfurt am Main, Germany, November 29–30.
43. I. E. Fox (2005). Numerical evaluation of the potential for fuel economy improvement due to boundary friction reduction within heavy-duty diesel engines, *Tribol. Int.*, 38(3), pp. 265–275.
44. U.S. Congress, Office of Technology Assessment (1991). *Improving Automotive Fuel Economy: New Standards, New Approaches*, OTA-E-504, U.S. Government Printing Office, Washington, DC.
45. A. E. Cleveland and I. N. Bishop (1960). *Several Possible Paths to Improved Part-Load Economy of Spark-Ignition Engines*, Society of Automotive Engineers, Paper 150A.
46. R. C. Rosenberg (1982). *General Friction Considerations for Engine Design*, Society of Automotive Engineers, Spec. Pub. 532, Paper No. 821576, p. 59.
47. R. A. Mufti and M. Priest (2005). The measurement of component friction losses in a fired engine, part 1 (experimental method), *Proceedings of World Tribology Congress III*, September 2005, Washington, DC,, Abstract no. WTC2005-64252.
48. Commercial software packages for engine and subsystem simulation: RINGPAK®, ENDYN®, VALDYN®, and PISDYN®, Ricardo Corp., Detroit, MI.
49. N. A. Hinein (1992). *Instantaneous Engine Frictional Torque, Its Components, and Piston Assembly Friction*, Final Report to the U.S. Department of Energy, Tribology Project, Contract No. 73072401 to Wayne State University, Detroit, MI.
50. PA. Gaydos and K. F. Dufrane (1992). *Studies of Dynamic Contact of Ceramics and Alloys for Advanced Heat Engines*, Oak Ridge National Laboratory Report ORNL/Sub/84-00216/1, available through Office of Scientific and Technical Information, P.O. Box 62, Oak Ridge, TN 37831.
51. L. L. Ting (1980). Lubricated piston rings and cylinder bore wear, in *ASME Wear Control Handbook*, eds. M. B. Peterson and W. O. Winer, ASME, New York, NY, pp. 609–665.
52. J. H. Fuchsluger and V. L. Vandusen (1980). Unlubricated piston rings, in *ASME Wear Control Handbook*, eds. M. B. Peterson and W. O. Winer, ASME, New York, NY, pp. 667–698.
53. P. C. Nautiyal, S. Singhal, and J. P. Sharma (1983). Friction and wear processes in piston rings, *Tribol. Int.*, 16(2), p. 43.
54. C. F. Taylor (1985). *The Internal Combustion Engine in Theory and Practice*, Vol. I, 2nd ed., M.I.T. Press, Cambridge, MA, p. 358.
55. P. J. Blau (2001). Simulation of cylinder bore surface finish parameters to improve laboratory-scale friction tests in new and used oil, in *Engine Systems: Lubricants, Components, Exhaust and Boosting System, Design and Simulation*, American Society of Mechanical Engineers, New York, NY, ASME ICE Vol. 37-3, pp. 57–63.
56. ASTM G181-04 (2004). Standard practice for conducting friction tests of piston ring and cylinder liner materials under lubricated conditions, in *Annual Book of Standards*, Vol. 03.02, ASTM International, West Conshohocken, PA.
57. R. C. Rosenberg (1982). *General Friction Considerations for Engine Design*, Society of Automotive Engineers, Warrendale, PA, Paper No. 821576.

58. J. A. Davis (1992). Friction and wear of internal combustion engine parts, in *ASM Handbook, Volume 18: Friction Lubrication, and Wear Technology*, 10th ed., ASM International, Materials Park, OH, pp. 553–562.

59. M. Hoshi (1984). Reducing friction losses in automobile engines, *Tribol. Int.*, 17(4), pp. 185–189.

60. J. T. Staron and P. A. Willermet (1983). *An Analysis of Valve Train Friction in Terms of Lubrication Principles*, Society of Automotive Engineers, Paper No. 830165.

61. J. C. Bell and T. A. Colgan (1991). Critical physical conditions in the lubrication of automotive valve train systems, *Tribol. Int.*, 24(2), pp. 77–84.

62. J. F. Braza, R. H. Licht, and E. Lilley (1992). *Ceramic Cam Roller Follower Simulation Tests and Evaluation*, STLE Preprint No. 92-AM-2F-1, Society of Tribology and Lubrication Engineers, Park Ridge, IL.

63. D. Dowson (1979). *History of Tribology*, Longman, London, p. 43.

64. Organization for Economic Cooperation and Development (1969). *Glossary of Terms and Definitions in the Field of Friction, Wear, and Lubrication (Tribology)*, International Research Group on Wear of Engineering Materials, Paris.

65. K. E. Drexler (1992). *Nanosystems*, Wiley, New York, NY, pp. 284–312.

66. T. A. Harris (1992). Friction and wear of rolling element bearings, in *ASM Handbook, Volume 18: Friction, Lubrication, and Wear Technology*, 10th ed., ASM International, Materials Park, OH, pp. 499–514.

67. W. J. Anderson (1964). Rolling element bearings, in *Advanced Bearing Technology*, eds. E. E. Bisson and W. J. Anderson, NASA Special Publication SP-38, U.S. Government Printing Office, Washington, DC (Chapter 6).

68. A. Palmgren (1959). *Ball and Roller Bearing Engineering*, SKF Industries, King of Prussia, PA.

69. L. B. Sibley (1980). Rolling bearings, in *ASME Wear Control Handbook*, eds. M. B. Peterson and W. O. Winer, ASME, New York, NY, pp. 699–726.

70. M. M. Khonsari and E. R. Booser (2001). *Applied Tribology—Bearing Design and Lubrication*, Wiley, New York, NY, p. 496.

71. M. J. Todd (1990). Modeling of ball bearings in spacecraft, *Tribol. Int.*, 23(2), pp. 123–128.

72. J. P. Ford (1969). Rolamite fundamentals, in *First Symposium on Rolamite*, ed. D. D. Eulert, University of New Mexico Press, Albuquerque, p. 9.

73. R. Pike and J. M. Conway-Jones (1992). Friction and wear of sliding bearings, in *ASM Handbook, Volume 18: Friction, Lubrication, and Wear Technology*, 10th ed., ASM International, Materials Park, OH, pp. 515–521.

74. N. S. Eiss (1984). Wear of non-metallic materials, in *CRC Handbook of Lubrication*, Vol. II, ed. R. Booser, CRC Press, Boca Raton, FL, p. 197.

75. R. G. Bayer (2002) *Wear Analysis for Engineers*, HNB Publishing, New York, NY, pp. 322–324.

76. G. C. Pratt (1967). Plastic-based bearings, in *Lubrication and Lubricants*, ed. E. R. Braithwaite, Elsevier, Amsterdam, pp. 376–426 (Chapter 8).

77. D. W. Dudley (1980). Gear wear, in *ASME Wear Control Handbook*, eds. M. B. Peterson and W. O. Winer, ASME, New York, NY, pp. 755–830.

78. K. L. Johnson and D. I. Spence (1991). Determination of gear tooth friction by disc machine, *Tribol. Int.*, 24(5), pp. 269–275.

79. W.-H. Chen and P. Tsai (1989). Finite element analysis of an involute gear drive considering friction effects, *J. Eng. Ind.*, 111, p. 94.

80. J. Kleemola and A. Lehtovaara (2007). Experimental evaluation of friction between contacting discs for the simulation of gear contact, *Tribotest*, 13, pp. 13–20.

81. K. E. Drexler (1992). *Nanosystems—Molecular Machinery, Manufacturing, and Computation*, Wiley, New York, NY, pp. 284–312.

82. STLE (1983). *A Glossary of Seal Terms*, Spec. Pub. SP-1, Society of Tribology and Lubrication Engineers, Park Ridge, IL.

83. J. F. Dray (1992). Friction and wear of seals, in *ASM Handbook, Volume 18: Friction, Lubrication, and Wear Technology*, 10th ed., ASM International, Materials Park, OH, pp. 546–552.

84. R. K. Flitney (1982). Reciprocating seals, *Tribol. Int.*, 15(4), pp. 219–226.

85. D. E. Johnson (1986). Rotary shaft seals, *Tribol. Int.*, 19(4), p. 170.

86. R. L. Johnson and K. Schoenherr (1980). Seal wear, in *ASME Wear Control Handbook*, eds. W. O. Winer and M. B. Peterson, American Society of Mechanical Engineers, New York, NY, pp. 727–753.

87. T. H. C. Childs (1983). Fine friction cutting: A useful wear process, *Tribol. Int.*, 16(2), pp. 67–84.

88. H. S. Hong and M. F. Ashby (1991). *Case Studies in the Application of Temperature Maps for Dry Sliding*, Cambridge University Report, CUED/C-MATS/TR. 186, February, pp. 9–16.

89. H. Ernst and M. E. Merchant (1940). Surface friction of clean metals—A basic factor in the metal cutting process, *Proceedings of M.I.T. Conference on Friction and Surface Finish*, M.I.T. Press, Cambridge, MA, pp. 76–101.

90. J. A. Schey (1983). *Tribology in Metalworking*, American Society for Metals, Metals Park, OH, pp. 574–588.

91. E. J. A. Armarego and R. H. Brown (1969). *The Machining of Metals*, Prentice Hall, Englewood Cliffs, NJ.

92. J. A. Williams and D. Tabor (1977). The role of lubricants in machining, *Wear*, 43, pp. 275–292.

93. H. S. Hong and M. F. Ashby (1991). *Case Studies in the Application of Temperature Maps for Dry Sliding*, Cambridge University Report, CUED/C-MATS/TR. 186, February, pp. 343–352.

94. H. S. Hong and M. F. Ashby (1991). *Case Studies in the Application of Temperature Maps for Dry Sliding*, Cambridge University Report, CUED/C-MATS/TR. 186, February, pp. 353–354.

95. J. G. Lenard (1992). Friction and forward slip in cold strip rolling, *Tribol. Trans.*, 35(3), pp. 423–428.

96. H. S. Hong and M. F. Ashby (1991). *Case Studies in the Application of Temperature Maps for Dry Sliding*, Cambridge University Report, CUED/C-MATS/TR. 186, February, pp. 268–312.

97. B. Avitzur (1992). Friction during metals forming, in *ASM Handbook, Volume 18: Friction, Lubrication, and Wear Technology*, 10th ed., ASM International, Materials Park, OH, pp. 59–69.

98. Committee on Friction Welding (1983). Friction welding, in *ASM Handbook*, Vol. 6, ASM International, Materials Park, OH, pp. 719–728.

99. American Welding Society (1968). Friction welding, in *Welding Handbook*, American Welding Society, New York, NY, pp. 239–261.

100. F. D. Duffin and A. S. Bahrani (1973). Frictional behaviour of mild steel in friction welding, *Wear*, 26, pp. 53–74.

101. ASM Committee on Friction Welding (1983). Friction welding, *ASM Handbook*, Vol. 6, ASM International, Materials Park, OH, pp. 719–728.

102. H. S. Kong and M. F. Ashby (1993). Case study in the application of temperature maps for dry sliding, in *Wear Mechanism Modeling*, Argonne National Laboratory, Final Subcontract Report, ANL/OTM/CR-3, Argonne, IL.

103. T. H. Hazlett (1967). Fundamentals of friction welding, *Metals Eng. Q.*, 7(1), pp. 1–7.

104. R. S. Mishra and M. W. Mahoney (2007). *Friction Stir Welding and Processing*, ASM International, Materials Park, OH, p. 360.

105. S. Aldajah, O. O. Ajayi, and G. R. Fenske (2006). Effects of friction stir processing on tribological performance of high carbon steel, *Proceedings of Seventh Annual U.A.E. University Research Conference*, pp. EN-9–EN-17.
106. J. Qu, H. Xu, Z. Feng, D. Frederick, B. Jolly, R. Battiste, P. Blau, and S. David (2008). Forming wear-resistant Al-Al$_2$O$_3$ surfaces using friction stir processing, Presented at STLE Annual Meeting, Cleveland, OH, May.
107. S. F. Miller, J. Tao, and A. J. Shih (2006). Friction drilling of cast metals, *Int. J. Machine Tools Manufacture*, 46(12–13), pp. 526–535.
108. S. F. Miller, P. J. Blau, and A. Shih (2005). Microstructural alterations associated with friction drilling of steel, aluminum, and titanium, *J. Mater. Eng. Perform.*, 14(5), pp. 647–653.
109. D. Dowson (2007). *Biotribology*, John Wiley and Sons Ltd, London, UK, p. 608.
110. O. S. Dinc, C. M. Ettles, S. J. Calabrese, and H. J. Scarton (1990). *Some Parameters Affecting Tactile Friction*, ASME Preprint 90-Trib-28, American Society of Mechanical Engineers, New York, NY, p. 6.
111. A. D. Roberts and C. A. Brackley (1992). Friction of surgeons' gloves. *J. Phys. D: Appl. Phys.*, 25, pp. A28–A32.
112. N. Gitis (2003). Tribometrology of skin, *Proceedings of STLE/ASME Joint International Tribology Conference*, Paper # 2003-TRIB-236, p. 8.
113. A. B. Cua, K. P. Wilhelm, and H. I. Maibach (1990). Friction properties of human skin: Relation to age, sex, and anatomical region, *Br. J. Dermatol.*, 123, pp. 473–479.
114. P. Elsner, D. Wilhelm and H. I. Maibach (1990). Friction properties of human forearm and vulvar skin: Influence of age and transepidermal water loss and capacitance, *Dermatologica*, 181, pp. 88–91.
115. M. G. Gee, P. Tomlins, A. Calver, R. H. Darling, and M. Rides (2005). A new friction measurement system for the frictional component of touch, *Wear*, 259(7–12), pp. 1437–1442.
116. P. Valint (1997). Posting to the EPFL Tribology Listserver, May 9.
117. H. D. Conway and M. Richman (1982). Effects of contact lens deformation on tear film pressures induced during blinking, *Am. J. Optom. Physiol.*, 59, p. 13.
118. A. Jacobson (2003). Biotribology: The tribology of living tissues, *Tribol. Lubric. Eng.*, 59(12), pp. 32–38.
119. J. A. Nairn and T.-B. Jiang (1995). Measurement of the friction and lubricity properties of contact lenses, *Proceedings of ANTEC*, Boston, MA, May 7–11, p. 5.
120. N. Nuño, R. Groppetti, and N. Senin (2006). Static coefficient of friction between stainless steel and PMMA used in cemented hip and knee implants, *Clin. Biomech.*, 21(9), pp. 956–962.
121. A. Unsworth (1978). The effects of lubrication in high joint prostheses, *Phys. Med. Biol.*, 23, pp. 253–268.
122. A. Unsworth (1991). Tribology of human and artificial joints, *J. Eng. Med.*, 205, pp. 163–172.
123. S. C. Scholes, A. Unsworth, R. M. Hall, and R. Scott (2000). The effects of material combination and lubricant on the friction of total hip prostheses, *Wear*, 241, pp. 209–213.
124. B. J. Hamrock and D. Dowson (1978). Elastohydrodynamic lubrication of elliptical contacts for materials with low elastic modulus. I. Fully flooded lubrication, *J. Lubric. Technol.*, 100, pp. 236–245.
125. I. M. Hutchings, ed. (2003). *Friction, Lubrication and Wear of Artificial Joints*, Prof. Engineering Pub., London, UK, p. 134.
126. D. W. Kelley (2000). Lubricious medical devices, U.S. Patent 6,071,266.
127. D. R. Dobies and A. Cohoon (2008). Case reports: Injecting lubricant into guiding catheter improves stent deployment, *J. Invasive Cardiol.*, online post, January 20, 2008, at www.invasivecardiology.com/article/5630.

128. D. Laroche, S. Delorme, T. Anderson, and R. di Raddo (2006). Computer prediction of friction in balloon angioplasty and stent implantation, in *Biomedical Simulation*, Lecture Notes in Computer Science series, Springer, Berlin, pp. 1–8.

129. *NYC Subway Tile Study* (2003). Voices of Safety International (VOSI), http://www.voicesofsafety.com.

130. ASTM C1028-07 (2007) Standard test method for determining the static coefficient of friction of ceramic tile and other like surfaces by the horizontal dynamometer pull-meter method, in *Annual Book of Standards*, Vol. 15.02, ASTM International, West Conshohocken, PA.

131. R. H. Knapp and X. Liu (2005). Cable vibration considering interlayer coulomb friction, *Int. J. Offshore Polar Eng.*, 15(3), pp. 229–234.

132. Anon. (not dated). Estimating tension when pulling cable into conduit, *Technical Talk*, Vol. 1, American Polywater, Stillwater, MN.

133. M. A. Urchegui, M. Hartelt, D. Klaffke, and X. Gomez (2007). Laboratory fretting tests with thin wire specimens, *Tribotest*, 13(1), pp. 67–81.

134. M. B. Karamis and B. Selcuk (1993). Analysis of the friction behaviour of bolted joints, *Wear*, 166, pp. 73–83.

135. R. E. Morf (1989). Flat belts shed the leather-strap image, *Machine Des.*, March 9, pp. 62–66.

136. G. Pelser (1989). Ball splines combine linear and rotary motion, *Machine Des.*, April 6, pp. 126–129.

137. H. Heshmat (1991). The rheology and hydrodynamics of powder lubrication, *Tribol. Trans.*, 34(3), pp. 433–439.

138. D. P. Krynine (1947). *Soil Mechanics*, McGraw-Hill, New York, NY.

139. M. J. Adams (1992). Friction of granular non-metals, in *Fundamentals of Friction*, eds. I. L. Singer and H. M. Pollock, Kluwer, Dordrecht, The Netherlands, pp. 183–207.

140. P. J. Blau (1993). Friction microprobe investigation of particle layer effects on sliding friction, *Wear*, 162–164, pp. 102–109.

141. P. J. Blau and C. E. Haberlin (1992). An investigation of the microfrictional behavior of C_{60} particle layers on aluminum, *Thin Solid Films*, 219, pp. 129–134.

142. D. Richard, I. Iordanoff, Y. Berthier, M. Renouf, and N. Fillot (2007). Friction coefficient as a macroscopic view of local dissipation, *J. Tribol.*, 129, pp. 829–835.

143. R. Holm (1958). *Electric Contacts Handbook*, 3rd ed., Springer-Verlag, Berlin.

144. M. Antler (1962). Wear, friction, and electrical noise phenomena in severe sliding systems, *A.S.L.E. Trans.*, 5, pp. 297–307.

145. M. Antler (1981). Sliding wear of metallic contacts, *IEEE Trans. Components, Hybrids, Manuf. Technol.*, CHMT-4(1), pp. 15–29.

146. R. A. Burton (1992). Friction and wear of electrical contacts, in *ASM Handbook, Volume 18: Friction, Lubrication, and Wear Technology*, 10th ed., ASM International, Materials Park, OH, pp. 682–284.

147. R. F. Hegel (1993). Hydroscopic effects on magnetic tape friction, *Tribol. Trans.*, 36(1), pp. 67–72.

148. B. Bhushan and S. Venkatesan (1993). Friction and wear studies of silicon in sliding contact with thin-film magnetic rigid disks, *J. Mater. Res.*, 8, pp. 1611–1628.

149. M. Scherge and S. Gorb (2001). *Biological Micro- and Nanotribology*, Springer-Verlag, Berlin, p. 304.

150. S. M. Hsu and Z. C. Ying (2002). *Nanotribology – Critical Assessment and Research Needs*, Kluwer Academic Pub., Norwell, MA, p. 460.

151. The International Forum on Nanotribology was established in 2001 and has a Web site http://www.nanotribology.org.

152. *Roller Coaster History*, at http://www.ultimaterollercoaster.com.

Index to Static and Kinetic Friction Coefficients

1 STATIC FRICTION COEFFICIENT DATA

1.1 METALS ON METALS

Aluminum on aluminum, 186
  in helium, 138
Copper on copper, 146, 186
  in helium, 138
  oxidized, 147
  sulfidized, 147
Copper on nickel, in helium, 138
Gold
  on palladium, 58
  on rhodium, 58
Gold on aluminum, in helium, 138
Gold on gold, 141, 186,
  in helium, 138
Iron
  on copper, in helium, 138
  on iron, 186
  on iron, in helium, 138
Lead on lead, 146
Metals on metals, 21, 143–144
  lightlyoxidized, 21
Molybdenum on molybdenum, 186
Nickel on copper, in helium, 138
Nickel on nickel, 186
  in helium, 138
Oxidized metals (*see* Metals, lightlyoxidized)
Silver
  on gold, 141
  on rhodium, 58
  on silver, 58, 141, 186
Steel on copper, 146
Steel on steel, 146
  oxidized, 147
  solid-lubricated, 250
  sulfidized, 147

1.2 CARBON MATERIALS AND CERAMICS ON OTHER MATERIALS

Carbon on carbon, 145
Diamond on diamond, 145

Glass (clean)
  on glass (clean), 21, 190
  on metal, 190
Glass (tempered)
  on aluminum (6061), 190
  on polymer (PTFE), 190
  on steel (1032), 190
Glass on metals and glass, 145
Graphite on graphite, 21

1.3 POLYMERS

Epoxy on steel (1018), 209
Nylon
  on nylon, lubricated, 195
  on steel, lubricated, 195
Polyethylene on steel (1018), 209
Polymers on metals and polymers, 145
Poly-tetrafluoroethylene
  on poly-tetrafluoroethylene, 21
  on steel (1018), 209
Polyurethane on steel (1018), 209

1.4 MISCELLANEOUS MATERIALS AND MIXED PAIRS

Brick on wood, 145
Cast iron on wood, 184
Cotton on cotton, 145
Ice
  on ice, 21, 145
  on waxes, 206
Leather
  on cast iron, 184
  on iron, 184
  on wood, 184
Mica on mica, 145
Paper on paper, 145
Silk on silk, 146
Ski wax on ice, 146
Solid lubricants, various, 251
Stone on stone, 184
Waxes on ice, 206
Wood on wood, 21, 146, 184

2.3 Polymers

2.4 Miscellaneous Materials and Mixed Pairs

2.5 Application-Specific Kinetic Friction Data

Subject Index

A

Abrasion
    ASTM G-65 dry sand test, 55
    friction models containing, 163
    friction with, 54
Abrasive papers, loading effects on friction, 163
Abrasive particles, role in transitions, 333
Additive package
    anti-wear, 233
    definition, 233
    formulations, 235–236
Additives, types, 234
    grease, 245
Adhesion
    bridges between asperities, 135
    comparison with adherence, 132
    models for friction, 150, 164
    relationship to friction, 10–11, 136–167,
        140–141
Adhesive bond formation, 124
Adhesive junction growth, 139
Adhesive junctions, polymers, 196
Adhesive transfer (*see* Transfer)
Adiabatic engines, ceramics for, 189
Adsorbed species, effects of oxygen and
        chlorine, 137–139
Air-conditioning systems, lubricants, 237
Aircraft brakes (*see* Brakes)
Airport runway, icy, 208
Alignment between surfaces, effects on
        running-in, 328
Amontons, G., law of friction, 7
Amusement park rides, 397–398
Angle of repose, 212
Anti-wear additives, 233
Archard, J. F., wear law, 10
Area of contact,
    modeling and size scales, 121–127
    normal force effects, 54
    polymer friction, 196
    real, 279–280, 309
    vibration effects, 309
Aristotle, 1
Arresting gear, aircraft carrier, 393
Articulating strut test, 95–96
Artificial joints, 390
    (*see also* Bio-implant friction)
Asbestos, brake linings, 352–353

Asperity
deformation, during sliding, 276–280
shapes, 163
truncation, 328
A-spots, 10, 202
ASTM
    Committee G2 on Wear and Erosion, 70
    friction standards, 69–72
    tabulation of standards, 111–112
Atomic force microscope (AFM), 74–75, 133
Atomic-scale friction, 73–75

B

Ball splines, 395
Band brakes, 32
Barus equation, 224
Base oils, 232–233
Basketball surfaces, 5
Bearings
    foil, 301
    journal, 35
    pivot, 30
    porous ceramic, 261
    porous metal, 260–261
    rolling element, 365–367
    sliding, 367–372
    types, 365–366
Belts, 31, 295
    friction testing, 69
    materials, 33
Bio-fuels, 236
Bio-implant friction, 5, 62, 79
Bio-implants (*see* Artifical joints)
Bismuth, as an EP additive, 246
Block-on-ring test, 61
Bolts, tightening and loosening, 273
Boric acid lubricants, 252–253
Boundary lubrication
    friction models, 229–230
    lubricant selection charts, 230–231
    piston ring model, 362
    Stribeck curve, 226
    wire-on-drum test, 81–82
Bowden, F. P., 11
Brakes
    aircraft, 346, 354
    drum, 347–348
    effectiveness, 353